VISCOSITY OF DENSE FLUIDS

VISCOSITY OF DENSE FLUIDS

K. Stephan

University of Stuttgart
Stuttgart, Federal Republic of Germany

and

K. Lucas

Gesamthochschule Duisburg
Duisburg, Federal Republic of Germany

SPRINGER SCIENCE+BUSINESS MEDIA, LLC

Library of Congress Cataloging in Publication Data

Stephan, Karl, 1930-
  Viscosity of dense fluids.

  Includes bibliographical references and index.
  1. Viscosity. 2. Fluids. I. Lucas, K., joint author. II. Title.
QC189.S84   532'.58   79-22620
ISBN 978-1-4757-6933-3   ISBN 978-1-4757-6931-9 (eBook)
DOI 10.1007/978-1-4757-6931-9

FOREWORD

The physical properties of fluids are perhaps among the most extensively investigated physical constants of any single group of materials. This is particularly true of the thermodynamic properties of pure substances since the condition of thermodynamic equilibrium provides the simplest considerations for experimental measurement as well as theoretical treatment. In the case of non-equilibrium transport properties, the situation is significantly complicated by the necessity of measurement of gradients in the experiment and the mathematical difficulties in handling non-equilibrium distribution functions in theoretical treatments. Hence, our knowledge of the transport properties of gases and liquids is perhaps one order of magnitude lower than for equilibrium thermodynamic properties. This situation is very much apparent when examining the available numerical data on the viscosity of fluids particularly at high pressures.

In this work, the authors have performed an outstanding contribution to the engineering literature by their critical evaluation of the pressure dependence of the available data on the viscosity of selected substances. The recommended values reported in the tables and figures also incorporate the saturated liquid and gas states as well as the data of the dilute gas in an attempt to integrate the present work with the recently published work by CINDAS/Purdue University on the viscosity of fluids at low pressures [166]. A deliberate effort was made to treat as many of the substances in the CINDAS volume as possible for which adequate high pressure data exist.

In these cases it was also attempted to establish internal consistency between the low pressure and saturation states data reported by CINDAS and the high pressure treatment presented herein. Unfortunately it was not possible to completely satisfy such internal consistency in all cases. Thus, these two works may well be considered complementary volumes particularly that there has been extensive cooperation between the two sets of authors.

While the primary goal of the present work is to present evaluated viscosity data at high pressures, the section on review of the state of theory and estimation techniques represents a significant contribution in its own right to the total volume. The skillfully prepared summary of these highly complex topics should prove to be a great assistance to those interested in having estimates of viscosity values for substances other than those reported herein. Similarly, the comprehensive list of references cited in both the textual part of the work as well as those reporting original data sources should provide the reader with a solid base for more extensive and in depth explorations.

The authors should be commended for their outstanding contribution in a most different and yet important area of thermophysics of vital interest to a wide spectrum of engineering applications.

Y. S. Touloukian, Director
Center for Information and Numerical
Data Analysis and Synthesis (CINDAS); and
Distinguished Atkins Professor of
Engineering, Purdue University

November 1978
West Lafayette, Indiana, USA

INTRODUCTION

This volume deals with the viscosity of gases and liquids at elevated pressures. Therefore, excluded from the discussion is the *dilute gas state* by which it is implied that the viscosity is independent of pressures. Defining the *dilute gas state* or alternatively the *states of elevated pressures*, in macroscopic terms, is not a simple task. No clear pressure limit can be given, but instead a bounding pressure-temperature curve, which is specific to each gas, must be specified as discussed by Childs and Hanley [1]. Thus the common assumption that at one bar pressure viscosity may be taken as that of a dilute gas, while correct in most cases, it is an incomplete statement. For example, for argon at 600 K, pressures up to 30 bar represent states that may still be considered in the dilute gas region, and for fluids with lower critical temperatures even higher pressures are allowed. On the other hand, even at very low pressures, a lower limit of dilution is reached, and the viscosity must be considered pressure dependent. This region, too, is not considered in this work. Therefore, the macroscopic pressure limits of coverage depend on the fluid and temperature considered. On the basis of molecular considerations the dilute gas region may best be defined as the state where binary collisions between molecules adequately define the transport properties of a gas. It is the region of the Boltzmann equation for the monatomic gas from which exact relations between the transport properties and the intermolecular pair potential may be found [2,3]. A vast literature exists concerning the viscosity of dilute gases, a comprehensive review of which may be found in [166] of the references to text.

The pressure dependence of a fluid is dramatic around its critical region, the derivative of viscosity with respect to pressure being infinite at the critical point itself. It is less pronounced far away from the critical point, yet by no means negligible. For most fluids reported in this volume the pressure effect represents an increase of viscosity between 10% and 20% per 100 bar in the liquid region far away from the critical point. In the gaseous region, the pressure dependence is a strong function of the temperature, and is influenced by the critical temperature. For temperatures high compared to the critical, and pressures which are not extremely high, the dilute gas state is approached and the pressure dependence fades away. This is the reason for the observed low pressure dependence at room temperature of the viscosity of fluids like helium, hydrogen, etc., which have low critical temperatures. On the basis of molecular considerations, the transport properties in the states considered in this volume are governed by higher order collisions and thus are beyond the scope of the original Boltzmann equation. The dynamics of many-body-collisions are not yet satisfactorily understood. Thus, contrary to the case of a monatomic dilute gas, no complete and rigorous theory has yet been developed for the dense fluid. Methods for the theoretical study of viscosity in dense fluids range from simplified models like Eyring's activated state theory and its successive modifications to the more formal but still idealized approaches like Enskog's hard sphere theory, and finally include the rigorous statistical mechanical theories in the form of the distribution function method or the time-correlation-function method. Some excellent reviews have been published on the subject [4-10]. In addition to the theoretical treatments, a large number of empirical estimation techniques for the viscosity of fluids under pressure may be found in the literature [11]. They range from purely empirical observations on the dependence of viscosity on temperature and density to the well-known group contribution methods that are recently studied successfully for equilibrium properties, and finally include correlations between viscosity and other fluid properties. A condensed, albeit incomplete summary of the more important developments in theory and estimation techniques is given in the introductory treatment of this volume.

The authors greatfully acknowledge the partial financial support provided by the Deutsche Forschungsgemenschaft (DFG) which made this work possible.

A number of individuals have contributed to the collection and evaluation of the data presented herein. In particular, the authors wish to acknowledge the contributions of J. Nagengast, Bochum and H. Dreyer, Berlin. Last, but not least, it is a pleasure to acknowledge the encouragement and support provided by TEPIAC, operated by the Center for Information and Numerical Data Analysis and Synthesis (CINDAS) of Purdue University, U.S.A. The former group provided bibliographic assistance and performed the final editorial functions and prepared the manuscript for publication.

Stuttgart, F.R.G., September 1978   Professor Dr.-Ing. K. Stephan

Duisburg, F.R.G., September 1978   Professor Dr.-Ing. K. Lucas

CONTENTS

NOTATION

b	$b = \frac{2}{3} \pi d^3$
d	hard sphere diameter
D	self-diffusion coefficient (Eq. 27)
E_o	activation energy of a molecule
ΔE_v	latent heat of vaporization
f	distribution function
$\bar{F}_{13}$	force exerted due to interaction of molecules 1 and 3 (with analogous meanings for other molecule pairs)
g(d)	equilibrium radial distribution function evaluated at distance d
g(R)	equilibrium radial distribution function (Eq. 5)
h	Planck's constant
J	microscopic stress tensor (Eq. 6, 7)
k	Boltzmann constant
m	molecular mass
M	molecular weight
n	number density
N	number of molecules (Eq. 18)
N_L	Loschmidt-number (Eq. 20)
p	equilibrium pressure
$\bar{P}_j$	the momentum of molecule j
P_{ix}	momentum of molecule i in x-direction
$\bar{\bar{P}}$	pressure tensor
q	partition function of a molecule per unit volume in the initial state
q*	same as q, except in activated state
r	distance of separation between two interacting molecules, (Eq. 14)
$\bar{r}$	the position considered in the system
$\bar{r}_j$	the position of molecule j
$\bar{r}_{kj}$	$\bar{r}_{kj} = \bar{r}_j - \bar{r}_k$
R	gas constant
T	absolute temperature
u	intermolecular potential (Eq. 14)
$\bar{u}$	the mass velocity
v	molar volume

V_f	free volume
z	compressibility factor
z^H	compressibility factor value for the hard sphere model

Greek letters

β	compressibility, (Eq. 37)
δ_{ij}	Kronecker delta
$\delta(x)$	the delta-function
ε	depth of potential-function well with the dimension of energy
$\bar{\bar{\varepsilon}}$	symmetric traceless part of the velocity gradient $\bar{\nabla}\bar{u}$, (Eq. 2)
ζ	friction coefficient related to intermolecular force field (Eq. 5)
θ	temperature function, (Eq. 34)
μ	viscosity
μ_0	viscosity of the hard sphere molecule gas model in the low density limit
μ_v	bulk viscosity (Eq. 2)
ξ	scaling factor for viscosity (Eq. 19, 20)
π	3.1416
ρ	density
σ	molecular parameter in the potential-function model with the dimension of distance
τ	time coordinate (Eq. 6, 7)
$\phi(r_{jk})$	potential function between molecule pairs j and k separated at distance r
χ	transmission coefficient (Eq. 23)
$\psi_2(R)$	a function represented by a differential equation with specified boundary conditions (Eq. 5)

Subscripts, superscripts and notations

$\bar{\bar{1}}$	unit tensor
$(\ )_R$	subscript R indicates a reduced (dimensionless) quantity. (See Eqs. 14-18)
$(\ )_{cr}$	subscript cr indicates values of the quantity at the critical point
c.p.	critical point
G	gas
L	liquid
n.b.p.	normal boiling point
n.m.p.	normal melting point
SL	saturated liquid
SV	saturated vapor

THEORY AND ESTIMATION

THEORY AND ESTIMATION

THEORY OF THE VISCOSITY OF DENSE GASES AND LIQUIDS

INTRODUCTION

Under the heading of theory, we wish to summarize briefly and without detailed reference to the complicated mathematics involved, those theories of momentum transfer in dense fluids which are based on molecular theory in a rigorous or at least modelized way. Various different approaches widely differing in rigour, mathematical complexity, and practical utility are considered and the presentation is grouped under the subheadings referred to as: rigorous statistical mechanical theories, corresponding states theories and model theories.

RIGOROUS STATISTICAL MECHANICAL THEORIES

General Formal Theory

Two different approaches are available to derive formal expressions for the viscosity of dense fluids by rigorous statistical mechanical theory, namely; the older distribution-function method and the more recent time-correlation-function method. The basic idea which leads to a connection of viscosity with molecular properties is the identification of the microscopic conservation equation for momentum with the analogous equation from continuum mechanics which contains viscosity as an empirical transport coefficient. Averaging by means of the non-equilibrium distribution function all quantities in the microscopic conservation equation for momentum yields the corresponding macroscopic conservation law. Its identification with the momentum conservation equation from continuum mechanics gives the stress tensor in terms of averaged microscopic quantities like the momentum, the locus and the force field of the individual molecules.

$$\bar{\bar{P}} = \left\langle \sum_{j=1}^{N} m \left(\frac{\bar{p}_j}{m} - \bar{u} \right) \left(\frac{\bar{p}_j}{m} - \bar{u} \right) \delta(\bar{r}_j - \bar{r}) - \frac{1}{2} \sum_{j \neq k} \sum \bar{r}_{jk} \left(\frac{\partial \phi(r_{jk})}{\partial \bar{r}_{jk}} \right) \delta(\bar{r}_j - \bar{r}) \right\rangle \tag{1}$$

where $\bar{\bar{P}}$ is the pressure tensor, m the molecular mass, $\bar{p}_j$ the momentum of molecule j, $\bar{u}$ the mass velocity, $\delta(x)$ the Delta-function, $\bar{r}_j$ the position of molecule j, $\bar{r}$ the position considered in the system and $\bar{r}_{kj} = \bar{r}_j - \bar{r}_k$. The term $\phi(r_{jk})$ is the pair potential, Eq. (1) implying the approximation of pairwise additivity.

The corresponding phenomenological form of the pressure tensor including the viscosity μ as an empirical coefficient, reads

3

$$\bar{\bar{P}} = [p - \left(\mu_v - \frac{2}{3}\mu\right) (\bar{\nabla} \cdot \bar{u})]\bar{\bar{I}} - 2\mu\bar{\bar{\varepsilon}}$$

with

$$\varepsilon_{ij} = \frac{1}{2}\left(\frac{\partial u_i}{\partial x_j} + \frac{\partial u_j}{\partial x_i}\right) - \frac{1}{3}(\bar{\nabla} \cdot \bar{u})\delta_{ij} \qquad (2)$$

where μ_v is the bulk viscosity, p the equilibrium pressure, $\bar{\bar{I}}$ the unit tensor, $\bar{\bar{\varepsilon}}$ the symmetric traceless part of the velocity gradient $\bar{\nabla}\cdot\bar{u}$, and δ_{ij} the Kronecker delta.

Proceeding further in line of the distribution function method, a kinetic equation has to be established for the non-equilibrium distribution function which is contained in the averaged microscopic quantities of the stress tensor [12]. Starting from the Liouville equation, which the N-body non-equilibrium distribution function must satisfy, it is possible to set up a hierarchy of equations for the lower order non-equilibrium functions, the BBGKY-hierarchy (Bogoliubov-Born-Green-Kirkwood-Yvon). If pairwise additivity of the intermolecular forces is assumed, such that the total potential energy of the intermolecular forces may be replaced by the sum of the potential energies of all molecular pairs, only the first and the second order distribution functions are important. The first two members of the hierarchy give coupling equations between the first order, the second order and the third order distributions.

$$\frac{\partial f_1}{\partial t} + \frac{\bar{p}_1}{m}\frac{\partial f_1}{\partial \bar{r}_1} = -\int F_{12}\left(\frac{\partial f_2}{\partial \bar{p}_1}\right)d\bar{r}_2 d\bar{p}_2 \qquad (3)$$

$$\frac{\partial f_2}{\partial t} + \frac{\bar{p}_1}{m}\left(\frac{\partial f_2}{\partial \bar{r}_1}\right) + \frac{\bar{p}_2}{m}\left(\frac{\partial f_2}{\partial \bar{r}_2}\right) + \bar{F}_{12}\left(\frac{\partial f_2}{\partial \bar{p}_1}\right) + \bar{F}_{21}\left(\frac{\partial f_2}{\partial \bar{p}_2}\right) = -\int\left[F_{13}\left(\frac{\partial f_3}{\partial \bar{p}_1}\right) + F_{23}\left(\frac{\partial f_3}{\partial \bar{p}_2}\right)\right]d\bar{r}_3 d\bar{p}_3 \qquad (4)$$

Here the index to the distribution function f indicates its order. $\bar{F}_{13}$ is the force exerted on molecule 1 due to its interaction with molecule 3, $\bar{F}_{23}$ has an analogous meaning.

In order to get a closed form kinetic equation for the second order non-equilibrium distribution function, some truncation or decoupling has to be effected in the hierarchy. This can only be done by introducing specific assumptions. For the dilute gas, the most important assumption is that of completely random molecular distribution, which transforms the coupling equation between the first order and the second order distribution function into a kinetic equation for the single particle distribution function, the Boltzmann equation (3).

For dense gases and liquids, Kirkwood's Brownian Motion Theory leads to Fokker-Planck-type equations for the time-smoothed first order and second order non-equilibrium distribution functions [13-15,17,18]. The equation for viscosity is [16,19]

$$\mu = \frac{nmkT}{2\zeta} + \frac{\pi\zeta}{15kT}n^2\int_0^\infty R^3\left(\frac{d\phi}{dR}\right)g(R)\psi_2(R)dR \qquad (5)$$

Here ζ is a friction coefficient related to the intermolecular force field, for which various theoretical expressions exist, all of which are not fully satisfactory. This quantity may also be obtained from experimental data for the selfdiffusion coefficient. The quantity ϕ is the pair potential, g(R) the equilibrium radial distribution function, $\psi_2(R)$ a function for which a differential equation along with boundary conditions has been specified. Comparison with experimental results reveals discrepancies in the order from one to several hundred percent depending on the values used for g(R) and ζ [16,19]. This, in addition to the complicated numerical evaluation, makes this approach unattractive for practical application.

Rice and Allnatt [20,21] have modified the Kirkwood theory. They have idealized the repulsive part of the intermolecular potential as a hard core. The change of the distribution function due to hardcore collisions is treated by means of an Enskog-type collision term, whereas the rate of change due to motion in the attractive fields of surrounding molecules between hard core collisions is treated by Kirkwood's approach. This theory, too, contains the friction coefficient ζ which must be obtained by some method. The resulting expressions are too complicated to be cited here. Many workers have discussed this approach [22-26]. Various other approaches to a theory of transport in dense media on the basis of distribution functions have been made [27-36]. Comparison of calculated viscosities to experimental data have been carried out. None of these approaches appears to be in good agreement with experiment for a large region of states.

Instead of non-equilibrium distribution functions, the viscosity can be related to the way in which spontaneous fluctuations regress in an equilibrium system. This idea leads to the time-correlation-function expression for viscosity, in which the viscosity is connected with the time-correlation-function of a dynamical phase function proportional to the microscopic stress tensor. The resulting expression for the viscosity of pure fluids [8] is

$$\mu = \frac{1}{VkT} \int_0^\infty \left\langle J^{xy}(0) J^{xy}(\tau) \right\rangle d\tau \tag{6}$$

with

$$J^{xy} = \sum_{i=1}^N \left(\frac{p_{ix} p_{iy}}{m} - \frac{1}{2} \sum_{j\neq1}^N r_{ij}^y \frac{d\phi}{dr_{ij}^x} \right) \tag{7}$$

where p_{ix} is the momentum of molecule i in x-direction. The angular brackets in Eq. (6) indicate averaging over an equilibrium ensemble.

The time-correlation function, which is the integrand in Eq. (6), measures the extent to which the value of a dynamical variable at a given time is affected by its value at some earlier time, and is therefore obviously a function of time. Various methods have been used to relate transport coefficients, especially the viscosity, to time-correlation functions [37-40]. Again the microscopic analog of the usual macroscopic momentum flux is used to identify the time-correlation function formula for viscosity. The resulting expression, Eq. (6), is valid irrespective of density. Special discussions have been given for the region close to the critical point [41, 42]. Of course, the time correlation expression for viscosity is purely formal. Solutions of the N-body problem are required before a rigorous evaluation can be completed. For dilute gases, where the main dynamical events are taken to be isolated binary encounters, the time-correlation-function expressions for the transport properties have been evaluated [37,43]. The results coincide with the results of the Chapman-Enskog solution of the Boltzmann equation. Calculations with the assumption of a hard sphere potential and a decaying exponential for the autocorrelation of the stress tensor lead to close agreement with Enskog's theory [44]. A general discussion on the connection between the kinetic approach and the time-correlation approach is given in [45]. A computer simulation solution of the time-correlation-function expression for viscosity by molecular dynamics has been given in [46]. Using the Lennard-Jones-potential, a comparison of the thus calculated values with experimental data for the viscosity of liquid argon reveal deviations in the order of 10%, which is by far better than the results of any other statistical calculation for viscosity in the liquid region without fitting of data. As molecular dynamics calculations are not practical as a tool to compute viscosity, theoretical solutions of the time-correlation-functions are needed. Such solutions for viscosity are discussed in [47-51].

A formal approach to the solution of the rigorous statistical mechanical formulae for the transport coefficients is the density expansion method. In equilibrium statistical mechanics, such a procedure yields the virial series. Similar efforts for the transport coefficients may either start from the BBGKY-hierarchy or from the time-correlation functions, whereby equivalent results are obtained. Due to successive correlated collisions, the density expansion is divergent, re-summation introducing logarithmic terms. The first density correction to the Chapman-Enskog dilute gas results requires the detailed analysis of triple collision events and is very involved [52-57].

The Enskog-Theory

Long before the general formal theory discussed above was developed, D. Enskog proposed a method for extending the Boltzmann equation to higher densities for the special case of a fluid consisting of hard spheres [9]. The assumption of hard spheres means that the forces between the molecules on collision are impulsive, i.e., the collision time goes to zero. Therefore, ternary and higher order collisions are neglected. Taking only static correlations into account and neglecting all successive binary collision events, by the assumption of molecular chaos for the momenta, leads to a modified Boltzmann equation for a dense fluid of hard spheres. It can be solved analogous to the case of a dilute gas, yielding the following expression for viscosity

$$\frac{\mu}{\mu_o} = \frac{1}{g(d)} + 0.800 \ bn + 0.761 \ b^2n^2g(d) \tag{8}$$

Here μ_o is the viscosity of the hard sphere gas in the low density limit, $g(d)$ is the equilibrium radial distribution function evaluated at a distance d, the hard sphere diameter $b = (2/3)\pi d^3$, and n is the number density. Equation (8) has been tested by comparison with molecular dynamics calculations [58,59]. As the hard sphere potential model has been used in these computations, the results serve to test primarily the molecular chaos assumption for momenta. It can be concluded that for viscosity the error of the Enskog formula for hard spheres lies within 20%. Good agreement is also found between the first density correction of the dilute gas viscosity taken from the Enskog equation and more accurate calculations for this quantity including triple collisions [60].

The Enskog formula for viscosity has also been used to predict the viscosity of dense real fluids. Various methods have been considered. From a theoretical viewpoint the most satisfactory approach is the one using high density-high temperature p-v-T data to obtain a temperature dependent value for the hard sphere diameter d [61]. Simple perturbation theory for the thermo-dynamic equation of state yields the compressibility factor Z as an expansion in terms of the reciprocal temperature around the hard sphere value Z^H

$$Z = Z^H + \frac{a_1(v)}{T} + O\left(\frac{1}{T^2}\right) \tag{9}$$

Experimental data for Z, plotted as isochores against 1/T, in fact give straight lines in the limit of high temperatures. Extrapolation to (1/T) → 0 an "experimental" value for Z^H is obtained, which is in good agreement with the theoretical equation of state for hard spheres, e.g., the Carnahan-Starling equation [62]. From interpretation of the slight curvature of the isochores at lower temperatures as the effect of a temperature dependent hard sphere diameter, a method may be developed to determine such values for d(T) from p-v-T data. In addition to that, g(d) is calcu-lated by the well-known hard sphere radial distribution function and μ_o from the well-known dilute gas viscosities of hard spheres. Computed viscosities for the heavy noble gases agree well with experimental data at high temperatures and high densities. For densities below the critical, the discrepancies become large. This is an expected result, as the hard sphere model appears to be realistic only for high densities and high temperatures. For more complicated fluids, this ap-proach gives less satisfactory results. Typical errors are between 10% at high densities and 40% at zero density [63]. Another method to determine the hard sphere diameter is fitting viscosity data at high densities and different temperatures to the theoretical Enskog curve. For monatomic substances, such d values agree with those from p-v-T data [64]. For simple polyatomic fluids, the viscosity is also predicted quite satisfactorily over a wide range of states by this method [63]. Another way to apply Enskog's theory to real fluids uses p-v-T data to determine an empiri-cal expression for the equilibrium radial distribution function

$$g(d) = \frac{1}{bn}\left[\frac{1}{nk}\left(\frac{\partial p}{\partial T}\right)_v - 1\right] \tag{10}$$

When d is chosen to fit the experimental dilute gas viscosity, reasonable agreement is obtained with experimental data for simple gases, for densities up to and somewhat above the critical, pro-vided that the temperature is sufficiently above the critical temperature [63,65].

The Enskog approach has been extended to a fluid of molecules interacting with a square-well potential [66-69]. This potential model takes attractive forces into consideration and at the same time retains the desirable feature of repulsive forces, now at two distinct separations. Again molecular chaos for the momenta is assumed. The resulting expression for viscosity is

$$\mu = \frac{5}{16\sigma^2}\left(\frac{mkT}{\pi}\right)^{1/2}\left\{\frac{\{1 + (2bn/5)[g(\sigma) + R^3 g(R\sigma)A]\}^2}{g(\sigma) + R^2 g(R\sigma)[B + (1/6)(\varepsilon/kT)^2]} + \frac{48}{25\pi}(bn)^2[g(\sigma) + R^4 g(R\sigma)B]\right\} \tag{11}$$

with

$$A = 1 - e^{-\varepsilon/kT} + \frac{\varepsilon}{2kT}\left\{1 + \frac{4}{\sqrt{\pi}} e^{\varepsilon/kT} \int_{(\varepsilon/kT)^{1/2}}^{\infty} e^{(-x^2)} x^2 dx\right\} \tag{12}$$

and

$$B = e^{\varepsilon/kT} - \frac{\varepsilon}{2kT} - 2 \int_0^\infty x^2 \left(x^2 + \frac{\varepsilon}{kT} \right)^{1/2} e^{-x^2} dx \tag{13}$$

Here m is the molecular mass, σ the distance for the hard sphere collisions, $R\sigma$ is the effective attraction distance, g the equilibrium radial distribution function, ε the depth of the well and R its extension. Comparison with experiment by using an accurate radial distribution function and potential parameters from second virial coefficient data gives a fair prediction of the density and temperature dependence of liquid viscosity. However, the absolute values may be substantially in error [70,71]. Generally, the square-well model is a significant improvement over the hard-sphere theory with a temperature dependent diameter which breaks down in the liquid region.

Another rigorous approach, which appears to be promising is that of perturbation theory. It has been quite successful for equilibrium properties. An application to transport properties is found in [72].

THEORY OF CORRESPONDING STATES

The formal derivation of the Theory of Corresponding States for the transport properties starts from the time-correlation function formulae [73]. The main assumption concerns the inter-molecular potential energy which is assumed to be of the form

$$u = f(r_{R1}, r_{R2}, \cdots r_{RN}) \tag{14}$$

where

$$r_R = r_i / \sigma$$

ε and σ are molecular parameters with the dimensions of energy and distance, respectively, and f is a universal function for all fluids considered. Equation (14) implies that the total potential energy may be characterized by two parameters only and that it is independent of molecular orientations. Furthermore, it is assumed that quantum effects can be neglected and that intra-molecular energy effects have no importance. Using the time-correlation function formula Eq. (6) for viscosity, both the canonical ensemble distribution function as well as the streaming operator governing the time dependence may be written in dimensionless form. The resulting equation for viscosity is

$$\mu_R = \mu_R(T_R, v_R) \tag{15}$$

with

$$\mu_R = \frac{\mu \sigma^2}{(m\varepsilon)^{1/2}} \tag{16}$$

$$T_R = \frac{kT}{\varepsilon} \tag{17}$$

$$v_R = \frac{v}{N\sigma^3} \tag{18}$$

As may be anticipated from the assumptions made, this simple formulation of the Corresponding States Principle works well for the heavy noble gases in all fluid states. Deviations are however encountered even for simple polyatomic fluids like O_2 and CH_4 [74,75], in the region of liquid densities. There, the transport properties are apparently more sensitive to the details of the intermolecular potential than equilibrium properties. Under the assumptions made, the potential parameters can be expressed in terms of the critical values of the fluid parameters, leading to the following formulation of the corresponding states principle for viscosity:

$$\mu\xi = f(T_N, p_N) \tag{19}$$

with

$$\xi = \frac{(T_{cr}R)^{1/6} N_L^{1/3}}{M^{1/2} p_{cr}^{2/3}} \tag{20}$$

$$T_N = \frac{T}{T_{cr}} \tag{21}$$

$$p_N = \frac{p}{p_{cr}} \tag{22}$$

where the subscript cr indicates the critical point. Such an equation has been extensively compared with data for dense gaseous viscosities.

Apart from gases with quantum effects, long chains, and strong dipole moments, this expression was found to predict gas viscosities with an uncertainty of about 6% [76]. Several extensions of the simple two parameter law of corresponding states have been proposed. Along the line of the statistical derivations, additional parameters taking into account specific molecular properties have been included. Such additional parameters have been used to take into account quantum effects [77], hindered rotation of the molecules [75], electrostatic properties [78], etc. An alternative concept of the Corresponding States Principle, which has an entirely phenomonological background, has also been developed [79]. Correspondence is found along single isotherms which are selected by the universal definition of characteristic points in the state diagram of a property, e.g., viscosity. Characteristic points can be defined along the critical isobar of viscosity at any desired temperature. The correspondence of different fluids along an isotherm through a characteristic point is remarkable. In fact, fluids like water, hydrogen and argon are found to fall on one line in a considerable region of states, which has not been found to apply for any other formulation of the Corresponding States Principle before.

MODEL THEORIES

Much work has been done on model theories for equilibrium properties, but relatively little on transport properties. The few important model theories of momentum transport in liquids may roughly be classified as activated state theories, free volume theories and combinations of these two. The model theories do not aim at giving a rigorous statistical description of the molecular momentum transport, but instead use a highly simplified and intuitive model of the process. As such, they lead to quite flexible formal structures for correlating viscosity data and predicting them from parameters which can be given a physical interpretation. Any conclusions derived from the degree of agreement with experimental data to the physical justification of the model should, however, be looked upon with caution. Instead, these semitheoretical approaches should be considered as a starting point to correlate viscosity data and predict them from other properties especially for complicated liquids.

Activated State Theories

Eyring has formulated a model theory for viscosity in liquids by the application of his reaction rate theory for chemical reactions to momentum transport [80-84]. According to this procedure, molecules are supposed to move within parallel liquid layers, where they change their sites by surpassing a certain energy barrier. The amount of this energy barrier is called activation energy. This is the energy that a molecule must acquire before it can jump over the potential barrier into another state. According to the reaction rate theory, this is a rate, i.e., the number of molecules crossing the energy barrier in unit time, and is given by the expression

$$r = \chi \left(\frac{kT}{h}\right) \frac{q^*}{q} e^{-E_o/kT} \tag{23}$$

Here E_o is the activation energy of a molecule. The quantities q^* and q are the partition functions per unit volume of molecules in the activated and the initial state, respectively. The only difference between these two partition functions is the fact that there is one translational degree of freedom less in the activated state than in the normal state [85]. Finally, χ is a transmission coefficient, which can be interpreted as the probability, for a molecule in the activated state to cross the potential barrier and jump into a new state. Since in the case of viscous flow, there is a shearing force acting on the molecules, the value of the potential barrier, i.e., the activation energy, will be diminished in the direction of the shear stress and will be increased in the opposite direction. As a result, there will be two different flow rates of molecules, giving a resulting difference of velocity between two adjacent liquid layers. With this difference in velocity and using Newton's law, one obtains

$$\mu = \frac{hN_L}{v}\left(\frac{q}{q^*}\right)\frac{1}{\chi} e^{-E_o/kT} \tag{24}$$

From this general form, a lot of practical viscosity equations were derived by making further simplifying assumptions. If one considers the partition function to be that of a single molecule in a free volume V_f, for which a simple expression is used, one obtains Eyring's equation

$$\mu = \left(\frac{N_L}{v}\right)^{2/3} \frac{2RT}{\chi\Delta E_v}(2\pi mkT)^{1/2} e^{E_o/kT} \tag{25}$$

Here E_o is the activation energy per mole and is frequently connected to the energy of vaporization ΔE_v. A similar equation has been derived by Weyman in a more statistical mechanical way [86,87].

Another viscosity equation is obtained by Eyring through the introduction of the chemical equilibrium constant. Assuming furthermore the transmission coefficient to be $\chi = 1$, one gets

$$\mu = \left(\frac{hN_L}{v}\right) e^{\Delta E_v / 2.45 RT} \tag{26}$$

Gold and Ogle [88] have tested this equation along with the similar Andrade equation and found deviations from experimental values between 50% and 100%. Clearly, this approach gives only an order of magnitude estimation.

Another equation for viscosity using the activated state model has been given by McLaughlin [89]. The liquid is considered to have a lattice structure, the transport phenomena being due to the existence of holes. The molecules in this model are surrounded by a spherical force field which is described by the Lennard-Jones potential. An equation for the self-diffusion coefficient as a function of the number of holes n, the distance of a molecule 'a' to a hole, and the rate r_o at which molecules attain the energy necessary to surpass the energy barrier is given by

$$D = nr_o(a^2/N_L) \tag{27}$$

This rate r_o is taken from the activated states theory of reaction rates. The ratio of the number of holes to the total number of molecules is given by the usual Boltzmann factor equation. The modified Stokes-Einstein relation between viscosity and self-diffusion coefficient is used to replace the diffusion coefficient by viscosity. For the free volume V_f which enters through the ratio of the partition functions in the activated states theory, the lattice model theory of Lennard-Jones and Devonshire [90] is invoked

$$V_f = 2\pi \sqrt{2G} \frac{v}{N_L} \tag{28}$$

with G being a tabulated function of temperature and volume. The energy barrier is considered to be a fraction R of the energy of the lattice system in which all molecules are in their equilibrium positions. This relation is

$$E_o = R(N_L)12\varepsilon \left[1.2045 \left(\frac{v_o}{v}\right)^2 - 0.5055 \left(\frac{v_o}{v}\right)^4 \right] \tag{29}$$

with ε as the Lennard-Jones potential energy parameter and $v_o = \sigma^3 N_L$, with σ as the Lennard-Jones potential distance parameter. The resulting final equation for viscosity is

$$\mu = \frac{(2\pi mkT)^{1/2}}{2\pi\sigma a^2} V_f^{1/3} e^{-\frac{\varepsilon \left(\frac{v_o}{v}\right)^4}{kT}} e^{\frac{E_o}{RT}} \tag{30}$$

with

$$a = \left[\sqrt{2} \frac{v}{N_L} \right]^{1/3} \tag{31}$$

from geometric considerations. The quantity R is defined by a temperature dependent relation (Eq. 32) which is supposed to be universally valid,

$$12R = 3.559 \left(\frac{T}{T_{cr}}\right) + 2.023 \tag{32}$$

Comparison of the predicted values from Eq. (30) with experimental data gives considerably better agreement in the case of some liquids between the melting point and the boiling point. Typical errors are between 10% and 20%, with larger errors for methane.

There is another equation due to Majumdar [91], which has a similar structure to that of McLaughlin [89]. The partition function is now replaced by that for the Barker tunnel model [92], which consists of a two-dimensional lattice of tunnels in which molecules are confined to positions anywhere in the tunnel. This partition function is given as

$$q = \left(\frac{2\pi mkT}{h^2}\right)^{3/2} e^{-F_1/kT} \; e^{-\frac{1}{2}\left(\frac{\phi_o}{N_L kT}\right)} A_f \tag{33}$$

Here F_1 is the free energy of a molecule, ϕ_o the energy of the system with all molecules in their equilibrium positions which is determined by the Lennard-Jones potential. The quantity A_f is a free area, tabulated as a function of temperature and volume. The final equation for the viscosity is

$$\mu = \frac{(2\pi mkT)^{1/2}}{2\pi\sigma a^2} A_f^{1/2} \; e^{\frac{\theta E_s}{kT}\left[\frac{v_s}{v} - \frac{1}{2n}\left(\frac{v_s}{v}\right)^3\right]} e^{\frac{\varepsilon}{kT}\left(\frac{v_o}{v}\right)^4} \tag{34}$$

Here v_s is the molar volume of the solid at the melting point, E_s the average potential energy of the molecules in their equilibrium positions in the solid, n is a constant and θ is a temperature function unique to the substance.

A further model based on the activated state theory is the well known liquid structure theory by Eyring et al. [93-97]. In this picture of a liquid, solid-like and gas-like structures are significant, leading to the following general expression for the viscosity.

$$\mu = \mu_s \frac{v_s}{v} + \mu_g\left(\frac{v - v_s}{v}\right) \tag{35}$$

where μ_s is the contribution of molecules with solid-like properties and μ_g the contributions due to molecules with gas-like properties. The contribution of the gas-like molecules to the viscosity of the liquid is generally considered to be given by the expression for a dilute system of hard spheres. To set up an equation for μ_s, the earlier model of the activated state in modified form is invoked, changes of sites now not being restricted any more to parallel liquid layers. Other relations have been derived from Eq. (35) using different partition functions to calculate the solid-like contribution. If one chooses a Lennard-Jones-Devonshire partition function, and considers the activation energy ϕ_o where all molecules are at their lattice sites and inversely proportional to the number of holes, the final viscosity equation reads

$$\mu = \frac{(\pi mkT)^{1/2}}{(v - v_s)}\left(\frac{6N_L V_f^{1/3}}{\chi Z}\right) e^{-\frac{a'v_s}{v - v_s}\left(\frac{Z}{2}\right)\frac{\phi_o}{kT}} + \frac{v - v_s}{v} \frac{5}{16d^2}\left(\frac{mkT}{\pi}\right)^{1/2} \tag{36}$$

Here V_f is the free volume, a' is a constant, Z is the number of nearest neighbours and d the hard sphere diameter. A test of the equation for hard sphere systems, which introduces some simplifications, reveals reasonable agreement without using data for liquid viscosity. Treating Eq. (36) as a formal structure with two empirical parameters, Hogenboom et al. [98] report good agreement with experimental viscosities in a temperature region of about 100 K. Eyring et al. [99] introduced an explicit expression for the intermolecular potential and introduced the pressure dependence through the dependence of v_s on pressure using the expression

$$v_s = v_{so}(1 - \beta \Delta p) \tag{37}$$

where v_{so} is the value of v_s at one bar pressure, β the compressibility of the solid-like molecules, and Δp the pressure difference between the actual pressure and one bar. Reasonable agreement is reported with experimental viscosities for four hydrocarbons up to 3600 bars. Eicher et al. [100] use a temperature dependent hard sphere diameter, fitted to dilute gas viscosities, and a Lennard-Jones potential. They furthermore treat the constants in the equation as adjustable parameters and then obtain good agreement with experiment for temperatures between the melting point and the boiling point. Further applications of the significant liquid structure theory to the calculation of viscosity may be found in [71,101-108].

Free Volume Theories

The term *Free Volume Theory* for transport processes should not be confused with its usual use for lattice or hole theories of equilibrium properties. Here it has a somewhat less rigorous significance. Batischinski [109] has given an empirical equation, which correlates the fluidity, i.e., the inverse of viscosity, with the molal volume of the liquid and a limiting volume v_o according to

$$\phi = \frac{1}{\mu} = A(v - v_o) \tag{38}$$

The difference $(v - v_o)$ is called *free volume*. The limiting volume v_o, which may be obtained by extrapolation of the linear plot to zero fluidity, has a value between solid and liquid volume at the melting point. Hildebrand [100-114] has investigated and discussed this correlation in considerable detail. He has verified its validity for several substances over a wide range of thermodynamic states. Correlations between the substance specific parameters and molecular mass have been found. Critical discussions, pointing out the limitations of the above correlation, by comparison with experimental data, have been made in [115-117]. Another empirical viscosity equation using the free volume concept was given by Doolittle [118-120]:

$$\mu = A \exp B/(v_f/v_o) \tag{39}$$

where A and B are parameters specific to the substance, v_f a free volume and v_o a reference volume, which is to be taken as the specific volume at absolute zero obtained by extrapolation. The connection to the actual experimental volume is given by

$$\frac{v_f}{v_o} = \frac{v - v_o}{v_o} \tag{40}$$

This equation has been modified in various ways in order to increase the region of validity [121-123]. For liquids, where density is a linear function of temperature, the following equation holds [121]

$$\ln \mu = A + \frac{B}{T - T_o} \qquad (41)$$

where T_o is a characteristic reference temperature to which some have attributed a fundamental significance. It is between absolute zero and the conventional glass-transition temperature and represents the limit in temperature under which there is only packing free volume and thus infinite viscosity. In [122], the pressure dependence of free volume is taken into account, yielding

$$\frac{f(p)}{[1 - a_s(T - T_o)]} = \frac{v^p(\log \mu - A)}{v_o(B + \log \mu - A)} \qquad (42)$$

where $f(p)$ is a function of pressure. A theoretical interpretation of the Doolittle equation has been given by Cohen and Turnball [124-126], who derived an equation for the self-diffusion coefficient under the assumption, that the diffusion process is determined by a statistical redistribution of the free volume.

Combined Model Theories

Macedo and Litovitz [127] have derived a hybrid equation for viscosity in which both mechanisms, the effect of the activation energy as well as that of the redistribution of the free volume are assumed to be simultaneously present. Viscosity is visualized to be inversely proportional to the product of two probabilities, namely, the probability of a molecule to surpass the potential barrier and the probability to find an adequate free volume. The final equation is

$$\mu = A \, e^{\left[\gamma \left(\frac{v_o}{v - v_o} \right) + \frac{E_o}{RT} \right]} \qquad (43)$$

where γ takes into consideration the overlap of free volumes and A is a substance specific parameter. A statistical derivation of this equation is given in [128]. In an empirical test, A, v_o and E_o are treated as adjustable parameters, and γ is set equal to unity. An indication that the activation energy must be taken as a function of density is given in [129].

Gubbins and Tham [130] have improved the Macedo-Litovitz-equation by taking into account a temperature dependence of v_o and a density dependence of E_o. Corresponding states relationships are given for both quantities.

EMPIRICAL ESTIMATION AND CORRELATION TECHNIQUES

INTRODUCTION

Under this heading, we wish to summarize some of those methods, which are basically empirical in origin and correlate viscosity data in terms of empirical curve-fitted equations in terms of molecular structures and in terms of different fluid properties.

EMPIRICAL CORRELATION EQUATIONS

The simplest representation of viscosity over a wide region of states is unquestionably achieved in terms of temperature and density. It is especially advantageous to plot the residual

viscosity, defined as the viscosity at some specified temperature and density minus its value at
the same temperature at zero density, against density. If the requirement of accuracy is not
stringent, it has been shown in a number of cases that the residual viscosity is only a function
of density over a large region of states

$$\mu(T,\rho) - \mu_o(T) = f(\rho) \tag{44}$$

This residual viscosity concept, which appears to have been developed from an analogous phenomenon
for thermal conductivity [131], has been discussed and used extensively for interpolation and extrapolation purposes [132-137]. In fact, it has been used for several substances in this volume
in order to interpolate and extrapolate a given data set. In principle, given the validity of
Eq. (44), one isotherm in the dense fluid region is sufficient, in combination with dilute gas
viscosities as a function of temperature, to obtain data for all fluid states for which p-v-T data
can be found. As p-v-T data are generally more readily available than dense fluid viscosity data,
this concept is extremely valuable in generating approximate viscosity values. Having stressed
the significance of the residual viscosity concept, its limitations must also be fully recognized.
It has become quite obvious that generally the concept is only approximately valid breaking down
especially at high densities and for wide temperature regions [138-139]. For low densities, too,
there is a temperature dependence, which however is not too important for practical purposes,
since here the dilute gas value is dominant anyhow. For some fluids, the residual viscosity concept is invalid in all region of states, a well-known example being p-hydrogen [140], and possibly
some other low boiling fluids as well.

In view of these factors, Eq. (44) has been supplemented by temperature dependent terms for
a more accurate representation of viscosity over a large region of states. Thus, a viscosity
equation for methane by [141] has the form

$$\mu(\rho,T) = \mu_o(T) + \mu_1(T)\rho + \Delta\mu(\rho,T) \tag{45}$$

where $\mu_o(T)$ is the dilute gas viscosity, $\mu_1(T)$ the first density correction and $\Delta\mu(\rho,T)$ a remainder.
The explicit equation has 12 empirical constants and fits the experimental data in a large region
of states within a few percent, well within the accuracy of the data. Additional empirical viscosity equations, containing a temperature dependent residual viscosity, are found in the literature
[142,143]. A discussion of the residual viscosity concept, based on statistical mechanics, is
given in [144].

The representation of viscosity in terms of density and temperature, although useful for
correlation purposes, is not very convenient in practical application. One needs in addition a
thermodynamic equation of state, from which density data have to be extracted iteratively for
given values of temperature and pressure which are the more widely used parameters. It would
therefore be highly desirable to have viscosity equations in terms of temperature and pressure.
An explicit equation for viscosity in the whole fluid range, as a function of temperature and
pressure, will not be possible because of infinite gradients at the critical point [135]. One is
therefore led to subdivide the total fluid region into various subregions such as the dilute gas,
the dense gas, and two liquid regions, one close to the critical temperature, another one at lower
temperatures. With the exception of the liquid region close to the critical temperature [145],

simple explicit equations may be given for the calculation of viscosity giving a representation
within experimental error [135,146,147]. A single, albeit implicit equation, for viscosity covering the whole fluid region in terms of temperature and pressure has yet to be given. Such equations of the form $p = p(T,\mu)$ have so far been established only for limited regions of states
[145,146].

GROUP CONTRIBUTION METHODS

Group contribution methods are capable of making approximate predictions of properties for
which only the molecular structure may be known. Since many molecules are made up of a number of
structural groups, this results in a considerable contraction of the formalism, and the potential
of group contribution methods for making predictions for large numbers of molecular systems is
great. In the recent literature on thermodynamic properties, these group contribution methods
have become quite popular [148,149] as a means for interpolating and extrapolating given data.

This method has also been applied to the viscosity of liquids [150-154]. A well-known
relationship, called Souder's method, gives the liquid viscosity as a function of density and a
single constant to be calculated from atomic and structural data [150]. In another investigation
[153], the two constants in the Andrade equation were correlated qualitatively with molecular
structure. Yet another equation uses density as well as temperature and one structural group constant to calculate liquid viscosity [151]. Further investigation on the structural dependence of
liquid viscosity for the higher hydrocarbons are reported in [152]. The predictions by a new
method for calculating viscosities of organic compounds based on their dependence on chemical constitution and structure were compared with an extensive body of experimental data and the agreement
appears to be quite good [154]. In all these works, valuable results of a qualitative nature have
been found. Generally an increase of molecular mass appears to increase the value of liquid viscosity and its pressure dependence. The effect of nondeformable structure, double and triple bonds,
branching and further structural characteristics of molecules on viscosity have been discussed in
the various references cited above.

CORRELATION OF VISCOSITY WITH VARIOUS THERMODYNAMIC AND TRANSPORT PROPERTIES

In numerous correlations of viscosity use is made of interrelationships between this quantity
and various thermodynamic and transport properties.

Among relations which relate viscosity to thermodynamic quantities is the well-known example
connecting the viscous energy in Eyring's reaction rate expression for liquid viscosity and the
internal energy of vaporization. This method of calculating viscosity is extended [155] to a
correlation between viscosity, specific volume and molal entropy of vaporization. An example for
the connection of liquid viscosity to the sonic velocity is given in [156], while the viscosity of
high pressure steam is correlated to the compressibility factor in [157]. In [158,159] relations
between viscosity and the isenthalpic Joule-Thomson coefficient are discussed theoretically. The
viscosity of liquid mixtures is correlated with thermodynamic excess quantities in [160].

Relations between viscosity and other transport properties may be rigorously found from
kinetic theory [2,3]. Such correlations include the Maxwell relation between viscosity and thermal

conductivity of a dilute monatomic gas, as well as the interrelations between viscosity, self-diffusion and isotopic thermal diffusion and between viscosity and diffusion coefficients of gaseous mixtures, discussed in [161,162]. A relation between the self-diffusion coefficient and viscosity for liquids is explored in [163]. An interrelation between diffusivity and solvent viscosity in dilute liquid solutions can be found in [164]. The universality of such relationships between the various transport properties is subject to the restrictions of irreversible thermodynamics [165], according to which transport properties belonging to fluxes of different tensorial rank do not generally interrelate.

REFERENCES TO TEXT

1. Childs, G.E. and Hanley, H.J.M., "Applicability of Dilute Gas Transport Property Tables to Real Gases," Cryogenics, 8, 94-7, 1968.

2. Hirschfelder, J.O., Curtiss, C.F., and Bird, R.B., "Molecular Theory of Gases and Liquids," John Wiley and Sons, Inc., 1964.

3. Chapman, S. and Cowling, T.G., "Mathematical Theory of Non-Uniform Gases," Cambridge University Press, 3rd Ed., 1970.

4. Gubbins, K.E., "Thermal Transport Coefficients for Simple Dense Fluids," in Statistical Mechanics-Volume 1, (Singer, K., Editor), Specialist Periodical Reports, The Chemical Society, 194-253, 1973.

5. Rice, S.A., Boon, J.P., and Davis, H.T., "Comments on the Experimental and Theoretical Study of Transport Phenomena in Simple Liquids," in Simple Dense Fluids (Frisch, H.L. and Salsburg, Z.W., Editors), Academic Press, 251-402, 1968.

6. Mazo, R.M., "Statistical Mechanical Theories of Transport Properties," in The International Encyclopedia of Physical Chemistry and Chemical Physics, Topic 9, Transport Phenomena, Pergamon Press, 1967.

7. Rice, S.A. and Gray, P., "The Statistical Mechanics of Simple Liquids," Interscience, 1965.

8. Steele, W.A., "Time-Correlations Functions" in Transport Phenomena in Fluids (Hanley, H.J.M., Editor), Dekker, 209-312, 1969.

9. Ernst, M.H., Haines, L.K., and Dorfman, J.R., "Theory of Transport Coefficients for Moderately Dense Gases," Rev. Mod. Phys., 41, 296-316, 1969.

10. Brush, S.G., "Theories of Liquid Viscosity," University of California, Lawrence Radiation Lab., Rept. No. UCRL-6400, 106 pp., 1961.

11. Reid, R.C., Prausnitz, J.M., and Sherwood, T.K., "The Properties of Gases and Liquids," McGraw Hill Book Co., Inc., New York, NY, 1977.

12. Liboff, R.L., "Introduction to the Theory of Kinetic Equations," Wiley, 1966.

13. Kirkwood, J.G., "The Statistical Mechanical Theory of Transport Processes. I. General Theory," J. Chem. Phys., 14, 180-201, 1946.

14. Chandrasekhar, S., "Stochastic Problems in Physics and Astronomy," Rev. Mod. Phys., 15, 1-89, 1943.

15. Wang, M.G. and Uhlenbeck, G.E., "On the Theory of the Brownian Motion II," Rev. Mod. Phys., 17, 323-42, 1945.

16. Kirkwood, J.G., Buff, F.P., and Green, M.S., "The Statistical Mechanical Theorie of Transport Processes. III. The Coefficients of Shear and Bulk Viscosity of Liquids," J. Chem. Phys., 17, 988-94, 1949.

17. Lebowitz, J. and Rubin, E., "Dynamical Study of Brownian Motion," Phys. Rev., 131, 2381-96, 1963.

18. Ross, J., "Statistical Mechanical Theory of Transport Processes. IX. Contribution to the Theory of Brownian Motion," J. Chem. Phys., 24, 375-80, 1956.

19. Zwanzig, R.W., Kirkwood, J.G., Stripp, K.F., and Oppenheim, I., "The Statistical Mechanical Theory of Transport Processes. VI. A Calculation of the Coefficients of the Shear and Bulk Viscosity of Liquids," J. Chem. Phys., 21, 2050-5, 1953.

20. Rice, S.A. and Allnatt, A.R., "On the Kinetic Theory of Dense Fluids. VI. Singlet Distribution Function for Rigid Spheres with an Attractive Potential," J. Chem. Phys., 34, 2144-55, 1961.

21. Allnatt, A.R. and Rice, S.A., "On the Kinetic Theory of Dense Fluids. VII. The Doublet Distribution Function for Rigid Spheres with an Attractive Potential," J. Chem. Phys., 34, 2156-65, 1961.

22. Schrodt, I.B., Kie, J.S., and Luks, K.D., "A Comparative Analysis of the Rice-Allnatt and Square-Well Liquid Transport Theories," Phys. Chem. Liquids, 2, 147-63, 1971.

23. Kie, J.S. and Luks, K.D., "Some Comments on the Rice-Allnatt Transport Theory," J. Phys. Chem., 76, 2133-7, 1972.

24. Collings, A.F. and Woolf, L.A., "Recalculation of the Friction Constant and Transport Coefficients of Liquid Argon from the Rice-Allnatt Theory," Aust. J. Chem., 24, 225-35, 1971.

25. Misguich, J. and Nicolis, G., "Generalized Rice-Allnatt Theory for Transport in Liquids," Mol. Phys., 24, 309-34, 1972.

26. Wei, C.C. and Davis, H.T., "Kinetic Theory of Dense Fluid Mixtures. II. Solution to the Singlet Distribution Functions for the Rice-Allnatt Model," J. Chem. Phys., 45, 2533-44, 1966.

27. Allen, P.M. and Cole, G.H.A., "Dense Fluid Non-Equilibria and the Prigogine Theory," Mol. Phys., 14, 413-24, 1968.

28. Allen, P.M. and Cole, G.H.A., "Singlet Distribution for Dense Fluid Non-Equilibria," Mol. Phys., 15, 549-55, 1968.

29. Misguich, J., "Kinetic Theory and Evaluation of the Coefficients of Thermal Transport in Liquids and Dense Systems," J. de Physique, 30, 221-42, 1969.

30. Allen, P.M. and Cole, G.H.A., "Pair Distribution and Dense Fluid Non-Equilibria," Mol. Phys., 15, 557-65, 1968.

31. Allen, P.M., "General Transport Equations in Dense Fluids," Physica, 52, 237-45, 1971.

32. Palyros, J.A., Davis, H.T., Misguich, J., and Nicolis, G., "Application of the Prigogine-Nicolis-Misguich Theory of Transport," J. Chem. Phys., 49, 4088-95, 1968.

33. Cowan, J.A. and Ball, R.N., "Temperature Dependence of Bulk Viscosity in Liquid Argon," Can. J. Phys., 50, 1881-6, 1972.

34. Rice, S.A. and Kirkwood, J.G., "On an Approximate Theory of Transport in Dense Media," J. Chem. Phys., 31, 901-8, 1959.

35. Prigogine, I., Nicolis, G., and Misguich, J., "Local Equilibrium Approach to Transport Processes in Dense Media," J. Chem. Phys., 43, 4516-21, 1965.

36. Forster, M.J. and Cole, G.H.A., "Singlet and Doublet Distribution for Classical Fluid Non-Equilibria," Mol. Phys., 20, 417-32, 1971.

37. Green, M.S., "Markoff Random Processes and the Statistical Mechanics of Time-Dependent Phenomena. II. Irreversible Processes in Fluids," J. Chem. Phys., 22, 398-413, 1954.

38. Mori, H., "Statistical-Mechanical Theory of Transport in Fluids," Phys. Rev., 112, 1829-42, 1958.

39. Kirkwood, J.G. and Fitts, D.D., "Statistical Mechanics of Transport Processes. XIV. Linear relations in Multi-Component Systems," J. Chem. Phys., 33, 1317-24, 1960.

40. Zwanzig, R., "Time Correlation in Functions and Transport Coefficients in Statistical Mechanics," Ann. Rev. Phys. Chem., 16, 67-102, 1965.

41. Kadanoff, L. and Swift, J., "Transport Coefficients Near the Liquid-Gas Critical Point," Phys. Rev., 166, 89-101, 1968.

42. Kawasaki, K., "Kinetic Equations and Time Correlation Functions of Critical Fluctuations," Ann. Phys., 61, 1-56, 1970.

43. Mori, H., "Time-Correlation Functions in the Statistical Mechanics of Transport Processes," Phys. Rev., 111, 694-706, 1958.

44. Wainwright, T., "Calculation of Hard-Sphere Viscosity by Means of Correlation Functions," J. Chem. Phys., 40, 2932-7, 1964.

45. Résibois, P., "On the Connection Between the Kinetic Approach and the Correlation-Function Method for Thermal Transport Coefficients," J. Chem. Phys., 41, 2979-92, 1964.

46. Gosling, E.M., "On the Calculation by Molecular Dynamics of the Shear Viscosity of a Simple Fluid," Mol. Phys., 26, 1475-84, 1973.

47. Forster, D., Martin, P.C., and Yip, S., "Moments of the Momentum Density Correlation Functions in Simple Liquids," Phys. Rev., 170, 155-9, 1968.

48. Forster, D., Martin, P.C., and Yip, S., "Moment Method Approximation for the Viscosity of Simple Liquids. Application to Argon," Phys. Rev., 170, 160-3, 1968.

49. Berne, J., Boon, J.P., and Rice, S.A., "On the Calculation of Autocorrelation Functions of Dynamical Variables," J. Chem. Phys., 45, 1086-96, 1966.

50. Boon, J.P., Berne, J., and Rice, S.A., "Reformulation of the Representation of Transport Coefficients Using the Autocorrelation-Function Formalism and the Linear Trajectory Approximation," J. Chem. Phys., 47, 2283-91, 1967.

51. Lucas, K. and Moser, B., "A Memory-Function Approach for the Viscosity of Simple Dense Fluids," Mol. Phys., 1979 (in press).

52. Zwanzig, R., "Method for Finding the Density Expansion of Transport Coefficients of Gases," Phys. Rev., 129, 486-94, 1963.

53. Ernst, M.H., Dorfman, J.R., and Cohen, E.G.D., "Transport Coefficients in Dense Gases. I. The Dilute and Moderately Dense Gas," Physica, 31, 493-521, 1965.

54. Kawasaki, K. and Oppenheim, I., "Correlation-Function Method for the Transport Coefficients of Dense Gases. I. First Density Correlation to the Shear Viscosity," Phys. Rev., 136A, 1519-34, 1964.

55. Weinstock, J., "Cluster Formulation of the Exact Equation for the Evolution of a Classical Many-Body System," Phys. Rev., 132, 454-69, 1963.

56. Ernst, M., Haines, L., and Dorfman, J.R., "Theory of Transport Coefficients for Moderately Dense Gases," Rev. Mod. Phys., 41, 296-316, 1969.

57. Sengers, J.V., Ernst, M., and Gillespie, D.T., "Three-Particle Collision Integrals for a Gas of Hard Spheres," J. Chem. Phys., 56, 5583-601, 1972.

58. Alder, B.J. and Wainwright, T.E., "Velocity Autocorrelations for Hard Spheres," Phys. Rev. Lett., 18, 988-90, 1967.

59. Alder, B.J., Gass, D.M., and Wainwright, T.E., "Studies in Molecular Dynamics. VIII. The Transport Coefficients for a Hard-Sphere Fluid," J. Chem. Phys., 53, 3813-26, 1970.

60. Henline, W.D. and Condiff, D.W., "Kinetic Theory of Moderately Dense Gases," J. Chem. Phys., 54, 5346-65, 1971.

61. Dymond, J.H. and Alder, B.J., "Van der Waals Theory of Transport in Dense Fluids," J. Chem. Phys., 45, 2061-8, 1966.

62. Carnahan, N.F. and Starling, K.E., "Equation of State for Nonattracting Rigid Spheres," J. Chem. Phys., 51, 635-6, 1969.

63. Lucas, K. and Ackmann, G., Unpublished Results.

64. Dymond, J.H., "Transport Properties in Dense Fluids," Proceedings of the Sixth ASME Symposium on Thermophysical Properties, Atlanta, GA, 143-7, 1973.

65. Hanley, J.H.M., McCarthy, R.D., and Cohen, E.G.D., "Analysis of the Transport Coefficients for Simple Dense Fluids. Application of the Modified Enskog Theory," Physica, 60, 322-56, 1972.

66. Longue-Higgins, H.C. and Valleau, J.P., "Transport Coefficients of Dense Fluids of Molecules Interacting According to a Square Well Potential," Mol. Phys., 1, 284-94, 1958.

67. Davis, H.T., Rice, S.A., and Sengers, J.V., "On the Kinetic Theory of Dense Fluids. IX. The Fluid of Rigid Spheres with a Square-Well Attraction," J. Chem. Phys., 35, 2210-33, 1961.

68. Davis, H.T. and Luks, K.D., "Transport Properties of a Dense Fluid of Molecular Interacting with a Square-Well Potential," J. Phys. Chem., 69, 869-80, 1965.

69. Luks, K.D., Miller, M.A., and Davis, H.T., "Transport Properties of a Dense Fluid of Molecules Interacting with a Square-Well Potential, Part II," AIChE J., 12, 1079-86, 1966.

70. Reed, T.M. and Gubbins, K.E., "Applied Statistical Mechanics," McGraw-Hill, New York, NY, 1973.

71. Herreman, W. and Grevendonk, W., "An Experimental Study on the Shear Viscosity of Liquid Neon," Cryogenics, 14, 395-8, 1974.

72. Mo, K.C., Gubbins, K.E., and Dufty, J.W., "Perturbation Theory for Dense Fluid Transport Properties," Proceedings of the Sixth ASME Symposium on Thermophysical Properties, Atlanta, GA, 158-67, 1973.

73. Helfand, E. and Rice, S.A., "Principle of Correspondings States for Transport Properties," J. Chem. Phys., 32, 1642-4, 1960.

74. Tham, M.J. and Gubbins, K.E., "Correspondence Principle for Transport Properties of Dense Fluids," Ind. Eng. Chem., Fundam., 8, 791-5, 1969.

75. Tham, M.J. and Gubbins, K.E., "Correspondence Principle for Transport Properties of Dense Fluids. Nonpolar Polyatomic Fluids," Ind. Eng. Chem., Fundam., 9, 63-70, 1970.

76. Lucas, K., "A Simple Technique for the Calculation of the Viscosity of Gases and Gas Mixtures," Chem.-Ing. Tech., 46, p. 157, 1974.

77. Rogers, J.D. and Brickwedde, F.G., "Comparison of Saturated Liquid Viscosities of Low Molecular Substances According to the Quantum Principle of Corresponding States," Physica, 32, 1001-8, 1966.

78. Stiel, L.I. and Thodos, G., "The Viscosity of Polar Gases at Normal Pressures," AIChE J., 8, 229-32, 1962.

79. Lucas, K., "Characteristic Points on Thermodynamic Surfaces and Their Use to Produce General Correlations Between State Quantities," Proceedings of the ASME Sixth Symposium on Thermophysical Properties, Atlanta, GA, 168-79, 1973.

80. Ewell, R.H. and Eyring, H., "Theory of the Viscosity of Liquids as a Function of Temperature and Pressure," J. Chem. Phys., 5, 726-36, 1937.

81. Powell, R.E., Roseveare, W.E., and Eyring, H., "Diffusion, Thermal Conductivity and Viscous Flow of Liquids," Ind. Eng. Chem., 33, 430-5, 1941.

82. Wynne-Jones, W.K.F. and Eyring, H., "The Absolute Rate of Reactions in Condensed Phases," J. Chem. Phys., 3, p. 492, 1935.

83. Eyring, H., "Viscosity, Plasticity and Diffusion as Examples of Absolute Reaction Rates," J. Chem. Phys., 4, 283-91, 1936.

84. Eyring, H., "The Activated Complex in Chemical Reactions," J. Chem. Phys., 3, 107-15, 1935.

85. Glasstone, S., Laidler, K.J., and Eyring, H., "The Theory of Rate Processes," McGraw-Hill, New York, NY, 485 pp., 1941.

86. Weymann, H.D., "Theoretical and Experimental Investigation of the Migration Theory of Viscous Flow," Kolloid-Z., 138, 41-56, 1954.

87. Weymann, H.D., "On the Hole Theory of Viscosity Compressibility and Expansivity of Liquids," Kolloid-Z., 181, 131-7, 1962.

88. Gold, P.E. and Ogle, G.J., "Estimating Thermophysical Properties of Liquids, Part 10 - Viscosity," Chem. Eng., 14, 121-3, 1969.

89. McLaughlin, E., "Viscosity and Self-Diffusion in Liquids," Trans. Faraday Soc., 55, 28-38, 1962.

90. Lennard-Jones, J. and Devonshire, A.F., "Critical Phenomena in Gases. I," Proc. R. Soc. London, 163A, 53-70, 1937.

91. Majamdar, D.K., "A Semi-Empirical Formula for the Viscosity of a 12:6 Liquid," J. Phys. Chem., 67, 1974-5, 1963.

92. Barker, J.A., "The Tunnel Theory of Fluids: The 12:6 Fluid," Proc. R. Soc. London, 259A, 442-57, 1961.

93. Eyring, H. and Ree, T., "Significant Liquid Structures. VI. The Vacancy Theory of Liquids," Proc. Mol. Ac. Sc., 47, 526-33, 1961.

94. Eyring, H., Henderson, D., Stover, B.J., and Eyring, E.M., "Statistical Mechanics and Dynamics," John Wiley Sons, Inc., New York, NY, 1964.

95. Eyring, H. and Ihon, M.S., "Significant Liquid Structures," John Wiley Sons, Inc., New York, NY, 1969.

96.   Ree, T., Ree, T.S., and Eyring, H., "Significant Structure Theory of Transport Phenomena,"
      J. Phys. Chem., 68, 3262-7, 1964.

97.   Faerber, G.L., Kim, S.W., and Eyring, H., "The Viscosities Flow and Glass Transition Temper-
      ature of Some Hydrocarbons," J. Phys. Chem., 74, 3510-8, 1970.

98.   Hogenboom, D.L., Webb, W., and Dixon, J.A., "Viscosity of Several Liquid Hydracarbons as a
      Function of Temperature, Pressure and Free Volume," J. Chem. Phys., 46, 2586-98, 1967.

99.   Ihon, M.S., Klotz, W.L., and Eyring, H., "Theoretical Calculation of the Pressure Dependence
      of Liquid Hydrocarbon Viscosities," J. Chem. Phys., 51, 3692-4, 1969.

100. Eicher, L.D. and Zwolinsky, B.J., "Molecular Structure and Shear Viscosity. Isomeric Hexanes,"
      J. Phys. Chem., 76, 3295-300, 1972.

101. Kim, K.C., Lu, W.C., Ree, T., and Eyring, H., "The Significant Structure Theory Applied to
      Liquid Oxygen," Proc. Nat. Acad. Sci., USA, 57, 861-7, 1967.

102. Ihon, M.S., Grosh, J., Ree, T., and Eyring, H., "The Significant Structure Theory Applied
      to Meta- and Para-Xylene," Proc. Nat. Acad. Sci., USA, 54, 1419-26, 1965.

103. Lu, W.C., Ree, T., Gerrad, V.G., and Eyring, H., "Significant Structure Theory Applied to
      Molten Salts," J. Chem. Phys., 49, 797-804, 1968.

104. Hsu, C.C., and Eyring, H., "Significant Liquid Structure Theory of Viscosity and Selfdiffusion
      of the Alkali Metals," Proc. Nat. Acad. Sci., USA, 69, 1342-5, 1972.

105. Carlson, C.M., Eyring, H., and Ree, T., "Significant Structures in Liquids. V. Thermodynamic
      and Transport Properties of Molten Metals," Proc. Nat. Acad. Sci., USA, 46, 649-59, 1960.

106. Thomson, T.R., Eyring, H., and Ree, T., "Thermodynamic and Physical Properties of Liquid
      Fluorine as Calculated by Significant Liquid Structure Theory of Liquids," J. Phys. Chem.,
      67, 2701-5, 1963.

107. Ma, S.M., Eyring, H., and Ihon, M.S., "The Significant Structure Theory Applied to Amorphous
      and Crystalline Polyethylene," Proc. Nat. Acad. Sci., USA, 71, 3096-100, 1974.

108. Leu, A.L., Ma, S.M., and Eyring, H., "Properties of Molten Magnesium Oxide," Proc. Nat. Acad.
      Sci., USA, 72, 1026-30, 1975.

109. Batschinski, A.J., "Investigation of the Internal Friction of Liquids. I," Z. Phys. Chem.,
      84, 643-705, 1913.

110. Hildebrand, J.H., "Kinetic Theory of Viscosity of Compressed Fluids," Proc. Nat. Acad. Sci.,
      USA, 72(5), 1970-2, 1975.

111. Hildebrand, J.H., "Motions of Molecules in Liquids: Viscosity and Diffusivity," Science,
      174, 490-2, 1971.

112. Hildebrand, J.H. and Lamoreaux, R.H., "Fluidity: A General Theory," Proc. Nat. Acad. Sci.,
      USA, 69(11), 3428-31, 1972.

113. Alder, B.J. and Hildebrand, J.H., "Activation Energy: Not Involved in Transport Processes
      in Liquids," Ind. Eng. Chem. Fundam., 12(3), 387-8, 1973.

114. Hildebrand, J.H. and Lamoreaux, R.H., "Fluidity and Liquid Structure," J. Phys. Chem., 77(11),
      1471-3, 1973.

115. Eicher, L.D. and Zwolinski, B.J., "Limitations of the Hildebrand-Batschinski Shear Viscosity
      Equation," Science, 77, p. 369, 1972.

116. Ertl, H. and Dullien, F.A.L., "Hildebrand's Equation for Viscosity and Diffusivity," J. Phys.
      Chem., 77(25), 3007-11, 1973.

117. Ertl, H. and Dullien, F.A.L., "Self-Diffusion and Viscosity of Some Liquids as a Function of
      Temperature," AIChE J., 19(6), 1215-23, 1973.

118. Doolittle, A.K., "Studies in Newtonian Flow. II. The Dependence of the Viscosity of Liquids
      on Free-Space," J. Appl. Phys., 22(12), 1471-5, 1951.

119. Doolittle, A.K., "Studies in Newtonian Flow. III. The Dependence of the Viscosity of
      Liquids on Molecular Weight and Free Space (in Homologous Series)," J. Appl. Phys., 23(2),
      236-9, 1952.

120. Doolittle, A.K. and Doolittle, D.B., "Studies in Newtonian Flow. V. Further Verification of the Free Space Viscosity Equation," J. Appl. Phys., 28(8), 901-5, 1957.

121. Barlow, J.A., Lamb, J., and Matheson, A.J., "Viscous Behaviour of Supercooled Liquids," Proc. R. Soc., London, 292A, 322-42, 1965.

122. Matheson, A.J., "Role of Free Volume in the Pressure Dependence of the Viscosity of Liquids," J. Chem. Phys., 44(2), 695-9, 1966.

123. Miller, A.A., "Free Volume and Viscosity of Liquid: Effects of Temperature," J. Phys. Chem., 67, 1031-5, 1963.

124. Cohen, M.H. and Turnbull, D., "Molecular Transport in Liquids and Glasses," J. Chem. Phys., 31(5), 1164-9, 1959.

125. Turnbull, D. and Cohen, M.H., "Free-Volume Model of the Amorphous Phase: Glass Transition," J. Chem. Phys., 34(1), 120-4, 1961.

126. Turnbull, D. and Cohen, M.H., "On the Free-Volume Model of the Liquid-Glass Transition," J. Chem. Phys., 52(6), 3038-41, 1970.

127. Macedo, P.B. and Litovitz, T.A., "On the Relative Roles of Free Volume and Activation Energy in the Viscosity of Liquids," J. Chem. Phys., 42(1), 245-56, 1965.

128. Chung, H.S., "On the Macedo-Litovitz Hybrid Equation for Liquid Viscosity," J. Chem. Phys., 44(4), 1362-4, 1966.

129. Brummer, S.B., "On the Relative Roles of Free Volume and Activation Energy in Transport Processes in Liquids: A Comment on the Paper of Macedo and Litovitz," J. Chem. Phys., 42, p. 4317, 1965.

130. Gubbins, K.E. and Tham, M.J., "Free Volume Theory for Viscosity of Simple Monopolar Liquids," AIChE J., 15(2), 264-71, 1969.

131. Abas-zade, A.K., "The Law of Heat Conduction of Liquids and Vapors," Zh. Eksp. Teor. Fiz., 23, 60-4, 1952.

132. Shimotake, H. and Thodos, G., "Viscosity: Reduced State Correlation for the Inert Gases," AIChE J., 4, 257-62, 1958.

133. Brebach, W.J. and Thodos, G., "Viscosity: Reduced State Correlation of Diatomic Gases," Ind. Eng. Chem., 50, 1095-100, 1958.

134. Groenier, W.S., and Thodos, G., "Viscosity and Thermal Conductivity of Ammonia in the Gaseous and Liquid States," J. Chem. Eng. Data, 6, 240-4, 1961.

135. Lucas, K. and Stephan, K., "Equations of State for the Transport Properties of Pure Liquid Materials," Chem.-Ing.-Tech., 45(5), 265-71, 1973.

136. Eakin, B.E. and Ellington, R.T., "Predicting the Viscosity of Pure Light Hydrocarbons," J. Petr. Technology, 15(2), 210-4, 1965.

137. Diller, D.E., Hanley, H.J.M., and Roder, H.M., "The Density and Temperature Dependence of the Viscosity and Thermal Conductivity of Dense Simple Fluids," Cryogenics, 286-94, 1970.

138. Rogers, J.D. and Brickwedde, F.G., "Excess Transport Properties of Light Molecules," AIChE J., 11(2), 304-10, 1965.

139. Kestin, J. and Wang, H.E., "On the Correlation of Experimental Viscosity Data," Physica, 24, 604-8, 1968.

140. Diller, D.E., "Measurements of the Viscosity of p-Hydrogen," J. Chem. Phys., 42, 2089-100, 1965.

141. Hanley, H.J.M., McCarty, R.D., and Haynes, W.M., "Equations for the Viscosity and Thermal Conductivity Coefficients of Methane," Cryogenics, 413-7, 1975.

142. Kessel'man, P.M. and Kamenetskii, V.R., "An Equation for Calculating the Viscosity of Compressed Gases," Teploenergetika, 14(9), 73-5, 1967.

143. Stein, W.A., "Equations for the Dynamic Viscosity and Thermal Conductivity of Pure Water," Waerme Stoffuebertrag., 2, 210-21, 1969.

144. Ashurst, W.T. and Hoover, W.G., "Dense Fluid Shear Viscosity and Thermal Conductivity - the Excess," AIChE J., 21(2), 410-1, 1975.

145. Lucas, K., "The Viscosity of Pure Liquids Near the Critical Temperature," Waerme
      Stoffuebertrag., 4, 236-43, 1971.

146. Tanishita, I., Nagashima, A., and Murai, Y., "Correlation of Viscosity, Thermal Conductivity
      and Prandtl Number for Water and Steam as a Function of Temperature and Pressure," Bull.
      JSME, 14(77), 1187-9, 1971.

147. Grigull, K., Mayinger, F., and Bach, J., "Viscosity, Thermal Conductivity and Prandtl Number
      of Water and Steam," Waerme Stoffuebertrag., 1, 15-34, 1968.

148. Nitta, T., Turek, E.A., Greenkorn, R.A., and Chao, K.C., "A Group Contribution Molecular
      Model of Liquids and Solutions," AIChE J., 23(2), 144-60, 1977.

149. Fredenslund, A., Jones, R.L., and Prausnitz, J.M., "Group-Contribution Estimation of Activity
      Coefficients in Non-Ideal Liquid Mixtures," AIChE J., 21, 1086-99, 1975.

150. Souders, M., "Viscosity and Chemical Constitution," J. Am. Chem. Soc., 60, 154-8, 1938.

151. Thomas, L.H., "The Dependence of the Viscosities of Liquids on Reduced Temperature and a
      Relation Between Viscosity, Density and Chemical Constitution," J. Chem. Soc. London, 573-9,
      1946.

152. Lowitz, D.A., Spencer, J.W., Webb, W., and Schiessler, R.W., "Temperature-Pressure-Structure
      Effects on the Viscosity of Several Higher Hydrocarbons," J. Chem. Phys., 30(1), 73-83, 1959.

153. Kierstead, H.A. and Turkevich, J., "Viscosity and Structure of Pure Hydrocarbons," J. Chem.
      Phys., 12(1), 24-7, 1944.

154. Van Velzen, D., Lopes Cardozo, R., and Langenkamp, H., "Liquid Viscosity and Chemical Con-
      stitution of Organic Compoints: A New Correlation and a Compilation of Literative Data,"
      Ind. Eng. Chem., Fundam., 11, 20-5, 1972.

155. Thomas, H., "Variation of the Viscosity of Liquids with Temperature and the Ratio of the
      Energy of Viscous Flow to the Energy of Vaporization," Chem. Eng. J., 11, 201-6, 1976.

156. Kreps, S.I. and Druin, M.L., "Prediction of Viscosity of Liquid Hydrocarbons," Ind. Eng.
      Chem., Fundam., 9(1), 79-63, 1970.

157. Sato, T. and Minamiyama, T., "Viscosity of Steam at High Temperatures and Pressures," Int.
      J. Heat Mass Transfer, 7, 199-209, 1964.

158. Grigull, K., Bach, J., and Reimann, M., "Properties of Water and Steam According to the
      1968 IFC-Formulation," Waerme-Stoffuebertrag., 1, 202-13, 1968.

159. Grigull, K., Reimann, M., and Bach, J., "Some Relationships Between Transport Coefficients
      and Thermodynamic Variables of State," Waerme-Stoffuebertrag., 3, 120-6, 1970.

160. Westmeier, S., "Relationships Between Viscosities and Thermodynamic Excess Quantities,"
      Wiss. Z. TH Leuna-Merseburg, 17(1), 99-103, 1975.

161. Holleran, E.M., "Interrelation of Viscosity, Self-Diffusion and Isotopic Thermal Diffusion,"
      J. Chem. Phys., 23(5), 847-53, 1955.

162. Weissman, S., "Estimation of Diffusion Coefficients from Viscosity Measurements: Polar and
      Polyatomic Gases," J. Chem. Phys., 40(11), 3397-406, 1964.

163. Lielmezs, J., "Relation Between the Self-Diffusion Coefficient, Density and Fluidity and
      the Temperature in Liquids," Z. Phys. Chem. Neue Folge, 91, 288-300, 1974.

164. Hayduk, W. and Cheng, S.C., "Review of Relation Between Diffusivity and Solvent Viscosity
      in Dilute Liquid Solutions," Chem. Eng. Sci., 26, 635-46, 1971.

165. Fitts, D.D., "Nonequilibrium Thermodynamics," McGraw-Hill, New York, NY, 1962.

166. Touloukian, Y.S., Saxena, S.C., and Hestermans, P., "Viscosity," Volume 11, in Thermophysical
      Properties of Matter - The TPRC Data Series (13 volumes), Plenum Publishing Corp., New York,
      NY, 1975.

NUMERICAL DATA

NUMERICAL DATA

DATA PRESENTATION AND RELATED GENERAL INFORMATION

SCOPE OF COVERAGE

Presented in this volume are data on the pressure and temperature dependence of viscosity for 50 pure fluids. The substances were selected based on scientific and technological interest as well as on availability of high pressure data.

To the extent that CINDAS has reported the viscosity data in the dilute gas region as well as for saturated vapor and saturated liquid conditions [166], this work concentrates on the pressure dependence of viscosity. The viscosity of fluids depends heavily on pressure in the viscinity of the critical point. To a lesser degree, the pressure dependence is appreciable in all other states as well, excluding the dilute gas region. Data on fluid mixtures have not been considered in this volume. While the actual contribution of this work is the pressure dependence of viscosity, the values in the dilute gas region, as well as those of the saturated vapor and the saturated liquid have been included in most cases. Consistency with the data of CINDAS [166] was aimed at, whenever possible, however this could not be achieved in all cases.

The fluids covered include 11 elements, 3 inorganic compounds and 36 organic compounds. Air is treated as a pure fluid. Only those sets of data which include a significant pressure range have been considered in the selection, thus limiting drastically the number of substances covered. It is hoped, though, that the great majority of pressure dependent data published up to now has been included.

Most experimental data have actually been reported as a function of pressure and only a minority of the cases give the density dependence as well. Thus, even when the density dependence was available in some original papers, the recommended values were always given as a function of temperature and pressure. This reflects the authors' view that temperature and pressure are the variables actually needed in most practical applications and that the recommended values generated from the original data are meant predominantly for practical use. Of course it must be realized that the density dependence correlation is the one which can directly be used to develop and check theoretical methods. The original data on density dependence, when available, was taken from the original publications. The viscosity-density-temperature plot was generated in many cases from a measured pressure dependence, using the thermodynamic equation of state data. The p-v-T data used

27

for this purpose are included in the references. This plot was used to extrapolate the available
body of data to higher temperatures and pressures using the approximate residual viscosity concept.
The extrapolated values may be identified in the diagrams for the recommended values.

PRESENTATION OF DATA

In the Material Index, the region of thermodynamic states in which data are available is
indicated by the following abbreviations

    L = Liquid
    G = Gas
    SL = Saturated Liquid
    SV = Saturated Vapor

One thus gets at a glance an overview of the distribution of data over the various thermodynamic
states.

The viscosity data and information for each pure fluid are presented in both tabular and
graphical form.

When more than one data source on the pressure dependence of the viscosity of a fluid was
available, generally data from one or several authors were selected to generate recommended values.
Even though criteria like experimental method, precision and abundance of data were applied, the
ultimate selection was made on a subjective basis considering the present authors' experience in
evaluating viscosity data. It is hoped that the selection was appropriate in most cases, though
errors in judgment can, of course, never be excluded in such a procedure. When the data of more
than one author had to be used to cover a large region of states, to assure consistency in the
overlapping region was an ever present problem. The approach used is discussed specifically for
each substance. In any case, the data of several authors were not averaged by some weighting
procedure, but instead a selection of one or several sets were made. This appeared to be a reasonable approach in view of the large discrepancies between some authors and the differences in
quality of experimental procedures used. The data of the selected authors were plotted against
pressure along measured isotherms on large scale diagrams. Quite frequently smoothing modifications were applied to the plot of the original data. They are discussed for each substance sepa-
rately. ·Using these large scale plots, the viscosity was plotted against temperature for even
pressures. The spacings of the isobars was such, that linear interpolation gave an error of less
than 1% with the exception of the critical region. Unfortunately, it was not convenient to delin-
eate in the tables of recommended values the region of extrapolation. Hence this is only shown in
the plots by dashed curves vs the solid lines for the regions where experimental measurements exist.

Whenever possible, a plot of residual viscosities against density was generated, as discussed
in the Section on Theory and Estimation and defined by the relation

$$\Delta\mu = \mu(T,\rho) - \mu(T,0) = f(\rho)$$

In such a plot, for most substances the residual viscosity $\Delta\mu$ is almost independent of temperature
in a large region of states. This behavior was used to extrapolate the given amount of data to
higher temperatures and pressures. Input data in this concept are dilute gas viscosity values as

taken from the literature, e.g., from the recommended values of [166], and p-v-T data. In most cases the limited availability of the p-v-T data did not allow an extrapolation over a wide region of states. The selections of these data are clearly indicated for each substance. Generally, the concept of residual viscosity becomes increasingly inaccurate at higher densities. Therefore it was not used above twice the critical density. At low densities, too, this concept does not hold in some cases, which however is less important because of the small relative contribution of the residual viscosity at low densities. More information about the merits and the validity of the residual viscosity concept may be found in the literature cited in the section on Theory and Estimation. The extrapolated values are included in the large scale μ-T-p plots by dashed lines, so that they may be readily identified. Their accuracy depends entirely on the validity of the concept for a specific substance as well as the reliability of the dilute gas viscosity values and the p-v-T data. It thus varies considerably from substance to substance and is not always easily assessed. Generally the accuracy of such extrapolated values should be a few percent less than that of the measured data.

From the large scale plots of isobars against temperature, tables were generated giving recommended viscosity values as a function of temperature and pressure, where viscosity is in the units of $10^{-5} Nsm^{-2}$, temperature is in degrees Kelvin and pressure in bar. Even entry values of temperature and pressure were selected, the spacing being such that generally linear interpolation resulted in an error well within the estimated uncertainty for the recommended values. In the critical region, such an effort was not considered to be reasonable as the experimental data there are less accurate anyhow and graphical interpolation between data points, even on large scale diagrams, introduces considerable uncertainties.

In the discussion for each fluid, the available experimental data and information on the pressure dependence of viscosity are thoroughly reviewed and assessed. Those sets of data which have not been selected in the generation of the recommended values were used to establish estimates of accuracy for the latter. When there were extensive data available from a single author, a selection of points representative for a statistical analysis was made. Corresponding recommended values were extracted from the diagrams. The percent deviation for each selected experimental point relative to the recommended value is computed according to the departure formula

$$x_i = \left(\frac{\mu_A - \mu_B}{\mu_A} \right) 100$$

where,

μ_A = author's values

μ_B = recommended values

x_i = percent deviation

The results of these evaluations are presented in tables. The deviations x_i were also used to produce departure plots, where by the above definition, departures are positive if the experimental values are greater than the recommended values and vice versa. To obtain an impression about the overall agreement between the recommended values and one specific set of data, an average deviation $\bar{x}$ with a corresponding standard deviation S was computed. The well-known definitions are

$$\bar{x} = \frac{\Sigma x_i}{n}$$

$$S = \frac{\sqrt{\sum_i (x_i - x)^2}}{n - 1}$$

where n is the number of experimental data points. These results are given for each set of data so treated in the summary tables of additional data.

In the majority of cases, the available experimental data were not adequate to allow a detailed comparison with the recommended values. In such cases, only the maximum deviation which was found in the course of the comparison of the single data points was listed. The data of some workers could not be compared at all with the recommended values, for a variety of reasons. Sometimes only density dependence was given, in other cases the data were not available in tabular form. Quite frequently, additional works on the pressure dependence of viscosity in different regions of state were available. Again, a comparison was not possible, however the references can be used by the reader to cover a larger region of states than actually presented in the table of recommended values.

In the discussion for each fluid an estimate of accuracy for the recommended values is given. It is mainly based on the comparison with the data points from other sources as well as on considerations of experimental precision. Evidently, the percent accuracy will be different in different regions of states. In the estimates of accuracy given here, the only differentiation made is between the region of the actual measurements and the region of extrapolation. The figures of reported accuracy are confirmed by about 90% of the data points of other authors with comparable precision. In the region of extrapolation, a slight reduction of accuracy was assumed, thus resulting in the reduced accuracy estimates for the extrapolated values.

SYMBOLS AND ABBREVIATIONS USED IN THE FIGURES AND TABLES

Most abbreviations and symbols used are those generally accepted in engineering and scientific practice and convention.

In this volume the word "data" is reserved for an experimentally determined quantity while quantities determined by calculation or estimation are referred to as values.

The notations "n.m.p.," "n.b.p.," and "c.p." refer to normal melting point, normal boiling point, and critical point, respectively. Numbers in square brackets in the discussion and those signified by the notation "Reference" on the departure plot correspond to the *References to Data Sources* listed at the end of this *Numerical Data* section.

In the deparature plots, curve numbers are surrounded either by circles or squares, the latter being used to indicate a single data point. Solid lines are used in the plot to connect experimental data points and dotted lines indicate calculated or correlated values. When the percent departure for any of the data points falls outside the range of the departure plot, the numerical value of the departure is correctly given at the data point with a vertical arrow pointing up or down from the data point to the given value to indicate the fact that the value is beyond the range of the plot.

Useful Physical Constants for Reported Substances

Name	Formula	Molecular Mass, g mol^{-1}	Melting (or Triple) Point, K	Normal Boiling Point, K	Critical Temp., K	Critical Pressure, bar	Dipole Moment, Debyes
Air	--	28.966	60	79b,82d	133	37.7	--
Ammonia	NH_3	17.031	195	240	405	112.8	1.47
Argon	Ar	39.948	84	87	151	48.7	0
Benzene	C_6H_6	78.113	279	353	563	49.0	0
Bromotrifluoromethane	$CBrF_3$	148.91	107.15	214	340	39.7	0.65
i-Butane	$i-C_4H_{10}$	58.124	114	262	408	36.5	0.132
n-Butane	$n-C_4H_{10}$	58.123	137	273	426	38.0	≤0.05
n-Butylacetate	$n-C_6H_{12}O_2$	116.16	195.25	399	579	31.4	1.84
Carbon Dioxide	CO_2	44.010	216(5 atm)	195	304	73.9	0
Carbon Monoxide	CO	28.010	68	81	134	35.0	0.112
Cyclohexane	C_6H_{12}	84.16	273.7	354	553	40.7	0
n-Decane	$n-C_{10}H_{22}$	142.286	243	447	619	21.1	0
n-Dodecane	$n-C_{12}H_{26}$	170.33	263.55	490	659	18.1	0
Ethane	C_2H_6	30.070	90	185	305	48.3	0
Ethanol	C_2H_6O	46.069	155.85	352	516	63.8	1.69
Ethylbenzene	C_8H_{10}	106.17	178.18	409	617	36.1	0.59
Ethylcyclohexane	C_8H_{16}	112.22	161.83	405	609	30.3	0
Ethylene	C_2H_4	28.054	104	170	283	50.6	0
Fluorine	F_2	37.997	54	85	144	52.2	0
Freon 12	CCl_2F_2	120.914	116	243	385	41.2	0.51
Freon 22	$CHClF_2$	86.469	113	233	369	49.8	1.42
Freon 113	$C_2Cl_3F_3$	187.376	238	321	487	34.1	--
Helium	He	4.003	0.95(26 atm)	4	5.2	2.27	0
n-Heptane	$n-C_7H_{16}$	100.203	183	371	540	27.4	0
n-Heptene	$n-C_7H_{14}$	98.19	154.15	367	537	28.4	0.34
n-Hexane	$n-C_6H_{14}$	86.177	178	342	508	30.1	0
n-Hexene	$n-C_6H_{12}$	84.16	133.33	337	504	30.7	0.45
Hydrogen	H_2	2.016	13.96	20	33	13.0	0
p-Hydrogen	$p-H_2$	2.016	13.81(99.8%)	20	33	12.9	0
Krypton	Kr	83.80	116	120	209	55.0	0
Methane	CH_4	16.043	90	112	190	46.1	0
Methanol	CH_4O	32.04	179.55	338	513	80.9	1.70
Methylcyclohexane	C_7H_{14}	98.19	146.56	374	572	34.7	0
Neon	Ne	20.179	25	27	44	27.5	0
Nitrogen	N_2	28.013	63	77	126	34.0	0
n-Nonane	$n-C_9H_{20}$	128.257	220	424	594	22.8	0
i-Octane	$i-C_8H_{18}$	114.22	163.65	391	560	24.8	0
n-Octane	$n-C_8H_{18}$	114.23	216	399	569	24.9	0
n-Octene	$n-C_8H_{16}$	112.22	171.42	394	567	26.2	0.34
Oxygen	O_2	31.999	55	90	155	50.4	0
i-Pentane	$i-C_5H_{12}$	72.15	113.45	301	460	33.7	0
n-Pentane	$n-C_5H_{12}$	72.150	144	309	470	33.8	0
Propane	C_3H_8	44.096	86	231	369	42.5	0.084
n-Propanol	$n-C_3H_8O$	60.10	146.65	370	537	51.7	1.66
i-Propanol	$i-C_3H_8O$	60.10	183.65	355	508	53.1	1.64
n-Propylacetate	$n-C_5H_{10}O_2$	102.13	178.15	375	549	33.3	1.84
Propylene	C_3H_6	42.080	88	225	365	46.0	0.366
Toluene	C_7H_8	92.140	178	384	594	41.1	0.36
n-Undecane	$n-C_{11}H_{24}$	156.30	247.562	469	640	19.4	0
Xenon	Xe	131.30	161	165	290	58.4	0

b Bubble point.
d Dew point.

CONVERSION FACTORS FOR UNITS OF VISCOSITY

The conversion factors for units of viscosity given in the table are based upon the following defined values and conversion factors given in NBS Special Publication 330, 1972:

Standard acceleration of free fall = 980.665 cm s^{-2}
1 in = 2.54 cm
1 lb = 453.59237 g

Conversion Factors for Units of Viscosity

MULTIPLY by appropriate factor to OBTAIN→	N s m⁻² (kg s⁻¹ m⁻¹)	Pa s (kg s⁻¹ m⁻¹)	Poise (dyne s cm⁻²) (g s⁻¹ cm⁻¹)	centipoise	micropoise	lb$_f$ s ft⁻²	poundal s ft⁻² (lb$_m$ s⁻¹ ft⁻¹)	lb$_m$ hr⁻¹ ft⁻¹	slug hr⁻¹ ft⁻¹
N s m⁻² (kg s⁻¹ m⁻¹)	1	1	10	1×10^3	1×10^7	2.08854×10^{-2}	0.671969	2.41909×10^3	75.1876
Pa s (kg s⁻¹ m⁻¹)	1	1	10	1×10^3	1×10^7	2.08854×10^{-2}	0.671969	2.41909×10^3	75.1876
Poise (dyne s cm⁻²) (g s⁻¹ cm⁻¹)	0.1	0.1	1	1×10^2	1×10^6	2.08854×10^{-3}	6.71969×10^{-2}	2.41909×10^2	7.51876
centipoise	1×10^{-3}	1×10^{-3}	1×10^{-2}	1	1×10^4	2.08854×10^{-5}	6.71969×10^{-4}	2.41909	7.51876×10^{-2}
micropoise	1×10^{-7}	1×10^{-7}	1×10^{-6}	1×10^{-4}	1	2.08854×10^{-9}	6.71969×10^{-8}	2.41909×10^{-4}	7.51876×10^{-6}
lb$_f$ s ft⁻²	47.8803	47.8803	4.78803×10^2	4.78803×10^4	4.78803×10^8	1	32.1740	1.15827×10^5	3.60000×10^3
poundal s ft⁻² (lb$_m$ s⁻¹ ft⁻¹)	1.48816	1.48816	14.8816	1.48816×10^3	1.48816×10^7	3.10810×10^{-2}	1	3.60000×10^3	1.11891×10^2
lb$_m$ hr⁻¹ ft⁻¹	4.13379×10^{-4}	4.13379×10^{-4}	4.13379×10^{-3}	0.413379	4.13379×10^3	8.63360×10^{-6}	2.77778×10^{-4}	1	3.10810×10^{-2}
slug hr⁻¹ ft⁻¹	1.33001×10^{-2}	1.33001×10^{-2}	0.133001	13.3001	1.33001×10^5	2.77778×10^{-4}	8.93724×10^{-3}	32.1740	1

CONVENTION FOR BIBLIOGRAPHIC CITATION

For the following types of documents the bibliographic information is cited in the sequences given below.

Journal Article

 a. Author(s) - The names and initials of all authors are given. The last name is written first, followed by initials.

 b. Title of the article - The title of a journal article is enclosed in quotation marks.

 c. Name of the Journal - The abbreviated name of the journal is given as used in *Chemical Abstracts*.

 d. Series, volume, and issue number - If the series is designated by a letter, no comma is used between the letter for series and the numeral for volume, and they are both in bold-face type. In case series is also designated by a numeral, a comma is used between the numeral for series and the numeral for volume, and only the numeral denoting volume is boldfaced. No comma is used between the numerals denoting volume and issue number. The numeral for issue number is enclosed in parentheses.

 e. Pages - The inclusive page numbers of the article.

 f. Year - The year of publication.

Report

    a. Author(s).

    b. Title of report - The title of a report is enclosed in quotation marks.

    c. Name of the sponsoring agency and report number.

    d. Part.

    e. Pages.

    f. Year.

Book

    a. Author(s).

    b. Title - The title of a book is underlined.

    c. Volume.

    d. Edition.

    e. Publisher.

    f. Location of the publisher.

    g. Pages.

    h. Year.

NUMERICAL DATA
ON VISCOSITY

AIR

The viscosity of air, considered here as a mixture of 21% oxygen, 78.1% nitrogen and 0.9% argon, is based on the experimental data of Lo et al. [133]. Their correlation was used to produce the recommended values presented on the next page. The values are plotted as isotherms against pressure in Figure 1, and as isobars against temperature in Figure 2.

Additional works on the pressure dependence of the viscosity of air are listed in the summary table below. An extensive comparison of these data with the recommended values was made and the agreement was found generally satisfactory, though it does not cover the entire region of states presented here. The departure plots given in Figure 3 present these comparisons. The uncertainty in the recommended values is estimated to be ± 3%.

Additional values on the pressure dependence of the viscosity of air, especially in the liquid region are given in [204], for temperatures from 75 K to 160 K up to 500 bars.

ADDITIONAL REFERENCES ON THE VISCOSITY OF AIR

Authors	Year	Ref. No.	Temperature K	Pressure bar	Method	Departure % (no. points)
Timrot and Serednitskayal	1975	191	300-570	1-120	Oscillating disk	1.5
Kurin and Golubev	1974	121	293-323	1-3200	Capillary tube	1.2
Latto and Saunders	1973	130	90-400	1-150	Capillary tube	1.3
Goring and Eagan	1971	64	423	37, 106	Capillary tube	1.0
Golubev and Petrov	1970	57	273-373	1-304	Capillary tube	1.11 ± 0.63 (37)
Kestin and Whitelaw	1964	110	298-523	1-142	Oscillating disk	1.98 ± 0.91 (36)
Filippova and Ishkin	1961	48	90-273	20-148	Capillary tube	-
Glaser and Gebhardt	1959	55	271-473	1-344	Falling ball	-
Kestin and Leidenfrost	1959	105	293-298	1-70	Oscillating disk	1.0 (13)
Kestin and Wang	1958	108	298	1-71	Oscillating disk	1.1 (10)
Makita	1957	139	298-473	1-785	Rolling ball	0.00 ± 3.25 (36)
Kestin and Pilarczyk	1954	107	294	1-71	Oscillating disk	-
Biles and Putnam	1952	16	297, 543	7-62	Capillary tube	0.9 (17)
Iwasaki	1951	90	323-423	1-203	Oscillating disk	0.28 ± 1.03 (27)
Kellström	1941	99	293	1-30	Rotating cylinder	0.6 (5)
Moulton and Beuschlein	1940	153	303	17-303	Capillary tube	7.0 (36)
Nasini and Pastones	1933	155	287	34-196	Capillary tube	3.5 (13)

VISCOSITY OF AIR

[μ, 10^{-6} N s m^{-2}]

T, K	Pressure, bar																
	1	20	37.7*	50	100	150	200	250	300	350	400	500	600	700	800	900	1000
200	13.4	13.9	14.5	15.2	18.5	22.9	27.5	32.0	36.7	41.3	45.9	55.1	64.3	72.8	82.6	92.0	–
220	14.5	14.9	15.5	16.0	18.5	21.8	25.5	29.2	33.0	36.7	40.4	47.9	55.3	62.0	69.1	75.1	80.7
240	15.5	15.9	16.4	16.8	18.7	21.2	24.1	27.2	30.1	33.1	36.3	42.4	48.5	54.0	59.7	64.5	68.6
260	16.5	16.9	17.3	17.5	19.1	21.1	23.4	26.0	28.5	31.0	33.8	38.8	44.0	48.8	53.5	58.0	61.8
280	17.5	17.8	18.1	18.4	19.7	21.3	23.2	25.4	27.6	29.9	32.2	36.7	41.2	45.5	49.9	53.8	57.5
300	18.4	18.7	19.0	19.2	20.3	21.8	23.4	25.4	27.2	29.3	31.3	35.3	39.2	43.2	47.2	50.8	54.3
320	19.3	19.6	19.9	20.0	21.1	22.3	23.8	25.4	27.1	28.9	30.7	34.3	37.8	41.5	45.0	48.5	51.7
340	20.2	20.5	20.7	20.9	21.8	23.0	24.3	25.7	27.2	28.8	30.3	33.7	36.8	40.1	43.4	46.5	49.6
360	21.1	21.3	21.5	21.7	22.6	23.7	24.8	26.0	27.4	28.8	30.2	33.2	36.1	39.0	42.0	44.9	47.8
380	21.9	22.1	22.3	22.5	23.3	24.3	25.3	26.5	27.7	29.0	30.2	32.9	35.6	38.2	40.9	43.6	46.3
400	22.7	22.9	23.1	23.3	24.1	25.0	25.9	26.9	28.0	29.2	30.3	32.8	35.2	37.5	40.1	42.5	45.0
450	24.7	24.9	25.1	25.3	25.9	26.7	27.4	28.3	29.1	30.0	31.0	32.8	34.8	36.7	38.7	40.8	43.0
500	26.6	26.8	27.0	27.1	27.6	28.3	28.9	29.7	30.3	31.0	31.9	33.4	35.0	36.7	38.3	40.2	42.1
600	30.1	30.3	30.4	30.5	31.0	31.4	31.9	32.4	33.0	33.5	34.1	35.3	36.4	37.7	39.0	40.4	41.8
700	33.4	33.5	33.6	33.7	34.0	34.4	34.7	35.1	35.5	36.0	36.5	37.4	38.3	39.3	40.4	41.5	42.6
800	36.3	36.4	36.5	36.5	36.8	37.1	37.4	37.7	38.1	38.4	38.8	39.6	40.5	41.3	42.2	43.1	44.0
900	39.0	39.1	39.2	39.2	39.5	39.7	40.0	40.3	40.6	40.9	41.3	42.0	42.7	43.5	44.2	45.0	45.8

*Critical pressure.

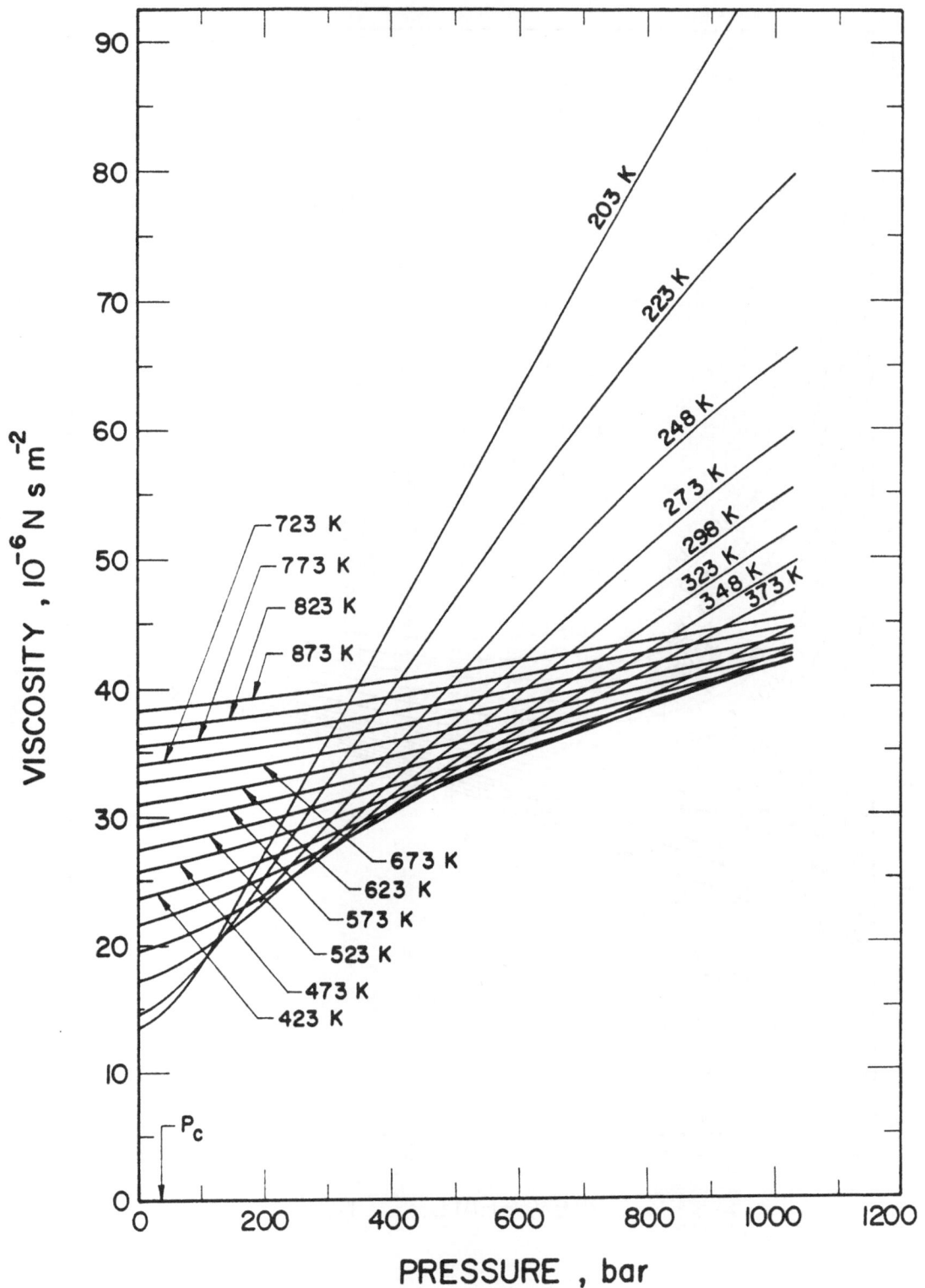

FIGURE I. VISCOSITY OF AIR [133].

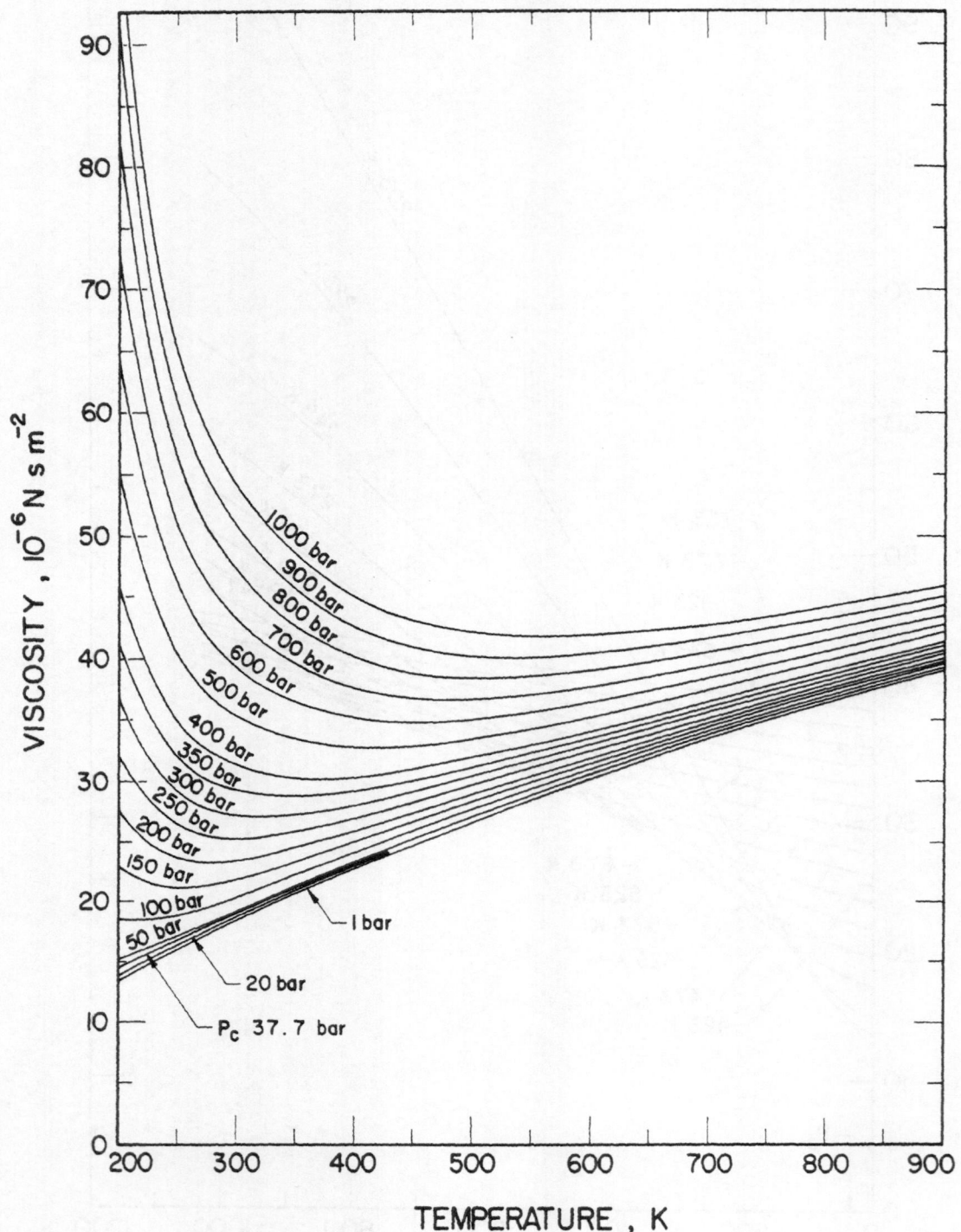

FIGURE 2. VISCOSITY OF AIR [133].

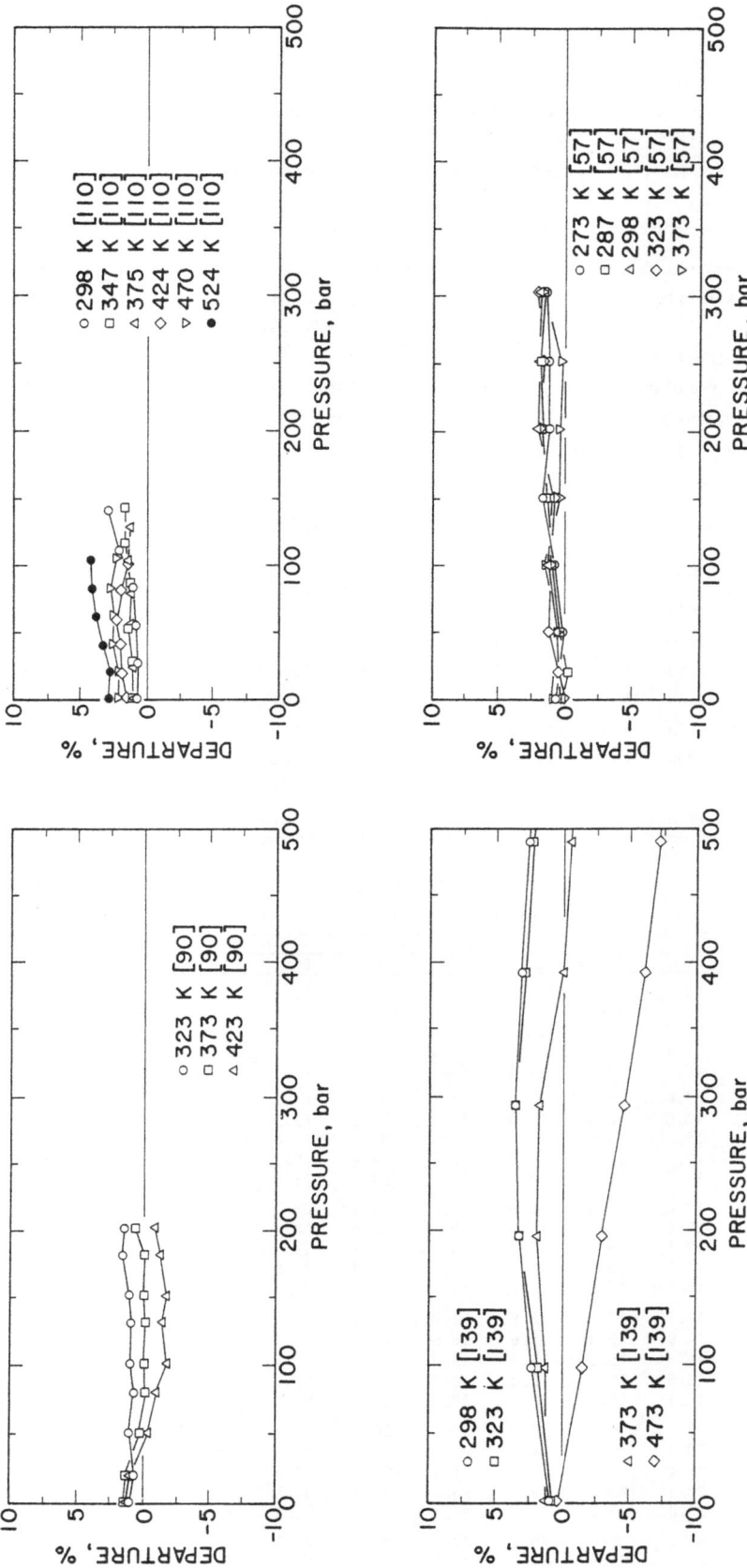

FIGURE 3. DEPARTURE PLOT ON THE VISCOSITY OF AIR.

AMMONIA

The data of Carmichael et al. [27] obtained by a rotating cylinder apparatus were used to generate the recommended values presented below. The measured isotherms are shown in Figure 1 as a function of pressure. Using the P-V-T data of [69], the 1 bar viscosity data of [19] with some extrapolation, which is in reasonable agreement with the formulation of [195], and the residual viscosity concept, it was possible to extrapolate the measured isotherms up to 600 bar, as well as from 478 K to 900 K up to 600 bar. The total amount of data is shown as isobars plotted against temperature in Figure 2, where the extrapolated values are indicated as dashed lines. The recommended values were read from Figure 2.

Additional works on the pressure dependence of the viscosity of ammonia are listed in the summary table below. Considerable discrepancies exist in the published viscosity data for this fluid. As seen from the departure plot, Figure 3, reasonable agreement is found with the data of Golubev et al. [116] in the dense gas region, while the deviations are higher in the liquid region. The disagreement with the data of Kiyama et al. [116] is excessive. Very good agreement is only found with data at lower pressures of Iwasaki et al. [91, 93]. In view of the large deviations between different investigations, any statement about accuracy must be provisional. The dense gas is estimated to have an accuracy of ± 7% in the region of measurements and ± 10% in the region of extrapolation.

ADDITIONAL REFERENCES ON THE VISCOSITY OF AMMONIA

Authors	Year	Ref. No.	Temperature K	Pressure bar	Method	Departure % (no. points)
Iwasaki and Takahashi	1968	93	298–408	1–93	Oscillating disk	1.2
Iwasaki	1964	91	293–303	1–25	Oscillating disk	0.3
Shimotake and Thodos	1963	180	373–473	17–473	Capillary tube	30.0
Makita	1955	137	323–573	1–101	Rolling ball	6.86 ± 8.07 (19)
Gobulev and Petrov	1953	57	303–523	1–811	Capillary tube	2.92 ± 3.74 (41)
Kiyama and Makita	1952	116	323–573	1–95	Rolling ball	6.86 ± 8.07 (19)
Carmichael and Sage	1952	28	278–378	9–345	Rolling ball	7.0
Stakelbeck	1933	183	253–353	1–25	Falling ball	52.0

VISCOSITY OF AMMONIA

$[\mu, \, 10^{-6} \, \text{N s m}^{-2}]$

T, K	1	50	100	112.8*	120	130	140	150	200	250	300	400	500	600
310	10.6	121	125	126	126	127	127	128	131	134	137	144	150	156
320	11.0	109	113	114	115	115	116	116	119	123	126	132	139	145
340	11.7	89.9	93.2	94.5	95.0	95.7	96.3	97.0	100	103	107	113	119	126
360	12.5	72.2	76.5	78.1	78.7	79.6	80.4	81.3	84.8	88.1	91.6	97.6	103	109
380	13.2	13.4	61.5	63.4	64.1	65.1	66.2	67.2	71.5	75.6	78.8	84.6	90.1	95.2
390	13.6	13.8	52.4	55.6	56.4	57.5	58.6	59.7	65.2	69.7	73.2	78.8	84.3	89.3
400	13.9	14.2	16.4	45.6	46.8	48.5	50.2	52.2	59.2	64.0	67.6	73.4	78.9	84.1
420	14.6	14.9	16.6	19.1	19.4	22.2	29.9	34.9	46.0	52.6	57.1	64.0	69.9	75.2
425	14.8	15.1	16.7	17.9	18.9	20.8	23.4	28.9	42.4	49.8	54.6	61.8	67.9	73.2
430	15.0	15.3	16.8	17.7	18.7	20.1	21.7	24.0	38.6	47.2	52.3	59.8	66.0	71.3
435	15.2	15.5	16.8	17.6	18.6	19.6	20.9	22.4	35.2	44.4	50.0	57.8	64.1	69.4
440	15.3	15.7	16.9	17.6	18.4	19.3	29.4	21.5	32.3	41.9	47.7	55.9	62.4	67.6
445	15.5	15.9	17.0	17.5	18.4	19.1	19.9	20.8	29.7	39.3	45.6	54.0	60.6	66.0
450	15.7	16.1	17.1	17.5	18.3	18.9	19.6	20.3	27.6	36.9	43.4	52.0	58.8	64.4
460	16.0	16.5	17.3	17.6	18.2	18.6	19.2	19.8	24.3	32.4	39.5	48.6	55.6	61.3
470	16.4	16.8	17.4	17.7	18.2	18.6	19.0	19.4	22.6	29.2	35.9	45.5	52.7	58.5
480	16.8	17.2	17.7	17.9	18.4	18.6	18.9	19.3	21.9	27.1	32.8	42.6	49.9	55.8
500	17.5	17.8	18.2	18.3	18.5	18.7	19.0	19.3	21.4	25.1	28.7	37.6	45.0	50.8
520	18.2	18.5	18.7	18.8	19.0	19.2	19.4	19.6	21.4	24.0	26.5	34.0	41.1	46.6
540	18.9	19.1	19.3	19.4	19.5	19.8	20.0	20.2	21.6	23.6	25.6	31.4	37.9	43.2
560	19.6	19.8	20.0	20.1	20.2	20.5	20.7	20.9	22.0	23.6	25.1	29.9	35.3	40.5
580	20.4	20.6	20.8	20.8	20.9	21.1	21.3	21.5	22.4	23.7	25.0	29.0	33.7	38.4
600	21.1	21.2	21.4	21.4	21.5	21.7	21.9	22.1	23.0	24.1	25.2	28.7	32.8	36.9
650	22.9	23.0	23.2	23.2	23.3	23.4	23.6	23.7	24.5	25.3	26.1	28.6	31.7	35.1
700	24.7	24.8	25.0	25.0	25.1	25.2	25.4	25.5	26.3	26.9	27.4	29.2	31.8	34.6
750	26.5	26.6	26.7	26.7	26.8	26.9	27.1	27.2	28.0	28.5	28.9	30.4	32.5	34.9
800	28.3	28.4	28.5	28.5	28.6	28.7	28.9	29.0	29.8	30.2	30.5	31.8	33.6	35.6
900	32.0	32.1	32.2	32.2	32.3	32.4	32.5	32.7	33.3	33.7	34.0	34.8	36.1	37.7

* Critical pressure.

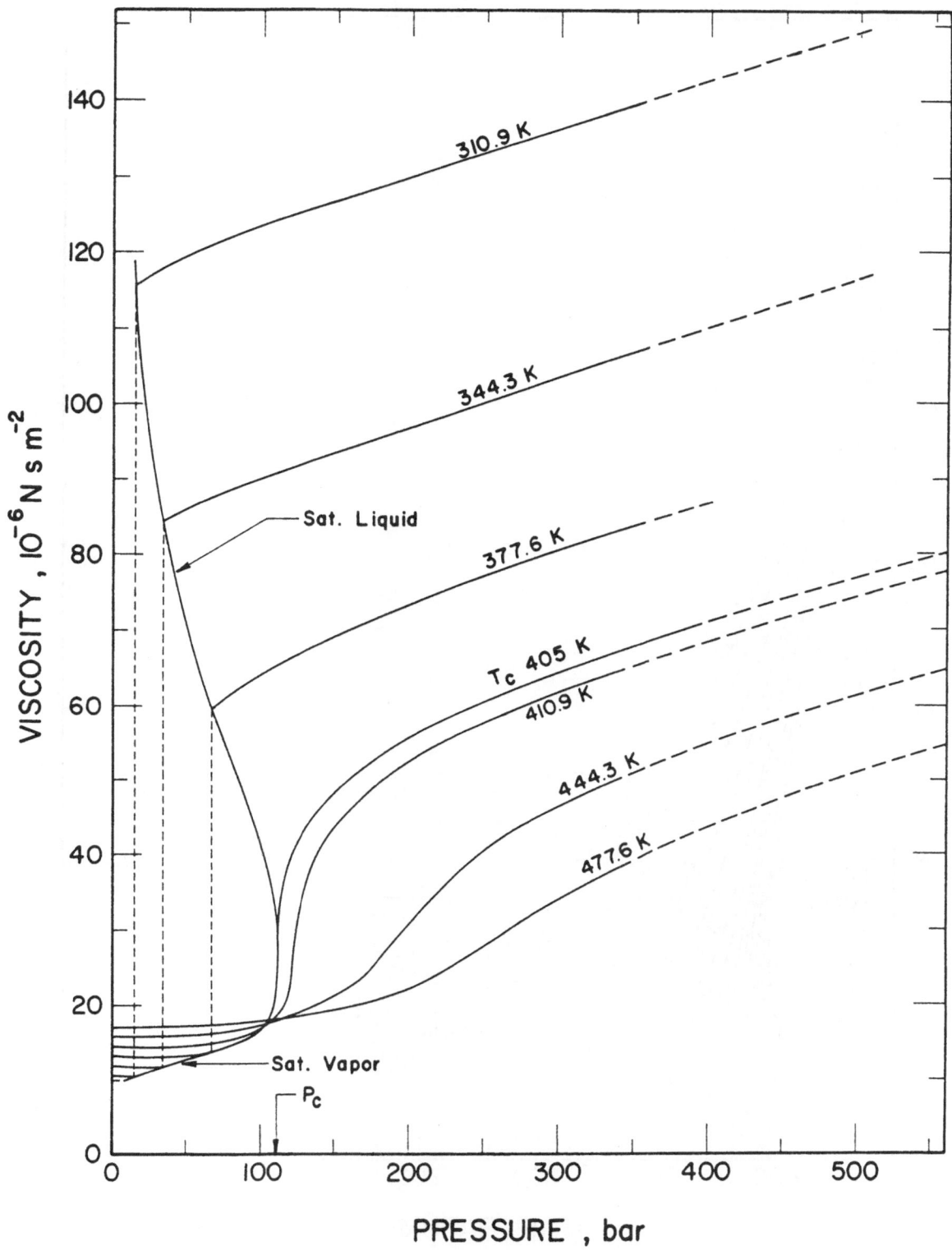

FIGURE 1. VISCOSITY OF AMMONIA [27].

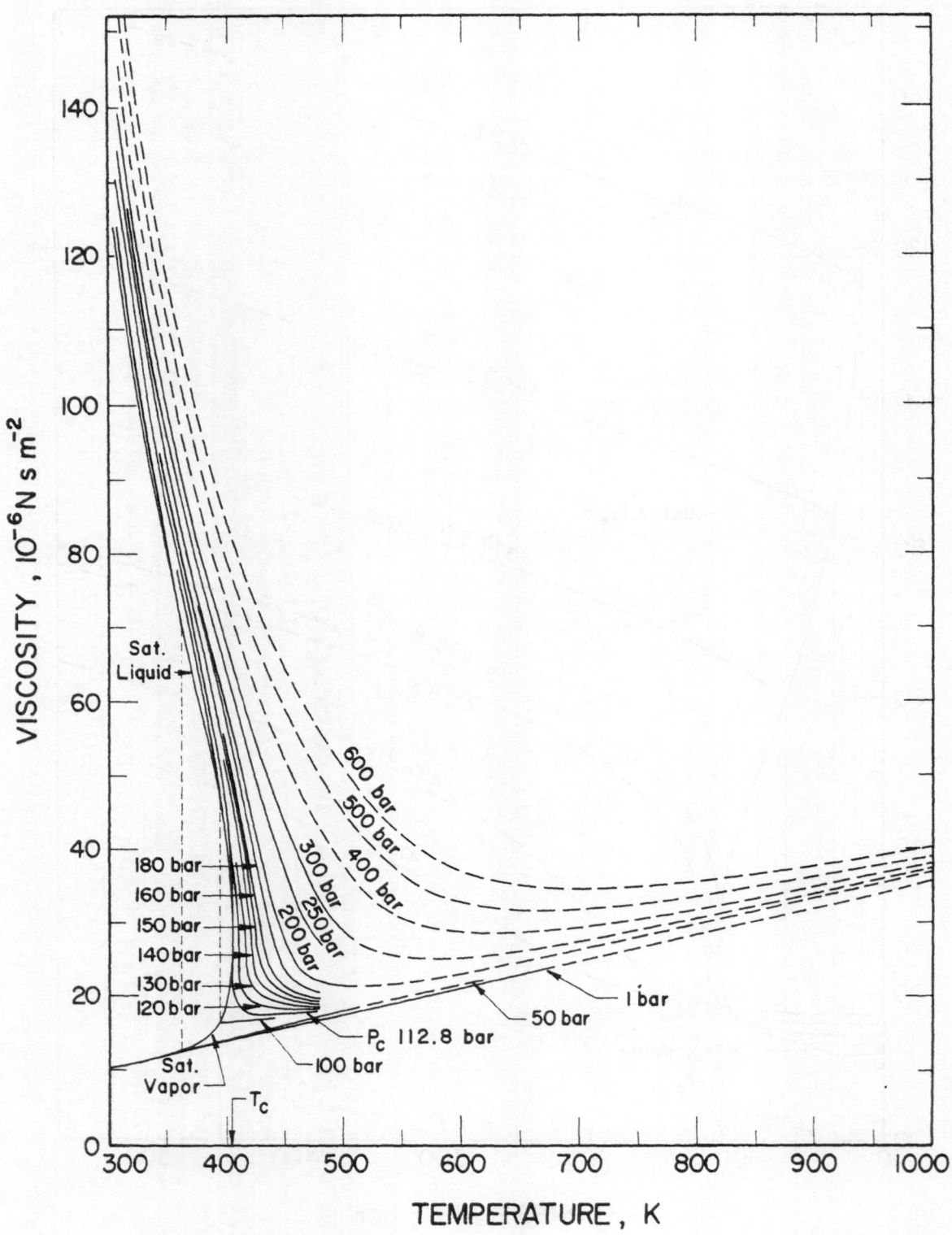

FIGURE 2. VISCOSITY OF AMMONIA [27].

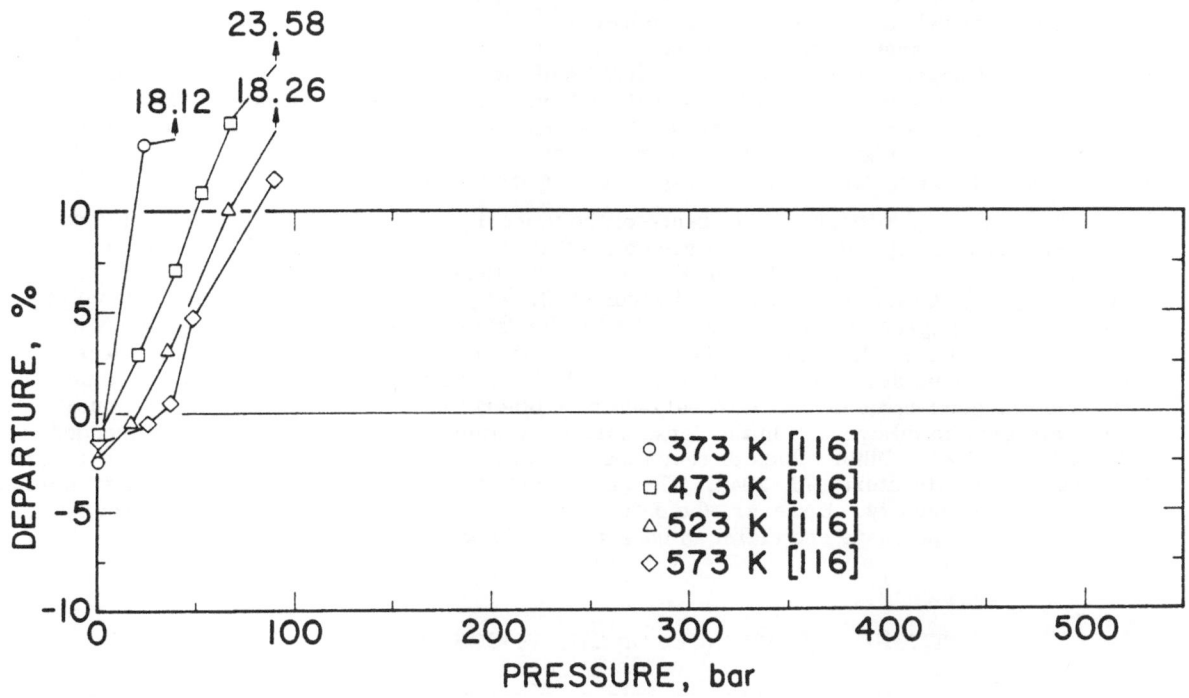

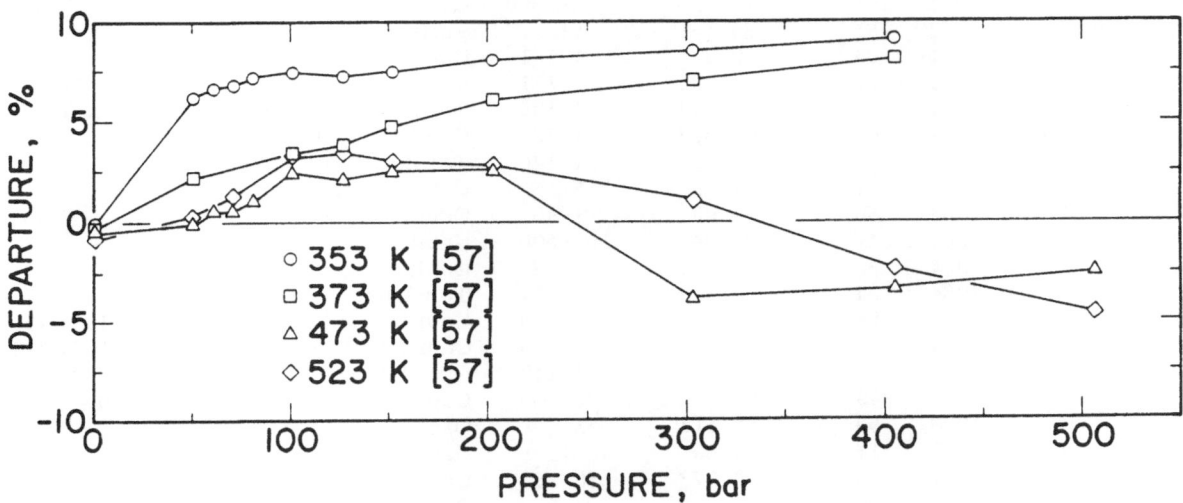

FIGURE 3. DEPARTURE PLOT ON THE VISCOSITY OF AMMONIA.

ARGON

A large amount of data exists for the viscosity of argon in all fluid states. The sources of data used to generate the recommended values are those of Andreyev et al. [8], Haynes [77] and Michels et al. [145]. These cover a temperature range from 108 K to 923 K, however, recommended values of up to 650 K only are tabulated on the next page. Figure 1 shows a selection of the actual data, where the smoothing modifications in the isotherms are about 1.5%. Using the P-V-T data of [148] the 1 bar data of [135] and the residual viscosity concept, it was possible to extrapolate the data of Haynes [77] in the temperature region of 170 K to 270 K up to 500 bar. Figure 2 shows isobars against temperature, the few extrapolated values being indicated by dashed lines. The saturated vapour data were taken from [195].

Additional works on the pressure dependence of the viscosity of argon are listed in the summary table below. An extensive comparison with the recommended values was made and the results are also shown in the summary table below as well as in Figure 3. Excellent agreement may be reported with the data of Kurin et al. [121], Gracki et al. [65] and Flynn et al. [49] where the deviations are seldom higher than 1%. Reasonably good agreement is seen to exist with the data of Kestin et al. [109], Golubev et al. [57], Reynes et al. [169], Kiyama et al. [116] and Makita [139], where the deviations amount to a few per cent. The agreement with the liquid state data of Hellemans et al. [81] and De Bock et al. [38] is quite unsatisfactory. In view of the good agreement of Haynes' data with others' in the gaseous region, his liquid state data are given preference in this work. In the dense gaseous region the accuracy of the recommended values are estimated to be ± 2%. Other tables of recommended values for the viscosity of gaseous and liquid argon may be found in the literature [76, 204]. The agreement of these values with the present recommendations is essentially satisfactory. However, the data in [76, 204], extend to lower temperatures and higher pressures in the liquid range and claim an uncertainty of about ± 5% and ± 2%, respectively.

ADDITIONAL REFERENCES ON THE VISCOSITY OF ARGON

Authors	Year	Ref. No.	Temperature K	Pressure bar	Method	Departure % (no. points)
Timrot et al.	1975	191	295–570	1–570	Oscillating-disk	1.8 (7)
Kurin and Golubev	1974	121	273–423	1–3800	Capillary tube	-0.07 ± 0.47 (20)
Slyusar and Rudenko	1972	181	85–151	SL	Falling cylinder	-
Kestin et al.	1971	106	298	1–101	Oscillating-disk	0.8
Gracki et al.	1969	65	173–298	7–171	Capillary tube	-0.60 ± 0.59 (29)
De Bock et al.	1967	39	89–140	1–196	Torsional crystal	7.87 ± 4.78 (23)
De Pippo and Kestin	1967	41	293–303	1–25	Oscillating-disk	0.4 (6)
De Bock et al.	1967	38	90	1–140	Torsional crystal	-
Boon et al.	1967	17	84–89	SL	Capillary tube	-
Van Itterbeek	1966	88	84–90	1–98	Oscillating-disk	- (11)
Martin et al.	1965	141	303–348	100–500	Capillary tube	-
Twasaki et al.	1964	91	293–303	1–53	Oscillating-disk	0.7 (14)
Lowry et al.	1964	134	102–128	51–507	Torsional crystal	18.6
Reynes and Thodos	1964	169	373–473	71–830	Capillary tube	3.26 ± 1.57 (21)
Reynes	1964	167	373–473	71–830	Capillary tube	3.26 ± 1.57 (21)
Kestin and Whitelaw	1963	109	296–537	1–142	Oscillation-disk	1.39 ± 6.28 (28)
Flynn et al.	1963	49	195–373	29–188	Capillary tube	-0.34 ± 0.44 (24)
Förster	1963	50	85–114	SL	Oscillating-disk	-
Boon and Thomaes	1963	18	84–89	SL	Capillary tube	-
Filippova and Ishkin	1961	48	90–273	36–153	Capillary tube	-
Kestin and Leidenfrost	1959	104	293–298	1–31	Oscillating-disk	-0.6 (15)
Kestin and Wang	1958	108	298	1–71	Oscillating-disk	0.7 (10)
Makita	1957	139	298–423	1–785	Rolling ball	0.16 ± 1.05 (30)
Zhdanova	1956	209	90–270	Isochores	Capillary tube	-
Robinson	1955	170	90–293	1–2027	Falling weight	-
Makita	1955	138	323–573	1–101	Rolling ball	0.16 ± 1.05 (30)
Kestin and Pilarczyk	1954	107	293	1–71	Oscillating-disk	-
Golubev and Petrov	1953	57	273–473	1–481	Capillary tube	2.89 ± 2.23 (33)
Kiyama and Makita	1952	116	323–573	1–97	Rolling ball	1.02 ± 2.53 (24)
Rudenko and Shubnikov	1934	172	84–87	SL	Capillary tube	-

VISCOSITY OF ARGON

[μ, 10^{-6} N s m^{-2}]

T, K	Pressure, bar																	
	1	20	30	40	48.7*	60	70	80	90	100	120	140	150	200	250	300	400	500
110	9.1	145.0	147.3	149.2	151.2	153.7	155.8	158.0	160.0	162.1	166.4	170.4	172.5	183.1	192.8	202.8	–	–
115	9.5	128.8	131.2	133.2	135.2	137.6	139.6	141.8	144.0	146.0	150.2	154.2	156.4	167.2	176.8	186.0	–	–
120	9.9	114.0	116.4	118.4	120.6	122.9	125.1	127.0	129.3	131.4	135.5	139.6	141.8	152.4	161.6	170.2	–	–
125	10.3	100.3	102.6	104.8	107.0	109.3	111.4	113.6	115.8	117.8	122.0	126.2	128.3	138.2	147.2	155.4	–	–
130	10.7	–	90.0	92.3	94.4	96.9	99.2	101.5	104.0	106.0	110.0	114.1	116.3	125.5	134.0	141.6	–	–
135	11.1	–	78.0	80.3	82.8	85.6	88.1	90.6	93.0	95.2	99.2	103.2	105.2	113.9	122.0	129.1	–	–
140	11.4	–	–	68.8	72.1	75.2	78.0	80.6	83.1	85.2	89.3	93.1	95.2	103.6	111.2	117.8	–	–
170	13.7	14.4	14.9	15.8	16.7	18.7	21.5	26.0	31.3	36.7	44.6	50.3	53.0	63.6	72.4	78.8	86.8	96.8
175	14.1	14.7	15.3	16.0	16.7	18.4	20.3	23.1	26.9	31.3	38.9	45.7	48.5	59.4	67.8	74.1	83.0	92.8
180	14.5	15.1	15.6	16.3	16.9	18.3	19.8	21.8	24.4	27.9	34.7	41.7	44.6	55.7	63.8	70.0	79.4	89.1
185	14.9	15.5	16.0	16.6	17.0	18.3	19.6	21.2	23.0	25.6	31.5	38.0	40.9	52.4	60.2	66.4	76.0	85.7
190	15.2	15.8	16.3	16.9	17.3	18.4	19.6	20.9	22.3	24.2	29.1	34.9	37.6	49.5	57.0	63.1	72.9	82.4
195	15.6	16.2	16.6	17.2	17.5	18.6	19.6	20.8	22.1	23.5	27.5	32.5	35.0	46.7	54.2	60.2	70.0	79.3
200	16.0	16.6	17.0	17.5	17.8	18.7	19.7	20.8	22.0	23.2	26.4	30.6	32.9	44.2	51.6	57.6	67.3	76.5
210	16.7	17.3	17.7	18.1	18.4	19.2	20.0	21.0	21.9	22.9	25.4	28.5	30.1	39.7	46.8	52.8	62.4	71.6
220	17.4	18.0	18.4	18.7	19.1	19.7	20.4	21.2	22.1	23.0	25.0	27.4	28.7	36.0	42.7	48.6	58.4	67.2
230	18.1	18.7	19.0	19.4	19.8	20.3	20.9	21.6	22.3	23.1	24.9	26.9	28.0	33.6	39.2	45.0	54.8	63.3
240	18.8	19.4	19.7	20.0	20.4	20.9	21.5	22.0	22.7	23.4	25.0	26.6	27.6	32.2	36.9	42.2	51.9	60.0
260	20.1	20.6	20.9	21.2	21.7	22.0	22.5	23.0	23.5	24.0	25.2	26.5	27.2	30.9	34.7	38.9	47.4	54.8
280	21.4	21.8	22.1	22.3	22.7	23.1	23.5	23.9	24.4	24.7	25.7	26.8	27.2	30.4	33.7	37.2	44.5	51.1
300	22.7	23.0	23.2	23.5	23.8	24.2	24.5	24.8	25.2	25.5	26.4	27.3	27.4	30.4	33.2	36.1	42.4	48.5
320	23.9	24.2	24.4	24.7	24.9	25.2	25.5	25.8	26.2	26.5	27.3	28.0	28.4	30.7	33.1	35.6	41.1	46.6
340	25.1	25.4	25.6	25.8	26.0	26.3	26.6	26.9	27.2	27.5	28.1	28.8	29.2	31.2	33.2	35.5	40.3	45.2
350	25.7	26.0	26.2	26.4	26.6	26.9	27.1	27.4	27.7	28.0	28.5	29.2	29.5	31.4	33.4	35.6	40.0	44.8
400	28.6	28.8	29.0	29.1	29.2	29.4	29.7	29.9	30.1	30.3	30.8	31.3	31.6	33.0	34.6	36.4	39.8	43.6
450	31.4	31.6	31.7	31.8	31.9	32.1	32.2	32.3	32.5	32.6	33.0	33.5	33.7	34.7	36.0	37.5	40.3	43.4
500	34.0	34.2	34.3	34.4	34.4	34.5	34.6	34.7	34.8	34.9	35.3	35.6	35.8	36.5	37.5	38.8	41.1	43.8
550	36.5	36.6	36.7	36.8	36.8	36.9	37.0	37.1	37.1	37.2	37.5	37.8	37.9	38.5	39.2	40.2	42.2	44.6
600	38.9	39.0	39.1	39.1	39.1	39.2	39.3	39.3	39.4	39.4	39.6	39.8	39.9	40.5	41.0	41.8	43.5	45.5
650	41.2	41.3	41.3	41.4	41.4	41.4	41.5	41.5	41.6	41.6	41.8	41.9	42.0	42.5	42.9	43.5	45.0	46.7

*Critical pressure.

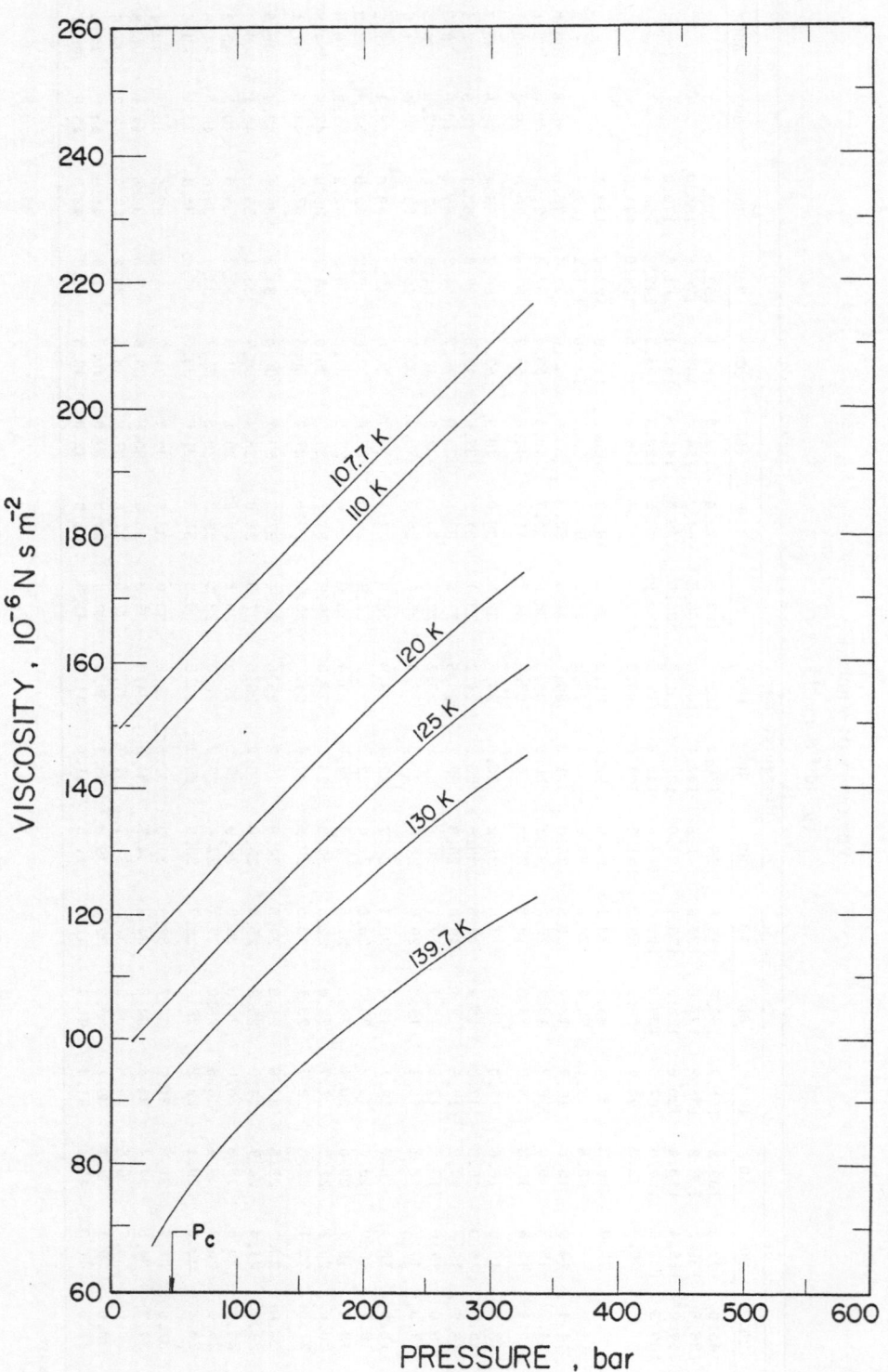

FIGURE IA. VISCOSITY OF ARGON [77].

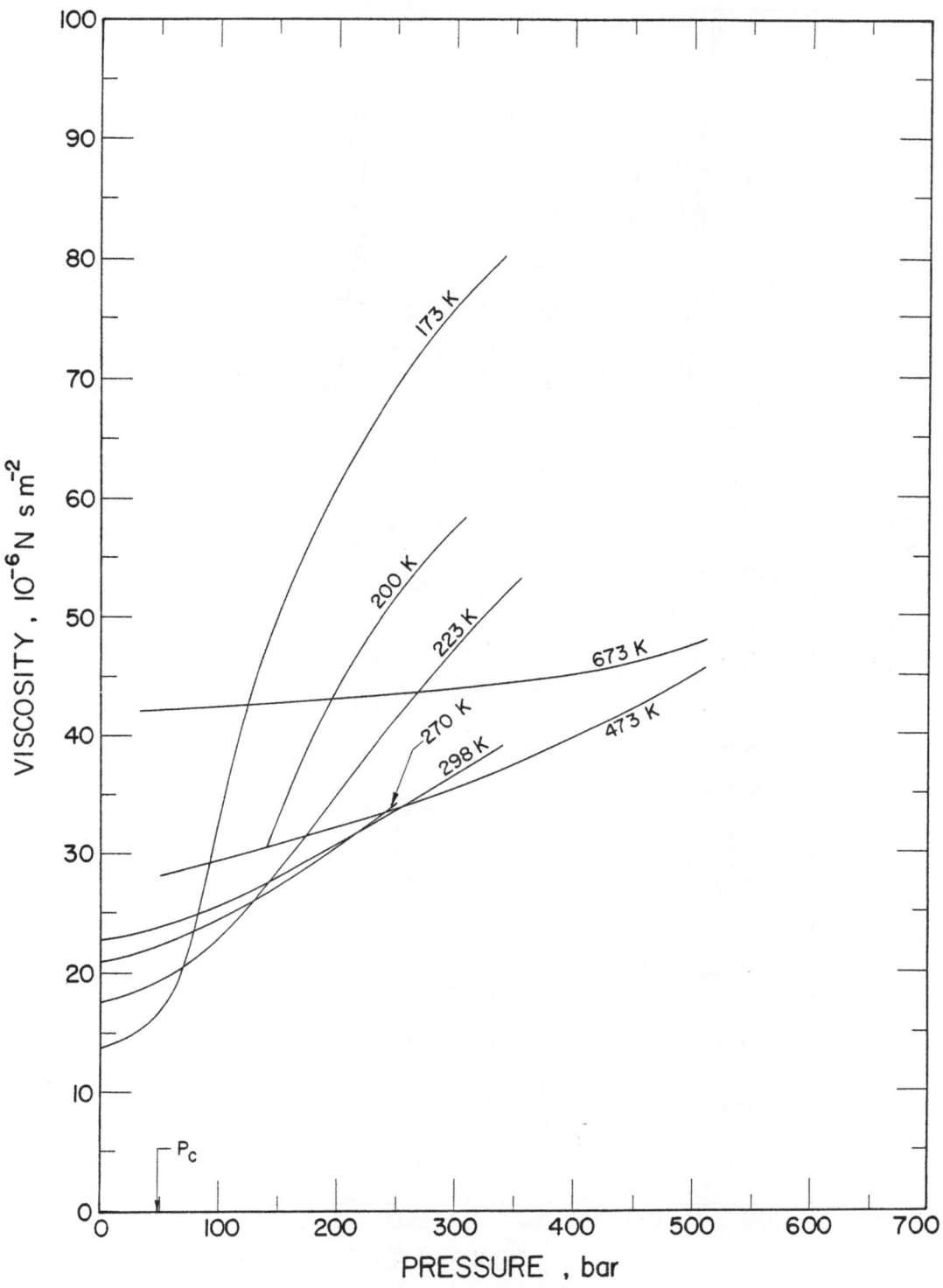

FIGURE IB. VISCOSITY OF ARGON [8, 77, 145].

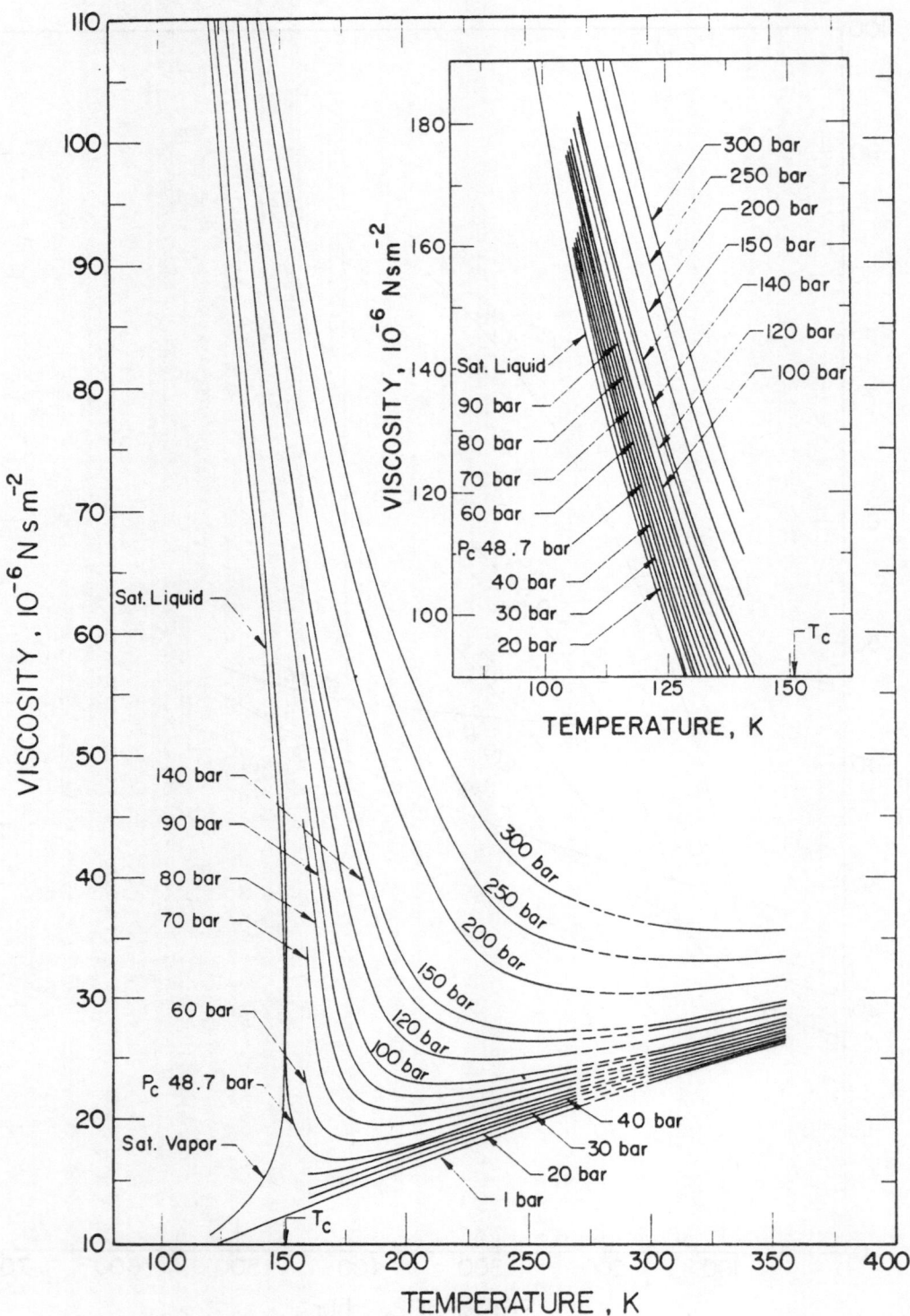

FIGURE 2A. VISCOSITY OF ARGON [77, 145].

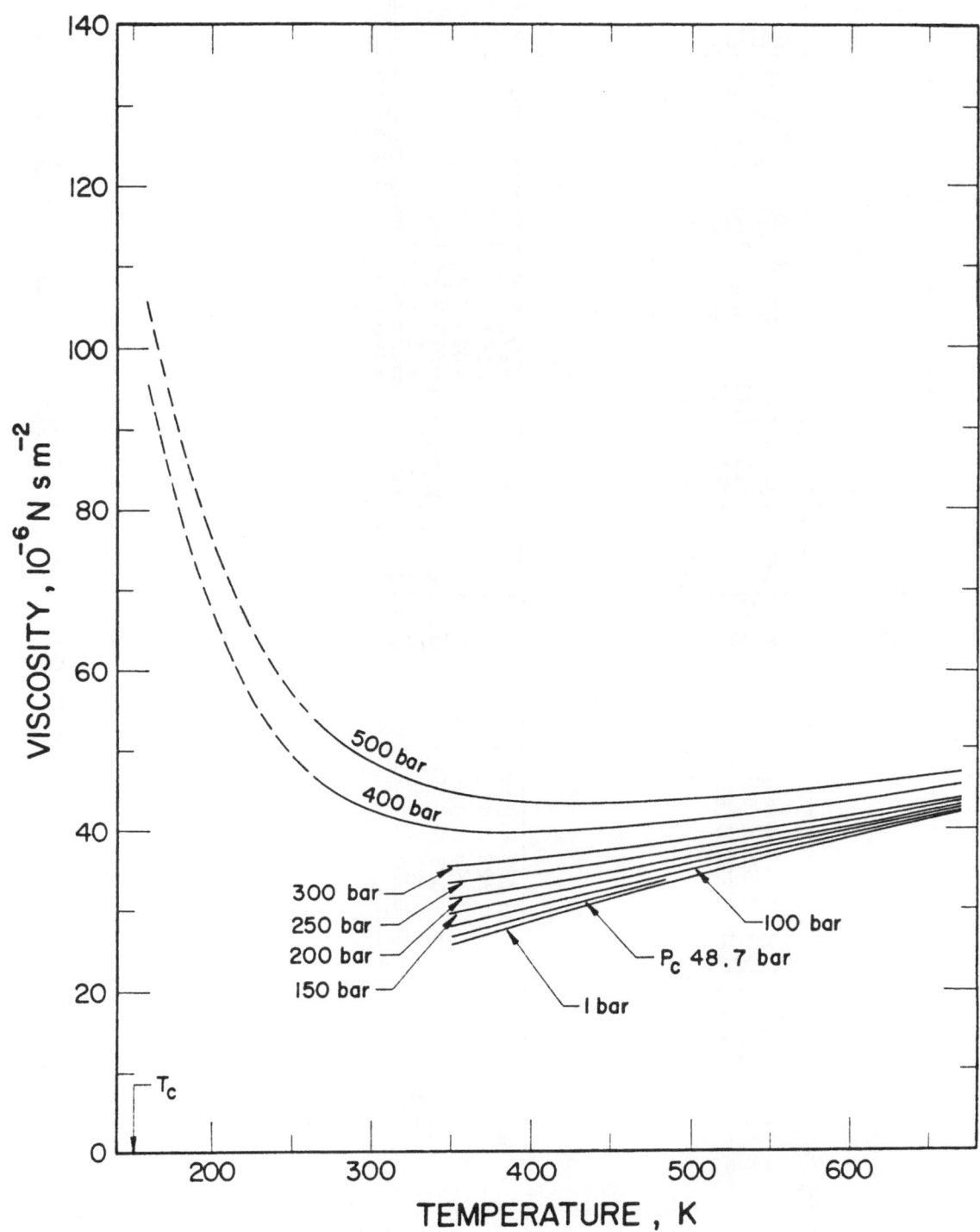

FIGURE 2B. VISCOSITY OF ARGON [8].

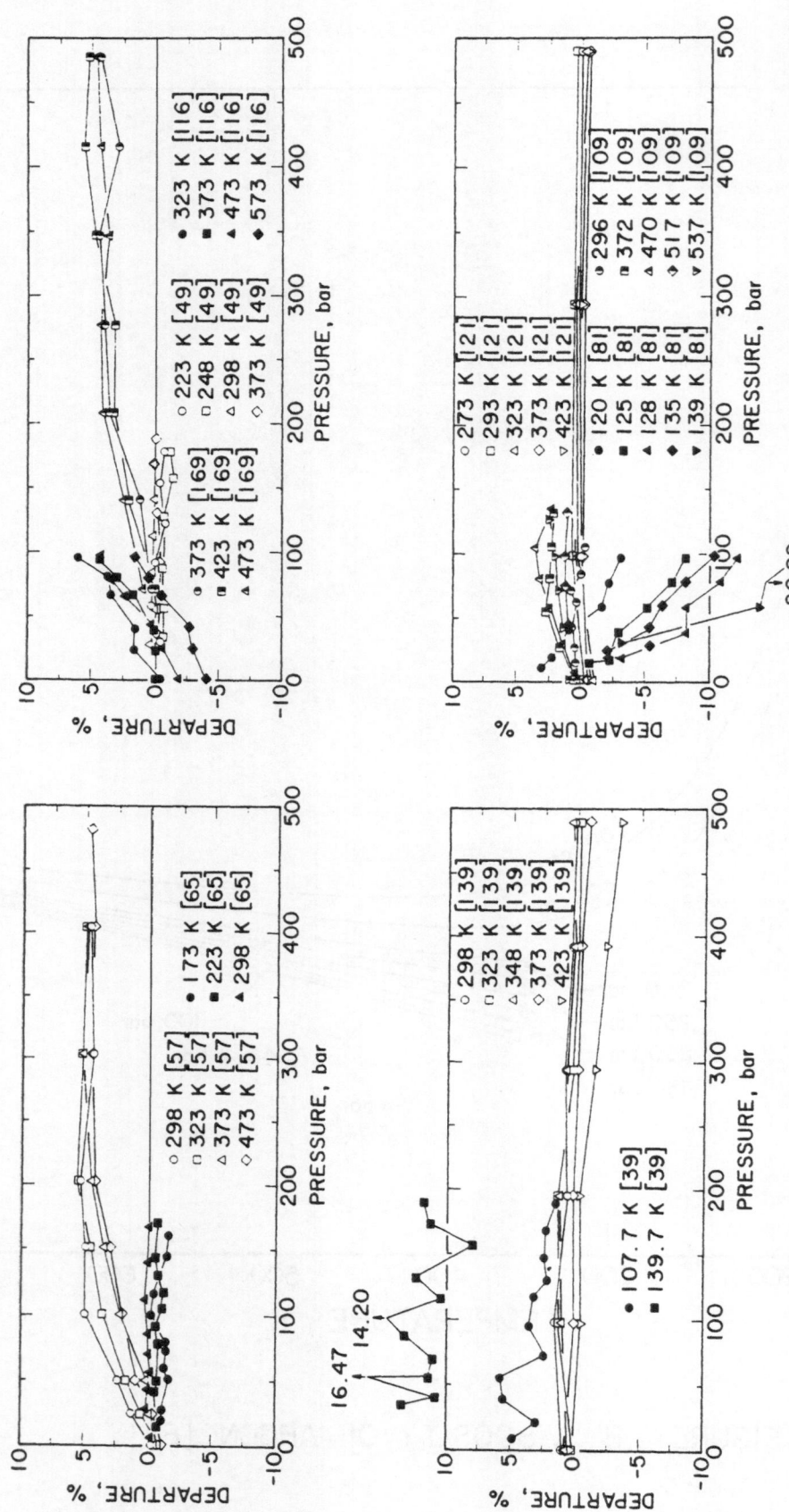

FIGURE 3. DEPARTURE PLOT ON THE VISCOSITY OF ARGON.

BENZENE

The data of Mamedov et al. [140] were used to generate the recommended values shown below. These data, obtained by a capillary tube viscosimeter, are plotted in Figures 1A and 1B as isotherms against pressure, where smoothing modifica- tions of a maximum of 1.1% had to be applied. In Figures 2A and 2B the data are plotted as isobars against temperature. The recommended values were read from Figures 2A and 2B.

Additional works on the pressure dependence of the viscosity of liquid benzene are listed in the summary table below. The agreement with the data of Collings et al. [34] is very satisfactory, though this could only be established over a narrow temperature range. Heterogeneous agreement is found with the data of Golubev et al. [57] as shown in the departure plot, Figure 3. In view of the confusing comparison with Golubev et al., the uncertainty in the recommended values is tentatively estimated to be ± 6% in the temperature range below 500 K. When this work was completed the formulation in [203] became known to the authors. These data include a larger region of states and tend to indicate that for temperatures above 500 K the recommended values given here may be too low, as borne out in the comparison with the Golubev's data. This is especially true at the lower pressures, where the values in [203] are higher than the recommended values by more than 10%.

ADDITIONAL REFERENCES ON THE VISCOSITY OF BENZENE

Authors	Year	Ref. No.	Temperature K	Pressure bar	Method	Departure % (no. points)
Collings	1971	34	303-323	1-1716	Torsional crystal	1.0
Golubev and Petrov	1970	57	368-521	1-659	Capillary tube	-0.62 ± 7.09 (20)
Heiks and Orban	1956	79	363-562	2-50	Falling weight	-
Kuss	1955	123	298-353	1-1961	Falling ball	-
Jobling and Lawrence	1951	94	303-363	98-441	Falling cylinder	3.7
Khalilov	1939	113	293-483	SL	Capillary tube	-
Bridgman	1926	21	303-348	1-11768	Falling weight	-

VISCOSITY OF BENZENE

$$[\mu, \ 10^{-6} \ N \ s \ m^{-2}]$$

T, K	Pressure, bar													
	1	20	40	49.0*	60	80	100	120	140	160	180	200	300	400
290	688	698	711	716	722	734	746	759	771	784	798	811	878	946
300	588	600	611	617	622	635	647	658	670	680	693	704	764	821
310	512	522	532	538	542	554	565	574	585	594	605	614	668	716
320	452	460	468	474	478	488	498	506	515	524	533	542	588	630
330	405	411	418	422	426	435	443	450	458	466	474	482	522	560
340	366	370	377	381	385	392	399	406	413	420	427	434	468	502
350	331	336	342	346	350	356	362	368	375	381	388	394	425	456
360	-	306	342	315	318	324	330	336	342	348	354	360	389	418
370	-	278	284	287	290	296	303	308	314	320	325	330	358	386
380	-	254	260	262	266	272	278	283	288	294	299	305	332	358
390	-	232	238	241	244	251	257	261	266	272	276	282	308	333
400	-	214	220	222	226	232	237	242	246	252	256	262	286	310
420	-	182	187	190	193	198	203	208	212	217	221	226	247	270
440	-	155	161	163	166	171	175	180	184	188	192	196	216	237
460	-	133	138	140	143	147	152	156	160	163	167	170	189	208
480	-	-	117	119	122	127	131	135	139	143	146	150	167	183
500	-	-	96.2	99.3	103	108	113	118	122	125	129	132	149	166
520	-	-	78.7	82.4	86.4	92.6	97.8	102	106	109	113	116	132	148
540	-	-	-	-	71.4	78.4	83.6	87.8	91.8	95.2	98.7	102	118	133
550	-	-	-	-	64.2	71.6	76.8	81.4	85.6	88.8	92.5	95.8	112	127

*Critical pressure.

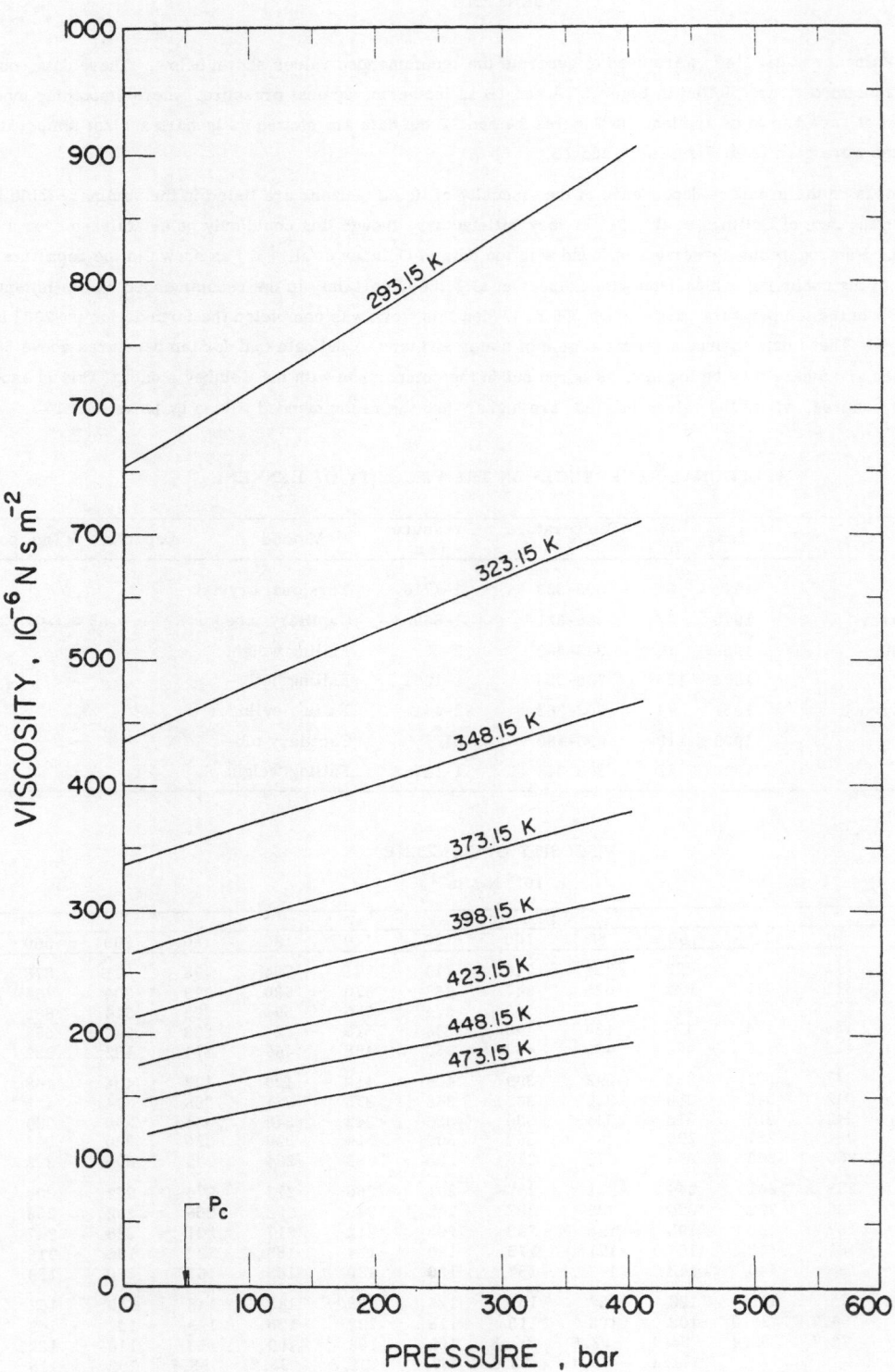

FIGURE I A. VISCOSITY OF BENZENE [140].

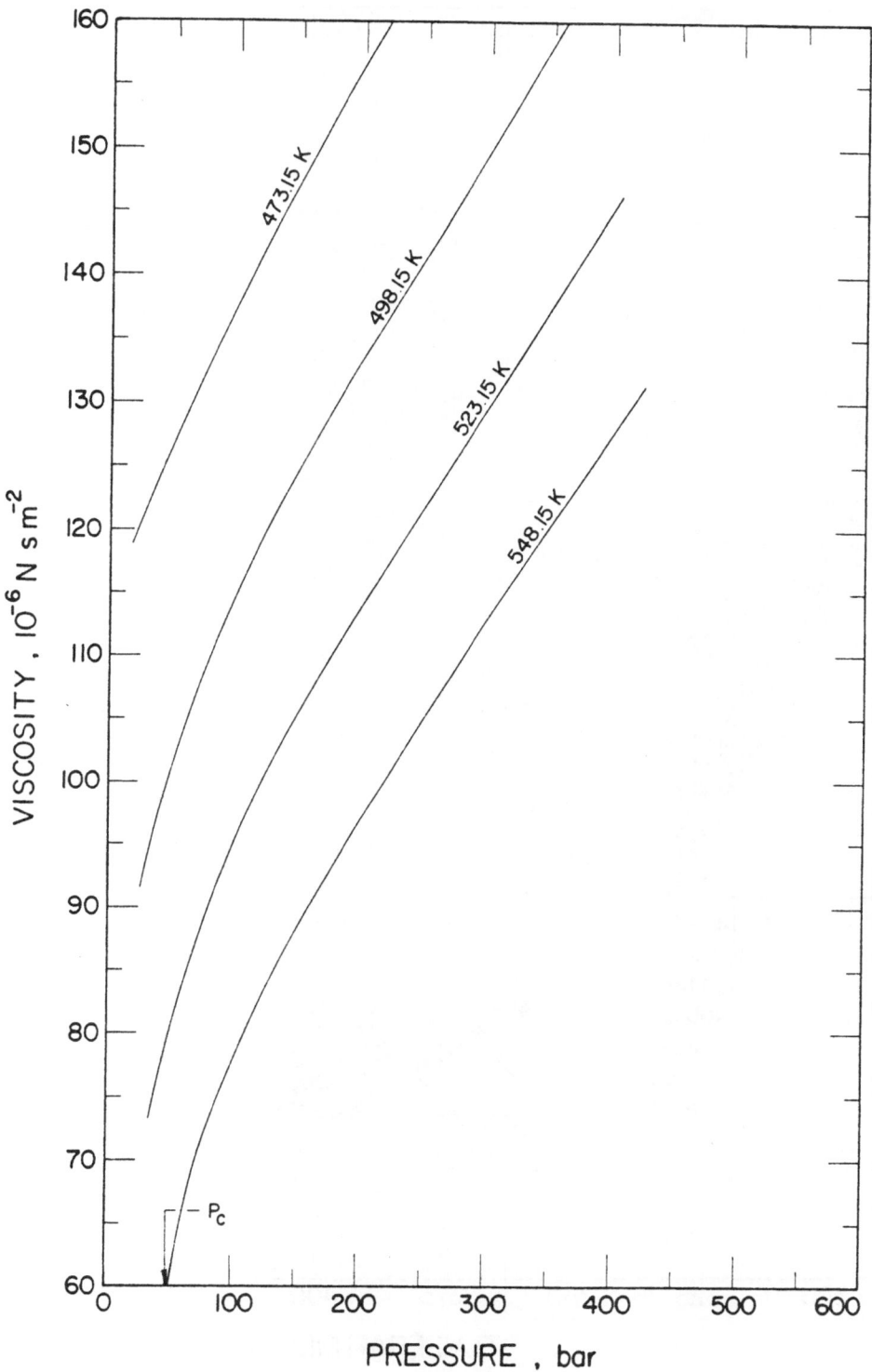

FIGURE 1B. VISCOSITY OF BENZENE [140].

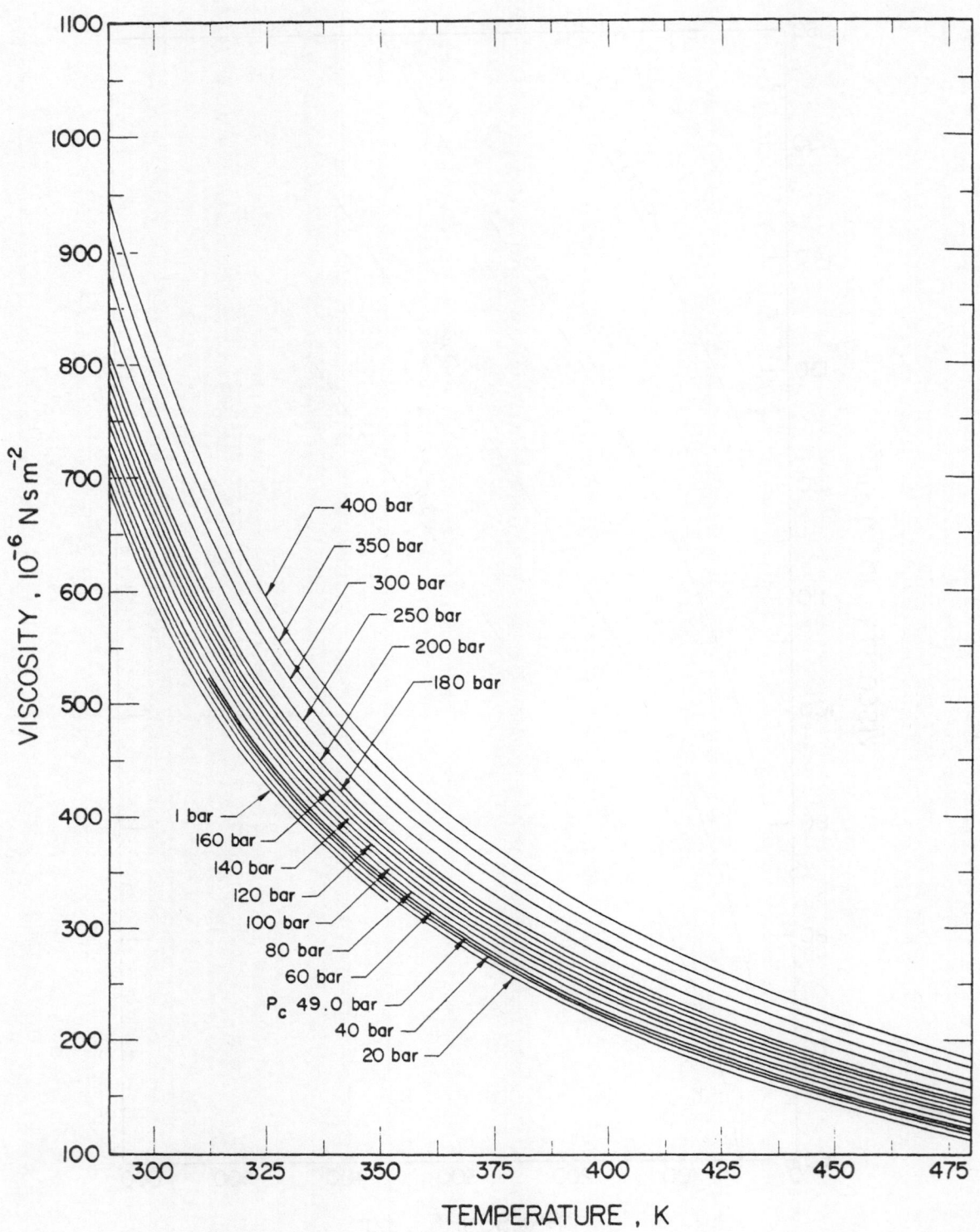

FIGURE 2A. VISCOSITY OF BENZENE [140].

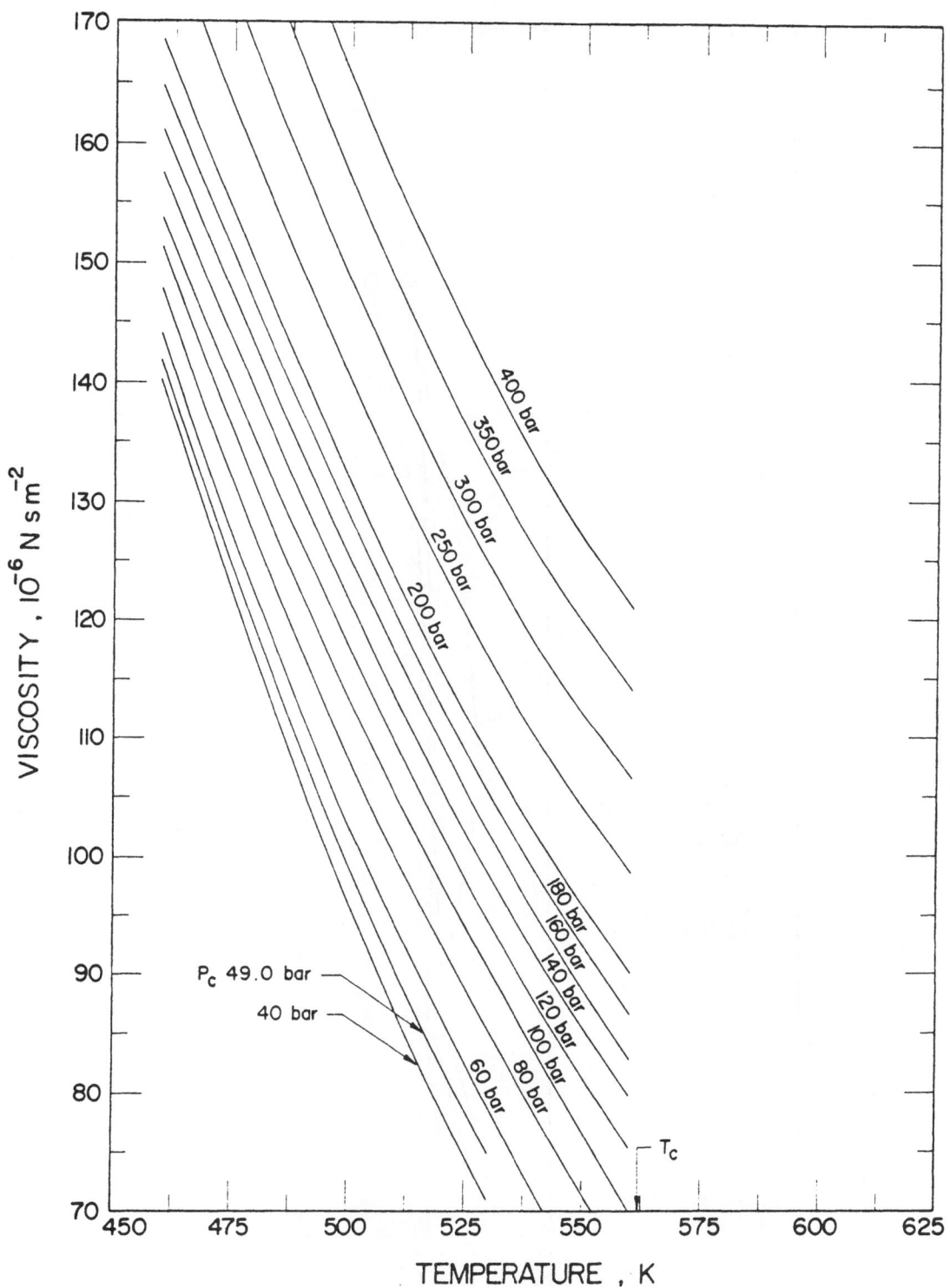

FIGURE 2B. VISCOSITY OF BENZENE [140].

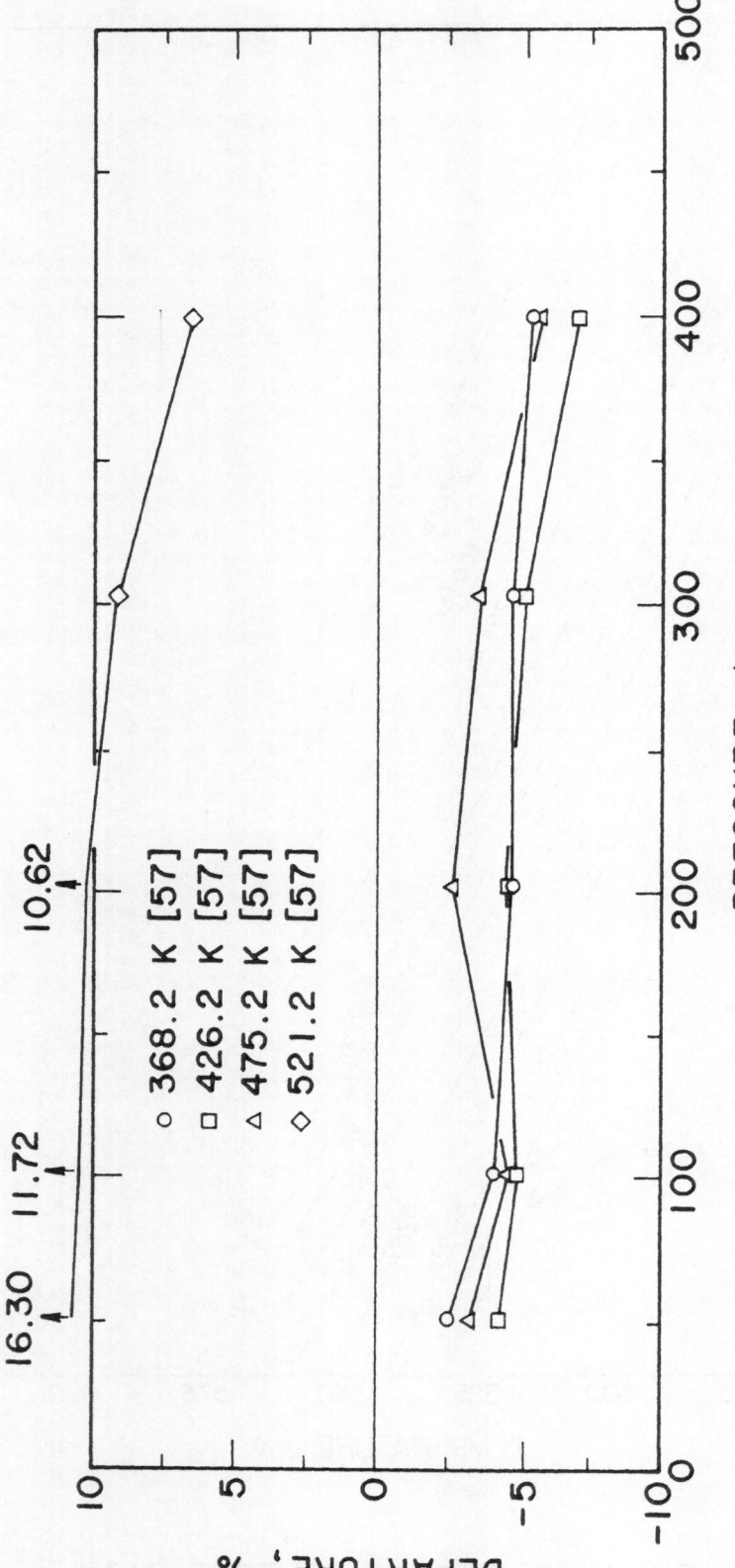

FIGURE 3. DEPARTURE PLOT ON THE VISCOSITY OF BENZENE.

BROMOTRIFLUOROMETHANE (R 13 B1)

The data of Karbanov and Geller [98], obtained by a capillary tube viscosimeter, were used to generate the recommended values tabulated below. The data are plotted as isotherms against pressure in Figure 1, where smoothing modifications of up to 1% have been applied. The data are plotted as isobars against temperature in Figure 2, where extrapolated values are indicated by dashed lines. Considerable deviations are found for the saturated liquid values reported in [195]. The limited data of Tsui [202], obtained by the rolling ball method with a small pressure range, exhibit deviations from the recommended values of 4 - 10%. The authors' statement of uncertainty is ± 1.5%.

VISCOSITY OF BROMOTRIFLUOROMETHANE (R 13 B1)

$[\mu, 10^{-6}\ N\ s\ m^{-2}]$

T, K	1	10	20	30	39.7*	50	60	70	80	90	100	120	140	160	180	200	300	400	500	600
290	14.5	16.5	155.0	159.0	162.7	166.9	169.7	174.0	177.0	180.8	184.6	189.8	196.0	202.4	207.7	212.4	238.2	263.5	289.2	314.4
295	14.7	16.4	145.5	149.8	153.5	158.2	161.4	165.5	168.8	172.5	176.4	182.0	188.2	194.5	200.0	204.8	230.6	255.8	280.8	305.4
300	15.0	16.4	136.0	140.4	144.4	149.4	153.0	157.0	160.7	164.4	168.2	174.3	180.5	186.8	192.3	197.4	223.2	248.0	272.4	296.4
305	15.2	16.4	126.4	131.2	135.2	140.7	144.7	148.8	152.8	156.4	160.2	166.7	173.0	179.1	184.8	190.2	216.0	240.4	264.4	287.6
310	15.4	16.4	17.9	122.0	126.2	132.0	136.4	140.7	144.8	148.4	152.4	159.2	165.5	171.6	177.4	183.0	208.8	232.8	256.3	278.9
315	15.7	16.5	17.9	112.6	117.0	123.2	128.1	132.6	137.0	140.6	144.7	151.7	158.2	164.4	170.2	176.0	201.6	225.5	248.4	270.4
320	15.9	16.7	18.0	103.4	108.0	114.4	119.8	124.6	129.3	133.0	137.2	144.5	151.0	157.2	163.2	169.0	194.8	218.4	240.7	262.0
325	16.2	16.9	18.0	94.0	98.7	105.7	111.5	116.8	121.8	125.6	130.0	137.4	144.0	150.1	156.2	162.2	187.9	211.2	233.2	254.0
330	16.4	17.1	18.2	20.6	89.3	97.0	103.2	109.0	114.4	118.4	122.8	130.4	137.1	143.3	149.4	155.6	181.2	204.2	225.8	146.0
335	16.6	17.4	18.3	20.2	18.3	88.0	94.8	101.3	107.0	111.2	115.8	123.7	130.4	136.8	142.9	149.2	174.7	197.5	218.8	238.6
345	17.1	17.8	18.7	20.0	24.2	67.3	77.5	86.3	92.8	98.0	102.8	110.9	118.0	124.4	130.8	137.2	162.5	184.8	205.4	224.6
350	17.4	18.0	19.0	20.2	23.2	53.6	68.6	78.9	86.0	91.6	96.7	104.9	112.2	118.8	125.2	131.5	156.8	178.8	199.2	218.2
355	17.6	18.2	19.1	20.3	23.0	35.6	59.6	71.7	79.2	85.6	90.8	99.2	106.6	113.4	120.0	126.2	151.4	173.2	193.3	212.2
360	17.8	18.4	19.4	20.5	22.8	30.7	50.2	64.6	72.6	79.8	85.0	93.8	101.4	108.4	115.0	121.2	146.4	168.0	188.0	206.5
365	18.1	18.7	19.5	20.6	22.7	28.3	41.3	57.5	66.4	74.2	79.6	88.8	96.6	103.8	110.3	116.5	141.6	163.1	182.8	201.3
370	18.3	18.9	19.8	20.8	22.6	26.8	35.6	50.4	60.8	68.9	74.4	84.1	92.0	99.4	106.0	112.0	137.1	158.4	178.0	196.4
375	18.6	19.1	20.0	21.0	22.6	26.0	32.5	44.0	55.6	64.0	69.7	79.7	87.8	95.2	101.7	108.0	132.9	154.0	173.5	191.8
380	18.8	19.4	20.2	21.2	22.6	25.4	30.6	39.6	51.2	59.5	65.2	75.5	83.7	91.2	97.8	104.0	128.9	150.0	169.2	187.5
385	19.0	19.6	20.4	21.7	22.6	25.0	29.2	36.7	47.2	55.3	61.2	71.6	80.0	87.5	94.1	100.2	125.2	146.0	165.2	183.3
390	19.3	19.8	20.6	21.5	22.7	24.7	28.3	34.6	43.8	51.6	57.6	67.8	76.2	84.0	90.6	96.7	121.5	142.4	161.5	179.4
395	19.6	20.0	20.8	21.7	22.8	24.5	27.6	33.1	41.2	48.4	54.2	64.4	72.8	80.6	87.3	93.3	118.9	138.8	157.8	175.8
400	19.8	20.2	21.0	21.9	22.9	24.4	27.2	32.0	39.0	45.8	51.3	61.3	69.6	77.6	84.2	90.2	114.9	135.6	154.6	172.4
405	20.0	20.5	21.2	22.0	23.0	24.4	26.9	31.2	37.2	43.6	48.8	58.4	66.7	74.6	81.3	87.2	112.0	132.5	151.4	169.2
410	20.3	20.7	21.4	22.2	23.1	24.5	26.8	30.6	35.8	41.6	46.4	55.8	64.0	71.8	78.5	84.4	103.1	129.6	148.4	166.2
415	20.5	20.9	21.6	22.4	23.3	24.6	26.8	30.2	34.6	39.8	44.5	53.4	61.3	69.1	75.8	81.8	106.4	126.8	145.6	163.3
420	20.8	21.1	21.8	22.6	23.5	24.7	26.8	29.9	33.6	38.4	42.7	51.1	58.9	66.6	73.4	79.2	104.0	124.2	142.8	160.5
425	21.0	21.4	22.0	22.8	23.6	24.9	26.9	29.7	32.8	37.0	41.0	49.0	56.6	64.1	71.0	76.8	101.4	121.7	140.3	157.8
430	21.2	21.6	22.2	22.9	23.9	25.1	27.1	29.6	32.3	36.0	39.5	47.0	54.4	61.8	68.6	74.5	99.1	119.3	137.8	155.3
435	21.5	21.8	22.4	23.1	24.1	25.4	27.4	29.6	32.0	34.8	38.0	45.0	52.3	59.6	66.4	72.4	96.8	117.0	135.4	152.8

*Critical pressure.

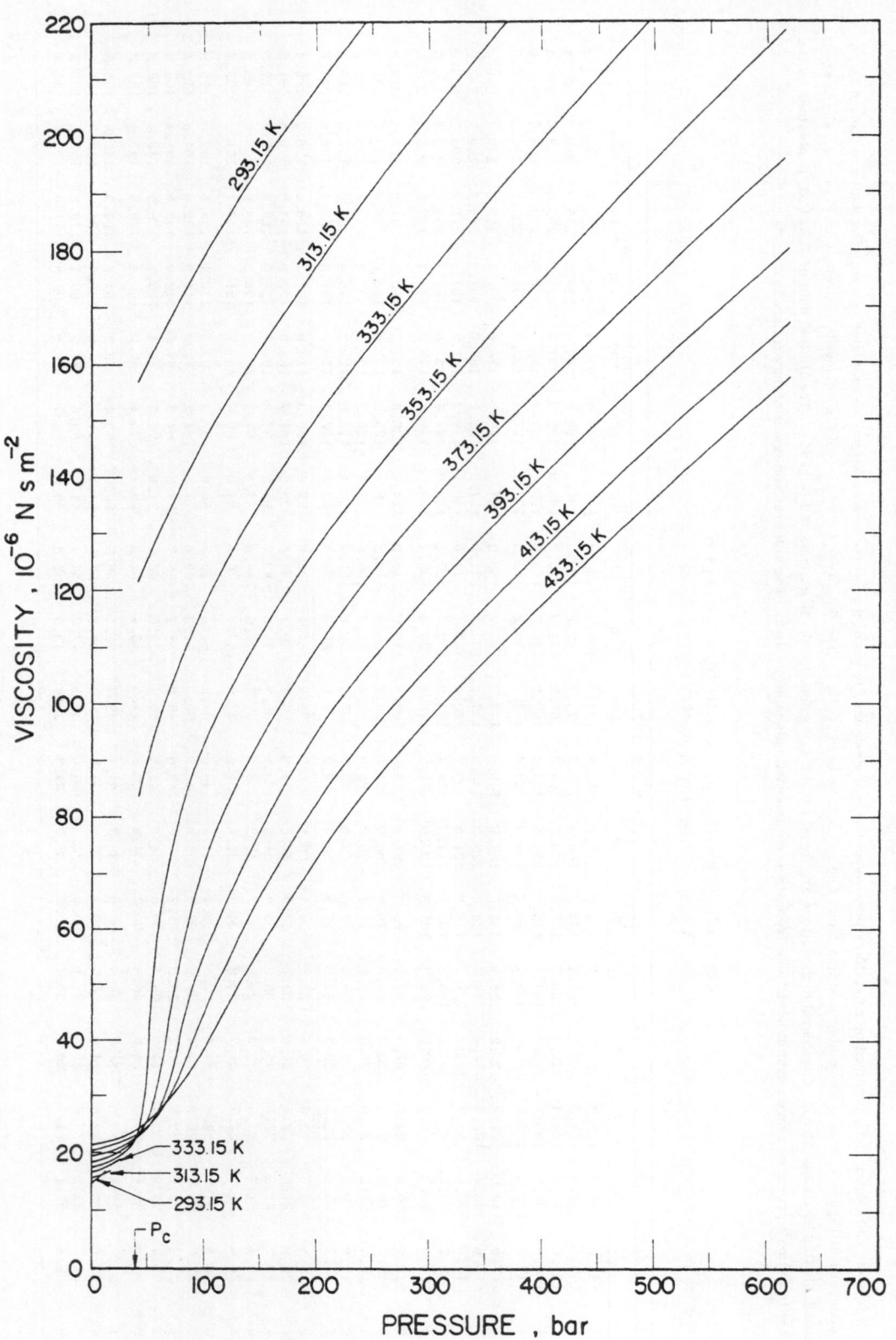

FIGURE I. VISCOSITY OF R 13BI [98].

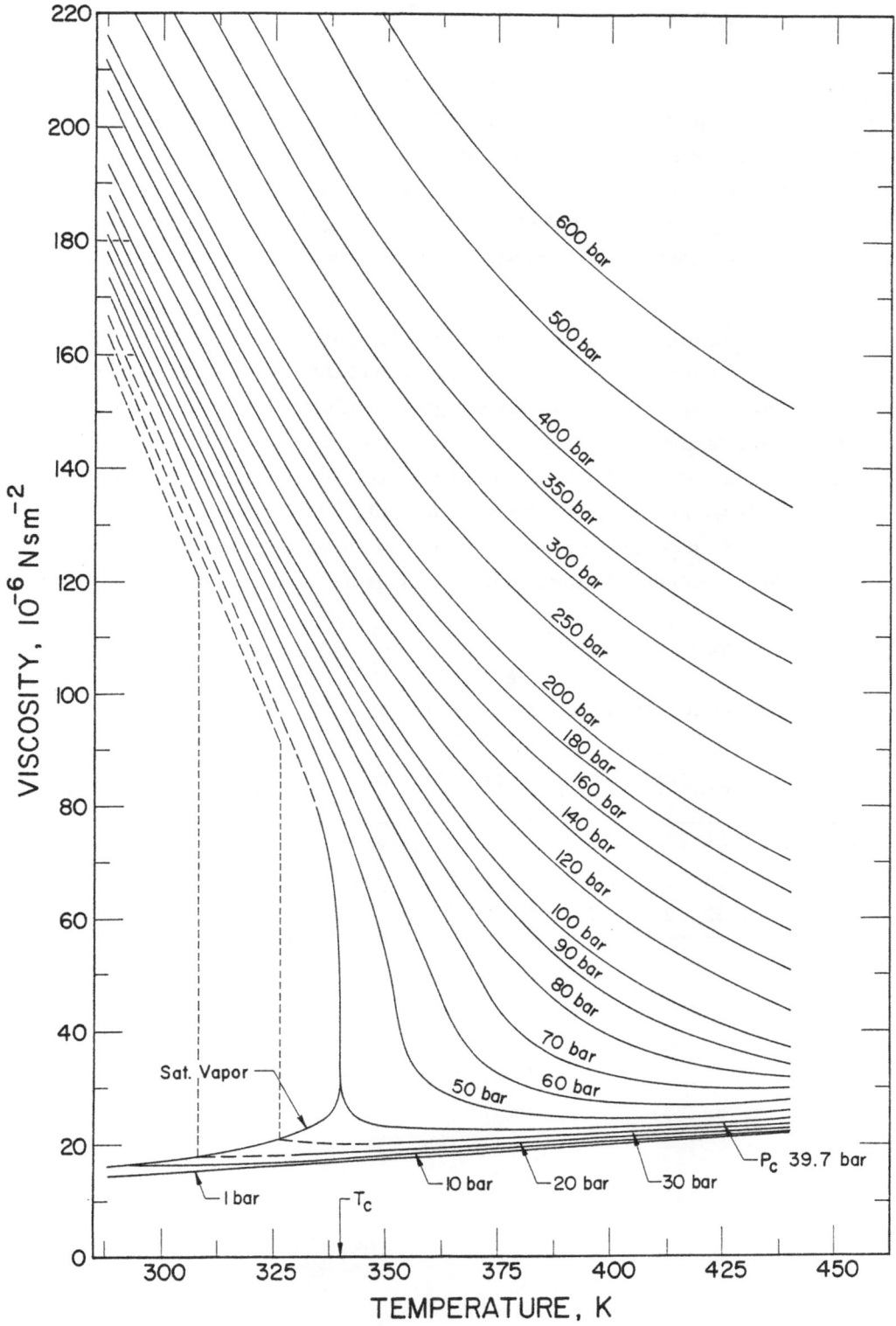

FIGURE 2. VISCOSITY OF R 13BI [98].

i-BUTANE

    The data of Gonzalez and Lee [63], obtained by a capillary tube viscosimeter, were used to gener-
ate the recommended values reported on the next page. The measured isotherms are plotted in Figure 1
as a function of pressure, where smoothing modifications of up to 3.1% had to be applied. Extrapolating
linearly the dilute gas values and using the P-V-T data of [143] as well as the residual viscosity concept,
an extrapolation from 510 K up to 850 K and 500 bar was possible. The data are shown as isobars against
temperature in Figure 2, from which the recommended values were read. Extrapolated values are indicated
by dashed lines.

    Additional works on the pressure dependence of the viscosity of i-butane are listed in the summary
table below. As seen from Figure 3, reasonably good agreement with the recommended values in the
region of experiments may be reported for the data of Agaev and Yusibova [4] which were also obtained by
a capillary tube viscosimeter and cover an even larger region of thermodynamic states. However, the
agreement with the data of Sage et al. [176], obtained by the rolling ball method, is quite unsatisfactory.
No comparison was possible with the data of Babb and Scott [10]. The uncertainty in the recommended
values is estimated to be ± 3% in the measured region. In view of some large discrepancies with the data
of Agaev and Yusibova the recommended values in the extrapolated region may have an uncertainty as high
as ± 10%.

ADDITIONAL REFERENCES ON THE VISCOSITY OF i-BUTANE

Authors	Year	Ref. No.	Temperature K	Pressure bar	Method	Departure % (no. points)
Agaev and Yusibova	1971	4	273-548	1-686	Capillary tube	-0.18 ± 3.58 (36)
Babb and Scott	1964	10	303	2000-8000	Rolling ball	– (4)
Sage et al.	1939	176	311-378	1-138	Rolling ball	-4.53 ± 6.3 (18)

VISCOSITY OF i-BUTANE

$[\mu \ 10^{-6} \ N \ s \ m^{-2}]$

T, K	|								Pressure, bar										
	1	20	30	36.5*	40	45	50	55	60	70	80	90	100	110	130	150	200	300	400	500
310	7.9	139.3	141.8	143.4	144.6	146.0	147.0	148.3	150.0	152.5	154.8	157.4	159.4	161.7	166.7	171.7	183.0	-	-	-
320	8.1	126.9	129.4	131.1	132.0	133.4	134.7	135.8	137.3	139.8	142.2	144.2	146.3	148.6	153.2	158.0	168.5	190.4	-	-
340	8.5	104.1	106.7	108.4	109.6	110.8	112.1	113.2	114.5	117.0	119.1	121.0	123.5	125.6	129.9	133.9	144.1	163.0	181.8	-
360	9.0	82.6	85.5	87.9	89.2	90.6	92.0	93.2	94.4	97.0	99.5	101.8	103.7	106.0	110.0	114.0	123.7	141.5	158.6	175.5
380	9.4	10.7	66.4	69.0	70.5	72.1	74.0	75.5	77.0	80.1	82.6	85.2	87.4	90.0	94.0	97.8	107.1	124.2	140.6	156.4
400	9.9	11.0	13.0	48.9	51.9	54.6	57.0	58.9	61.3	65.4	68.3	71.1	73.5	76.0	80.3	84.3	93.6	110.5	126.2	141.0
420	10.3	11.3	12.4	13.9	16.0	20.6	37.8	42.5	46.9	50.7	54.5	57.9	60.7	63.4	68.5	72.5	82.0	98.8	113.7	127.8
422	10.4	11.3	12.4	13.8	15.6	19.1	34.0	40.5	45.4	49.2	53.2	56.5	59.5	62.3	67.4	71.5	81.0	97.7	112.5	126.5
424	10.4	11.4	12.4	13.7	15.3	18.1	29.5	38.5	43.9	47.7	51.9	55.3	58.3	61.0	66.3	70.4	80.0	96.5	111.4	125.3
426	10.5	11.4	12.4	13.6	15.0	17.4	24.0	36.5	42.2	46.2	50.5	54.0	57.1	60.0	65.1	69.3	79.0	95.5	110.3	124.1
428	10.5	11.4	12.4	13.5	14.8	16.9	21.6	34.0	40.6	44.8	49.2	52.8	56.0	58.8	64.1	68.3	77.9	94.5	109.2	123.0
430	10.5	11.5	12.4	13.5	14.6	16.5	20.4	31.5	38.9	43.3	47.9	51.5	54.7	57.7	63.0	67.2	76.9	93.5	108.1	121.9
435	10.6	11.5	12.4	13.4	14.3	15.8	18.6	24.9	34.5	39.7	44.5	48.6	52.0	55.0	60.5	64.7	74.5	91.0	105.5	119.1
440	10.7	11.6	12.4	13.4	14.0	15.3	17.5	21.4	29.0	35.9	41.2	45.5	49.1	52.4	57.9	62.4	72.0	88.5	103.0	116.5
445	10.9	11.7	12.5	13.3	13.9	15.1	16.8	19.6	24.6	32.3	38.0	42.7	46.5	49.8	55.5	60.1	69.9	86.2	100.6	114.0
450	11.0	11.8	12.5	13.2	13.8	14.9	16.4	18.4	22.3	28.5	34.6	39.8	43.9	47.2	53.0	57.9	67.7	84.0	98.2	111.5
455	11.0	11.9	12.6	13.2	13.7	14.8	15.9	17.7	20.6	25.4	31.5	36.9	41.5	44.8	50.9	55.8	65.7	81.9	96.0	109.3
460	11.1	12.0	12.6	13.2	13.7	14.6	15.6	17.1	19.3	23.4	28.9	34.3	39.0	42.5	48.8	53.8	63.7	79.9	93.9	107.0
470	11.4	12.1	12.8	13.2	13.7	14.5	15.1	16.3	17.6	20.9	25.4	30.1	34.8	38.5	44.9	50.0	60.0	75.9	89.9	103.0
480	11.5	12.4	12.9	13.3	13.7	14.4	14.9	15.7	16.7	19.4	23.0	27.0	31.2	34.9	41.4	46.4	56.5	72.3	86.2	99.1
490	11.8	12.5	13.0	13.4	13.8	14.4	14.7	15.5	16.3	18.5	21.4	25.0	28.4	32.0	38.2	43.2	53.4	68.9	82.6	95.4
500	12.0	12.7	13.2	13.5	13.9	14.4	14.6	15.4	16.1	17.9	20.3	23.3	26.1	29.5	35.4	40.3	50.6	65.9	79.4	91.9
510	12.2	12.9	13.3	13.5	14.0	14.4	14.6	15.3	16.0	17.5	19.5	21.9	24.5	27.5	33.0	37.8	48.1	63.2	76.3	88.6
520	12.4	13.0	13.5	13.6	14.1	14.5	14.8	15.3	15.9	17.3	19.0	20.9	23.2	25.9	30.9	35.5	45.8			
540	12.9	13.5	13.8	14.0	14.3	14.6	15.0	15.3	15.8	16.9	18.3	19.9	21.6	23.6	27.7	31.8	41.9			
560	13.2	13.8	14.1	14.3	14.5	14.9	15.2	15.5	15.9	16.7	17.9	19.3	20.6	22.3	25.8	29.2	38.5			
580	13.6	14.1	14.4	14.6	14.8	15.2	15.5	15.6	15.9	16.6	17.6	18.8	20.0	21.5	24.4	27.3	35.7			
600	14.0	14.5	14.8	15.0	15.2	15.4	15.7	15.9	16.2	16.9	17.8	18.7	19.7	20.9	23.4	25.9	33.4			
620	14.4	14.9	15.2	15.4	15.6	15.8	16.0	16.2	16.5	17.1	17.9	18.7	19.5	20.6	22.7	24.9	31.5			
640	14.8	15.3	15.5	15.7	15.9	16.1	16.4	16.5	16.8	17.4	18.0	18.7	19.4	20.3	22.2	24.1	30.0			
650	15.0	15.5	15.7	15.9	16.1	16.3	16.5	16.7	16.9	17.5	18.0	18.7	19.4	20.3	22.0	23.8	29.3			
700	15.9	16.3	16.6	16.7	16.9	17.1	17.3	17.5	17.7	18.2	18.6	19.1	19.5	20.2	21.6	23.0	27.1			
750	16.7	17.1	17.4	17.5	17.6	17.8	18.0	18.2	18.4	18.8	19.1	19.5	19.9	20.5	21.6	22.8	26.0			
800	17.5	17.9	18.1	18.2	18.3	18.5	18.7	18.9	19.0	19.4	19.7	20.1	20.4	20.9	21.8	22.7	25.4			
850	18.4	18.7	18.8	18.9	19.0	19.2	19.4	19.6	19.7	20.0	20.3	20.6	20.9	21.3	22.2	23.0	25.2			

*Critical pressure.

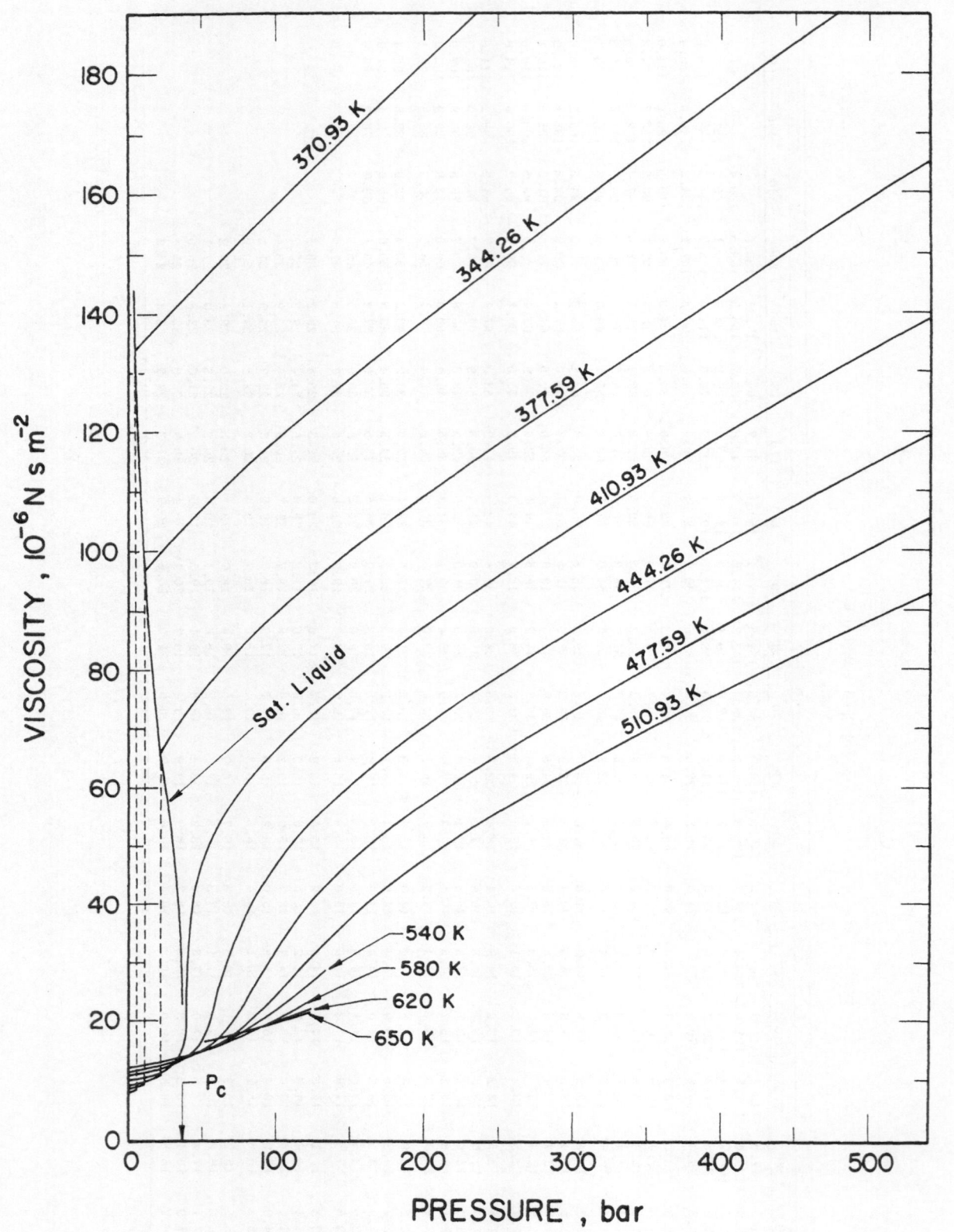

FIGURE 1. VISCOSITY OF *i*-BUTANE [63].

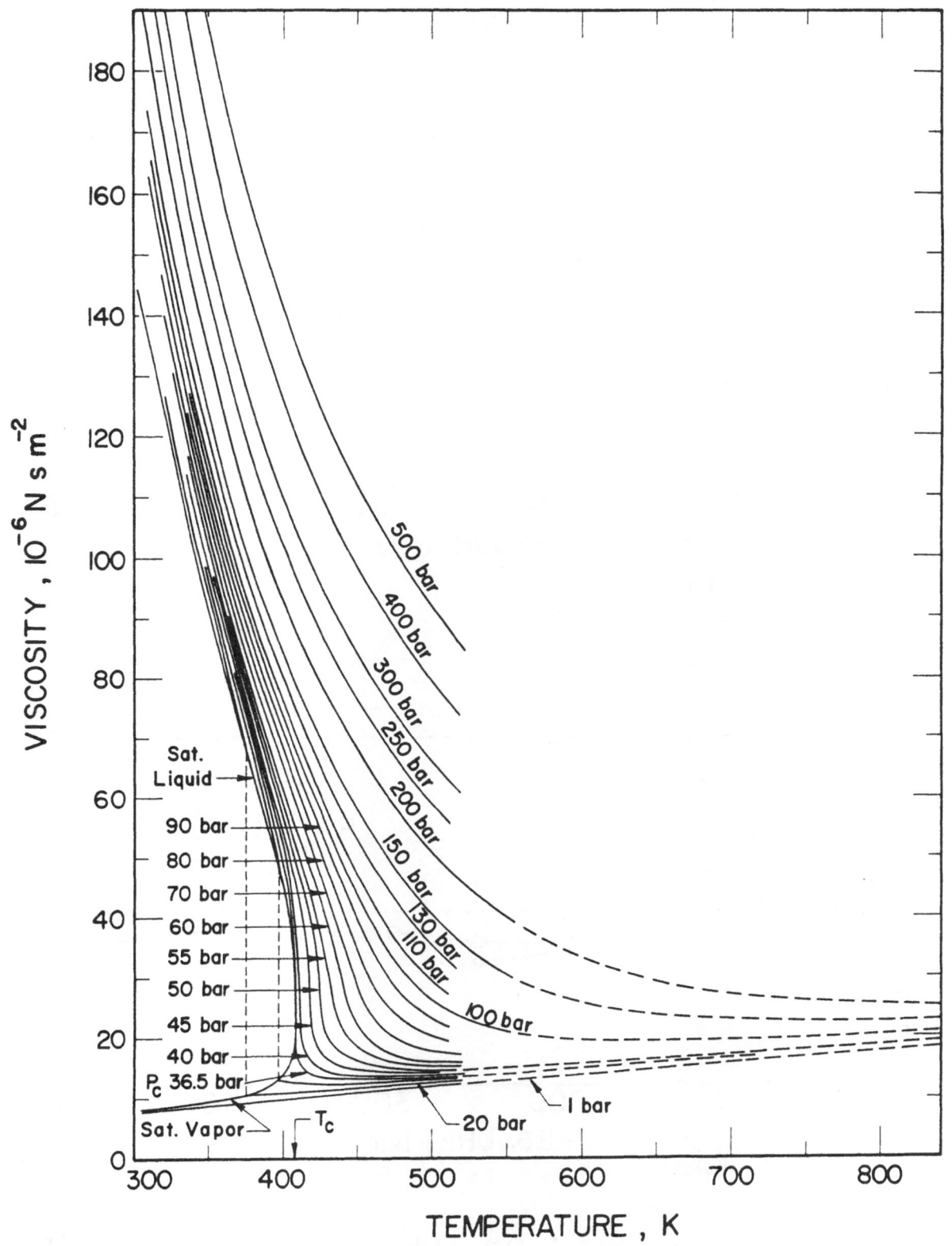

FIGURE 2. VISCOSITY OF i-BUTANE [63].

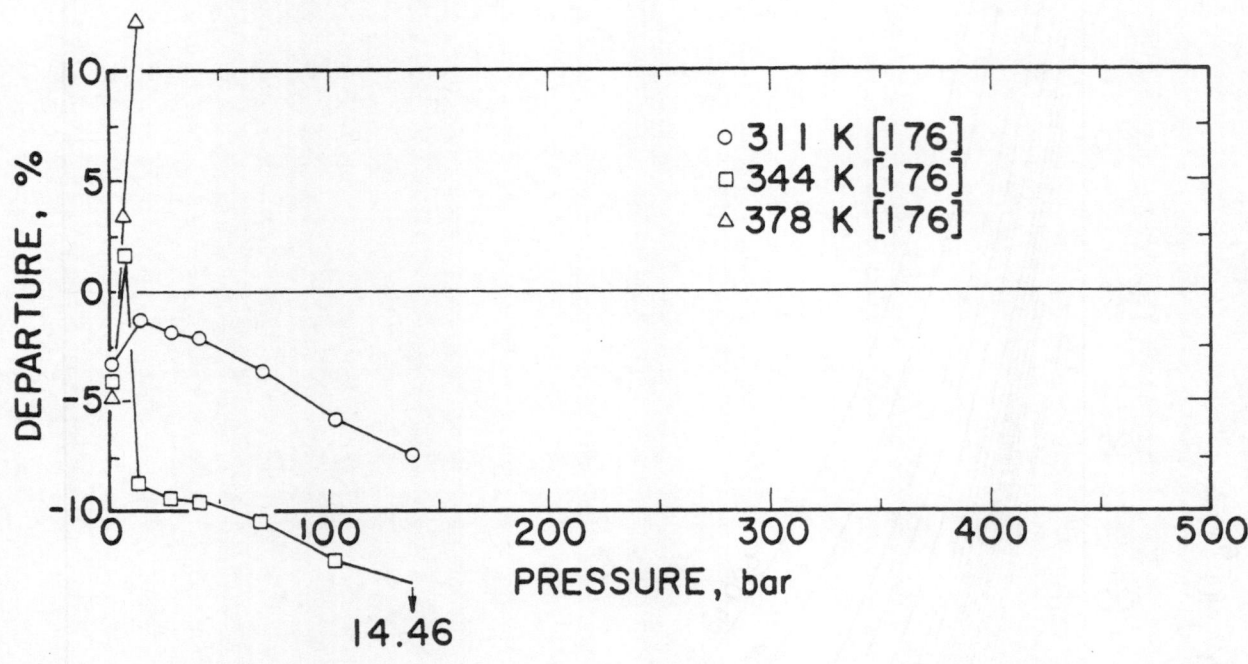

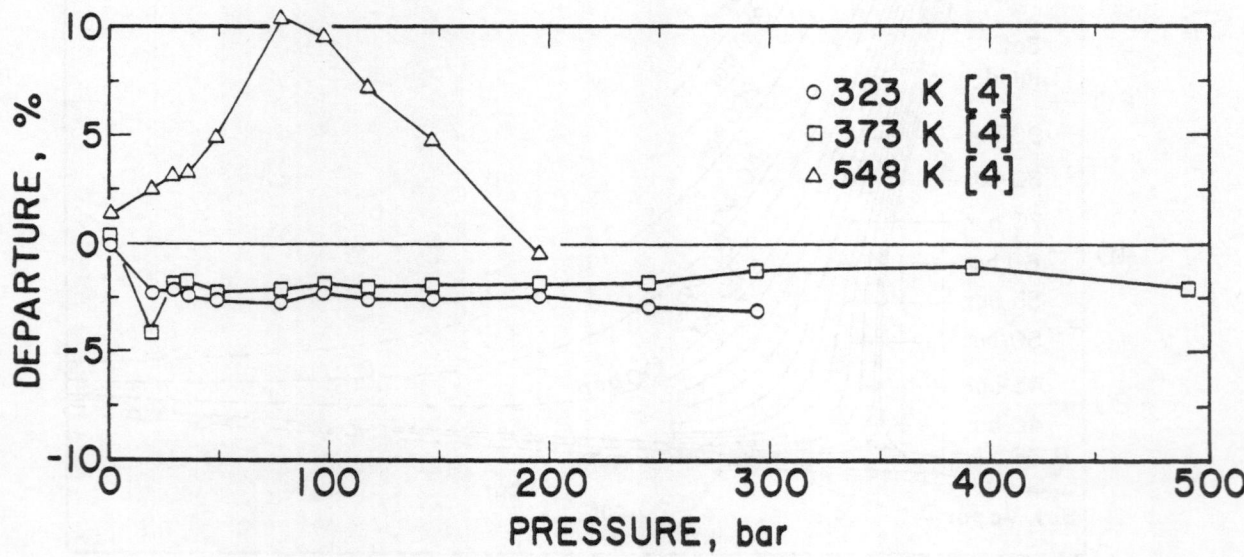

FIGURE 3. DEPARTURE PLOT ON THE VISCOSITY OF i - BUTANE.

n-BUTANE

The data of Dolan et al. [42] were used to generate the main body of the recommended values, presented on the next page. These data which have been in part extrapolated by the authors [42] from a smaller set of measurements, are presented in Figure 1 as isotherms against pressure. The precision of the measured data, which were obtained by a capillary tube viscosimeter, is claimed to be within ± 0.5%, while the uncertainty in all of the data presented is estimated in [42] to be ± 2% or better. Included in Figure 1 is the isotherm T = 277.59 K as measured by Carmichael and Sage [30]. Figure 2 shows a plot of viscosity against temperature for isobars, were the low temperature isotherm of [30] is again included. Extrapolating linearly the dilute gas viscosities and using the P-V-T data of [143] as well as the residual viscosity concept, it was possible to extrapolate from 510 K to 850 K up to 200 bar. The recommended values were read from Figure 2.

Additional works on the pressure dependence of the viscosity of n-butane are listed in the summary table below. As seen from the departure plots of Figure 3, the agreement with the data of Carmichael and Sage [30] is quite satisfactory. As already noted by these authors, the data of Dolan et al. [42] are a little higher at atmospheric pressure. Large deviations are found for the data of Sage et al. [176]. The view is taken that these older data, obtained by the rolling ball method, do not have the same accuracy as the more recent data. Good agreement is obtained with the highly accurate data of Kestin. The uncertainty in the recommended values is estimated to be ± 4% in the region reported by Figure 1 and ± 7% for the extrapolated values.

ADDITIONAL REFERENCES ON THE VISCOSITY OF n-BUTANE

Authors	Year	Ref. No.	Temperature K	Pressure bar	Method	Departure % (no. points)
Kestin and Yata	1968	112	293-303	1-25	Oscillating disk	3.7
Babb and Scott	1964	10	303	2000-10000	Rolling ball	- (5)
Carmichael and Sage	1963	30	278-433	1-345	Rotating cylinder	-0.59 ± 1.45 (17)
Starling et al.	1962	184	425-427	30-46	Capillary tube	-
Swift et al.	1960	190	293-373	SL	Falling body	-
Sage et al.	1939	176	311-378	1-138	Rolling ball	-0.95 ± 6.01 (33)

VISCOSITY OF n-BUTANE

$[\mu, 10^{-6} \text{ N s m}^{-2}]$

T, K	Pressure, bar																			
	1	20	30	38.0*	40	45	50	60	70	80	100	120	130	150	200	300	400	500	600	700
280	7.4	190.0	193.3	195.8	196.6	198.3	200.0	202.0	204.0	206.0	210.0	214.4	216.6	221.0	232.0	252.0	-	-	-	-
300	7.8	159.2	161.1	162.6	163.1	164.2	165.3	167.3	169.3	171.3	175.3	178.9	180.8	184.4	190.8	212.0	228.0	245.6	263.3	279.4
320	8.2	133.3	135.1	136.6	137.1	138.2	139.3	141.3	143.3	145.3	149.3	152.7	154.5	157.9	165.0	183.8	200.0	215.7	231.0	246.4
340	8.7	110.6	112.7	114.3	114.8	115.9	117.0	119.0	121.0	123.0	127.0	130.5	132.2	135.7	143.9	161.1	176.3	190.9	205.4	220.1
360	9.2	91.0	93.2	95.0	95.5	96.6	97.7	99.7	101.7	103.7	107.7	111.3	113.1	116.7	125.5	141.9	156.3	170.3	184.3	197.9
380	9.7	73.5	75.9	77.9	78.5	79.9	81.3	82.8	85.3	87.8	91.4	95.3	97.5	100.7	109.5	125.6	140.0	153.0	166.0	179.1
400	10.1	11.0	58.6	62.0	62.6	64.2	65.7	68.0	70.6	73.5	77.4	81.6	83.8	87.3	96.0	111.8	126.0	138.6	150.7	162.7
450	11.4	12.0	12.9	14.1	14.7	16.5	20.0	26.6	36.6	41.7	48.6	54.2	56.9	61.1	70.5	85.5	98.6	110.0	120.5	130.6
452	11.4	12.0	13.0	14.1	14.6	16.2	19.4	23.8	34.1	40.0	47.2	53.1	55.9	60.0	69.5	84.6	97.5	109.0	119.3	129.4
454	11.4	12.0	13.0	14.1	14.6	16.0	18.9	22.6	32.2	38.5	46.3	52.3	55.0	59.2	68.7	83.8	96.7	108.1	118.4	128.4
456	11.4	12.1	13.0	14.0	14.6	15.8	18.5	21.7	29.2	36.7	45.1	51.3	54.0	58.2	67.8	82.8	95.7	107.1	117.3	127.3
458	11.5	12.1	13.0	14.0	14.5	15.6	18.2	20.9	27.1	35.3	44.0	50.5	53.2	57.4	67.0	82.0	94.7	106.1	116.5	126.4
460	11.5	12.1	13.0	14.0	14.5	15.5	17.9	20.2	25.1	33.4	42.8	49.3	52.1	56.5	66.0	81.0	93.9	105.2	115.4	125.3
465	11.6	12.2	13.0	13.9	14.4	15.3	17.3	19.2	22.8	29.1	40.1	47.0	49.9	54.2	63.9	78.8	91.4	102.7	112.8	122.7
470	11.8	12.3	13.1	13.9	14.3	15.1	16.8	18.5	21.5	26.2	37.7	44.8	47.7	52.1	61.8	76.7	89.2	100.5	110.5	120.1
475	11.9	12.4	13.1	13.9	14.3	14.9	16.5	17.9	20.7	24.5	35.1	42.7	45.5	50.1	59.8	74.7	87.1	98.2	108.1	117.9
480	12.0	12.5	13.2	13.9	14.3	14.8	16.1	17.5	20.0	23.3	33.1	40.7	43.5	48.2	57.9	72.7	85.0	96.0	106.0	115.4
490	12.1	12.7	13.2	13.9	14.3	14.7	15.8	16.9	19.0	21.8	29.4	36.8	39.7	44.6	54.3	69.1	81.2	92.0	101.9	111.1
500	12.3	12.8	13.4	13.9	14.3	14.7	15.4	16.6	18.4	20.7	26.5	33.4	36.2	41.3	51.2	65.8	77.7	88.4	97.9	107.0
520	12.7	13.2	13.6	13.9	14.2	14.6	15.1	16.0	17.5	19.3	23.3	28.5	31.1	35.5	45.4	60.0	71.5	82.1	91.0	99.8
540	13.2	13.7	13.9	14.1	14.3	14.6	15.0	16.1	17.1	18.5	21.6	25.5	27.4	31.3	40.9					
560	13.5	14.0	14.2	14.4	14.5	14.9	15.2	16.2	17.0	18.2	20.7	23.6	25.1	28.0	37.2					
580	14.0	14.4	14.5	14.7	14.8	15.1	15.4	16.2	17.0	18.0	20.0	22.4	23.7	26.1	34.2					
600	14.5	14.8	15.0	15.1	15.2	15.4	15.6	16.4	17.2	17.9	19.5	21.6	22.7	24.8	31.7					
620	14.8	15.2	15.3	15.5	15.6	15.8	16.0	16.7	17.4	18.0	19.4	21.3	22.2	24.1	29.8					
640	15.3	15.7	15.8	16.0	16.1	16.2	16.4	17.0	17.6	18.2	19.4	21.0	21.8	23.4	28.2					
660	15.7	16.0	16.2	16.3	16.4	16.5	16.7	17.2	17.8	18.3	19.4	20.8	21.6	23.0	27.1					
680	16.0	16.4	16.5	16.7	16.8	16.9	17.1	17.6	18.1	18.5	19.5	20.8	21.5	22.8	26.2					
700	16.5	16.8	17.0	16.7	17.2	17.3	17.4	17.9	18.4	18.8	19.8	21.0	21.5	22.7	25.8					
750	17.4	17.7	17.8	17.9	18.0	18.1	18.2	18.6	18.9	19.3	20.0	21.0	21.6	22.6	25.1					
800	18.2	18.4	18.5*	18.6	18.7	18.8	18.9	19.2	19.6	19.9	20.6	21.4	21.9	22.7	24.8					
850	18.9	19.1	19.2	19.3	19.4	19.5	19.6	19.9	20.2	20.4	21.0	21.8	22.2	23.0	24.8					

*Critical pressure.

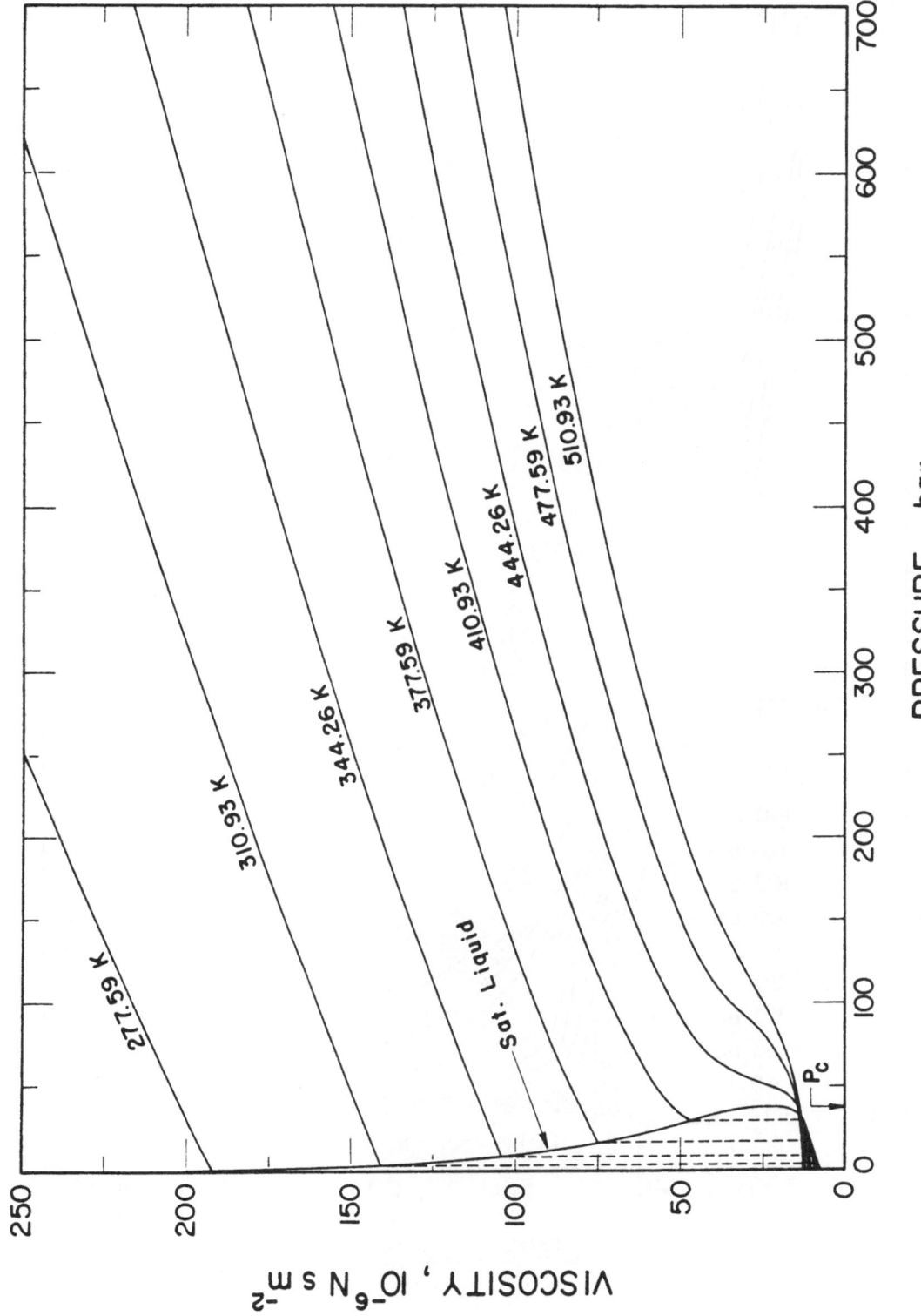

FIGURE 1. VISCOSITY OF n-BUTANE [42].

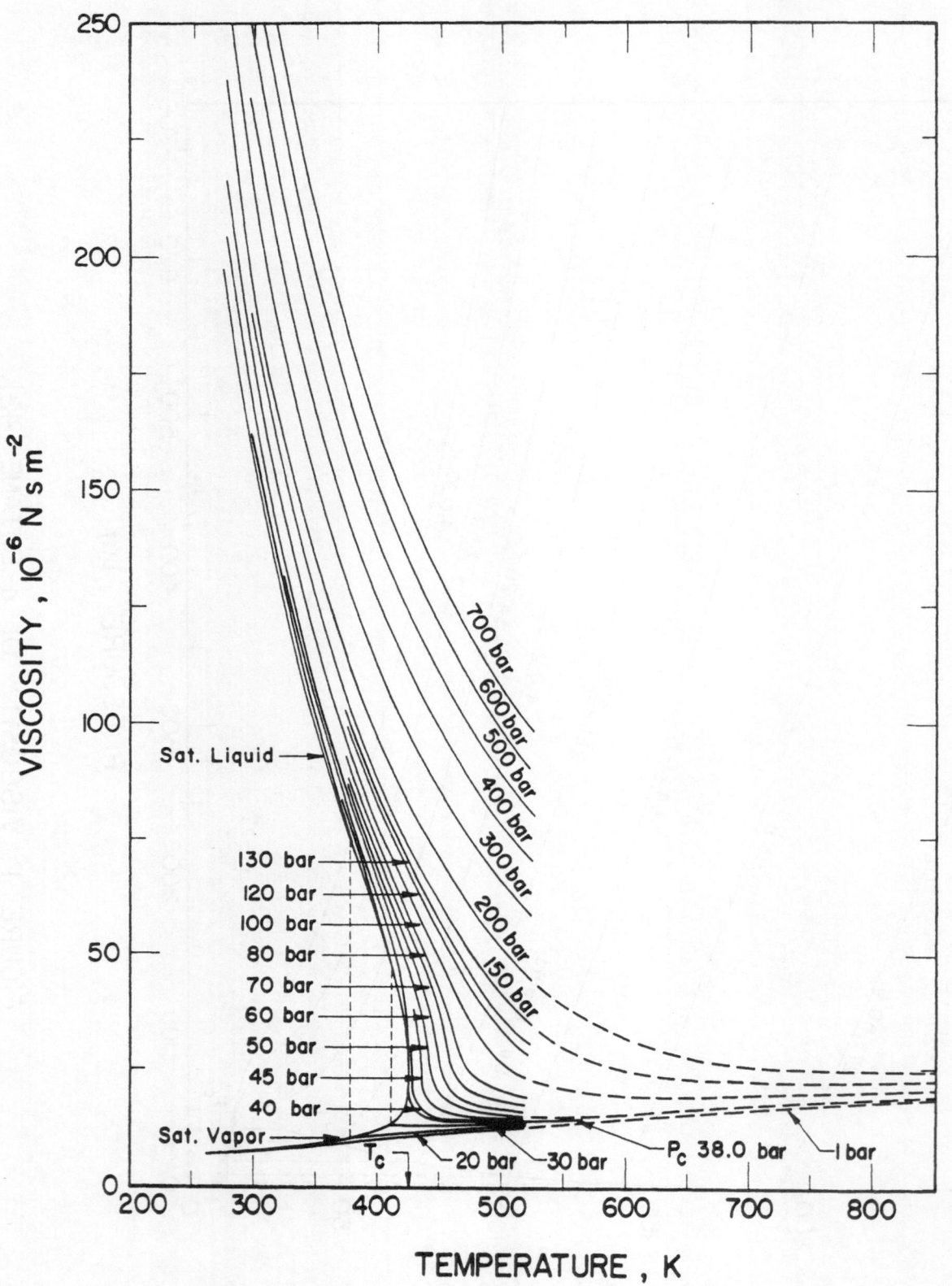

FIGURE 2. VISCOSITY OF n-BUTANE [42].

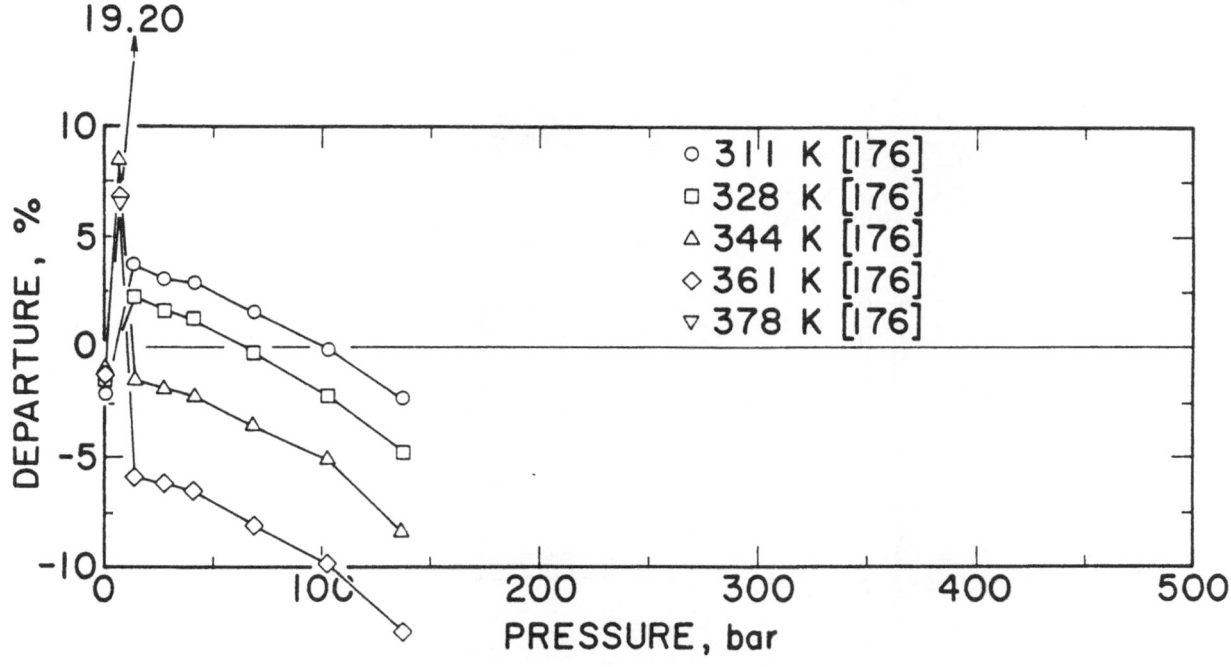

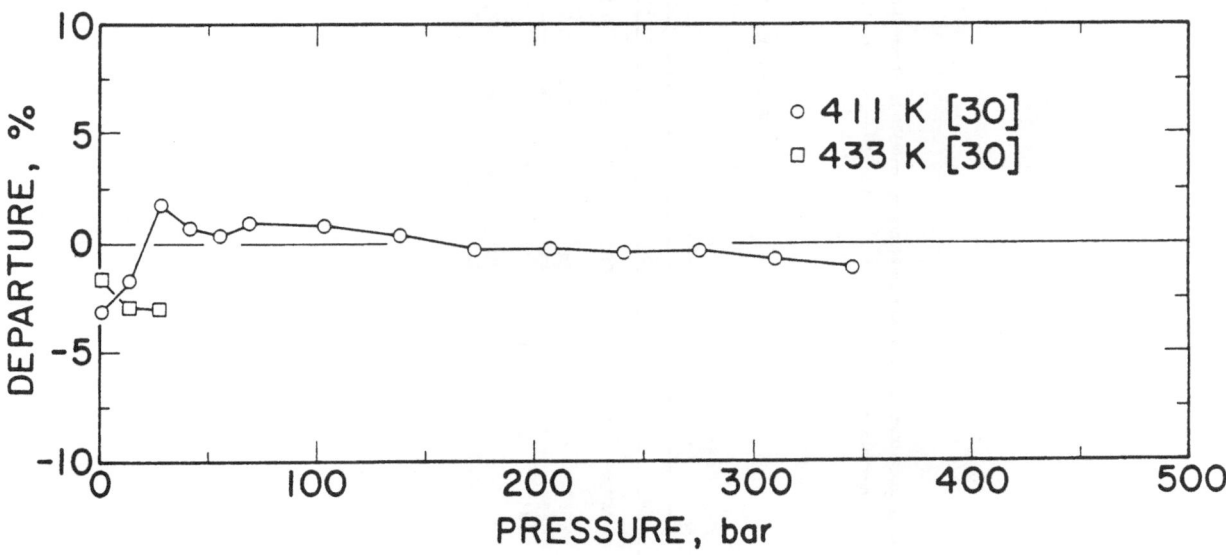

FIGURE 3. DEPARTURE PLOT ON THE VISCOSITY OF n – BUTANE.

BUTYLACETATE

The only available set of data is that by Guselnov and Kadscharov [70]. It was used to generate the recommended values tabulated below. These data were obtained with a capillary tube viscosimeter, the uncertainty is stated to be 1.1%. The data are plotted as isotherms against pressure in Figure 1, and as isobars against temperature in Figure 2.

VISCOSITY OF BUTYLACETATE
$[\mu, 10^{-6} \text{ N s m}^{-2}]$

T, K	\multicolumn{20}{c}{Pressure, bar}

T, K	1	5	10	20	30	40	50	60	70	80	90	100	150	200	250	300	350	400	450	500
290	755	759	763	767	775	782	790	797	805	812	820	827	862	900	935	970	1005	1040	1075	–
300	667	671	675	680	687	695	702	709	716	723	730	737	775	807	840	875	900	930	970	1000
320	517	520	523	527	533	539	545	550	555	560	565	570	600	627	655	680	705	730	757	785
340	417	420	424	427	431	436	440	444	448	452	456	460	487	507	525	547	570	585	605	625
360	340	342	345	347	352	357	362	365	368	371	374	377	400	415	432	450	467	485	500	515
380							300	303	307	310	314	320	335	350	365	380	395	410	422	437
400							255	258	262	265	269	272	285	297	310	322	337	350	377	375
420							220	223	226	229	232	235	245	255	267	280	292	305	377	327
440							190	192	195	197	200	202	212	222	235	245	255	270	277	287
460							165	168	171	174	177	180	187	197	210	220	230	240	250	257
480							145	147	150	152	155	157	167	177	187	200	207	217	227	235
500							132	134	136	138	140	142	152	165	172	185	192	202	210	222

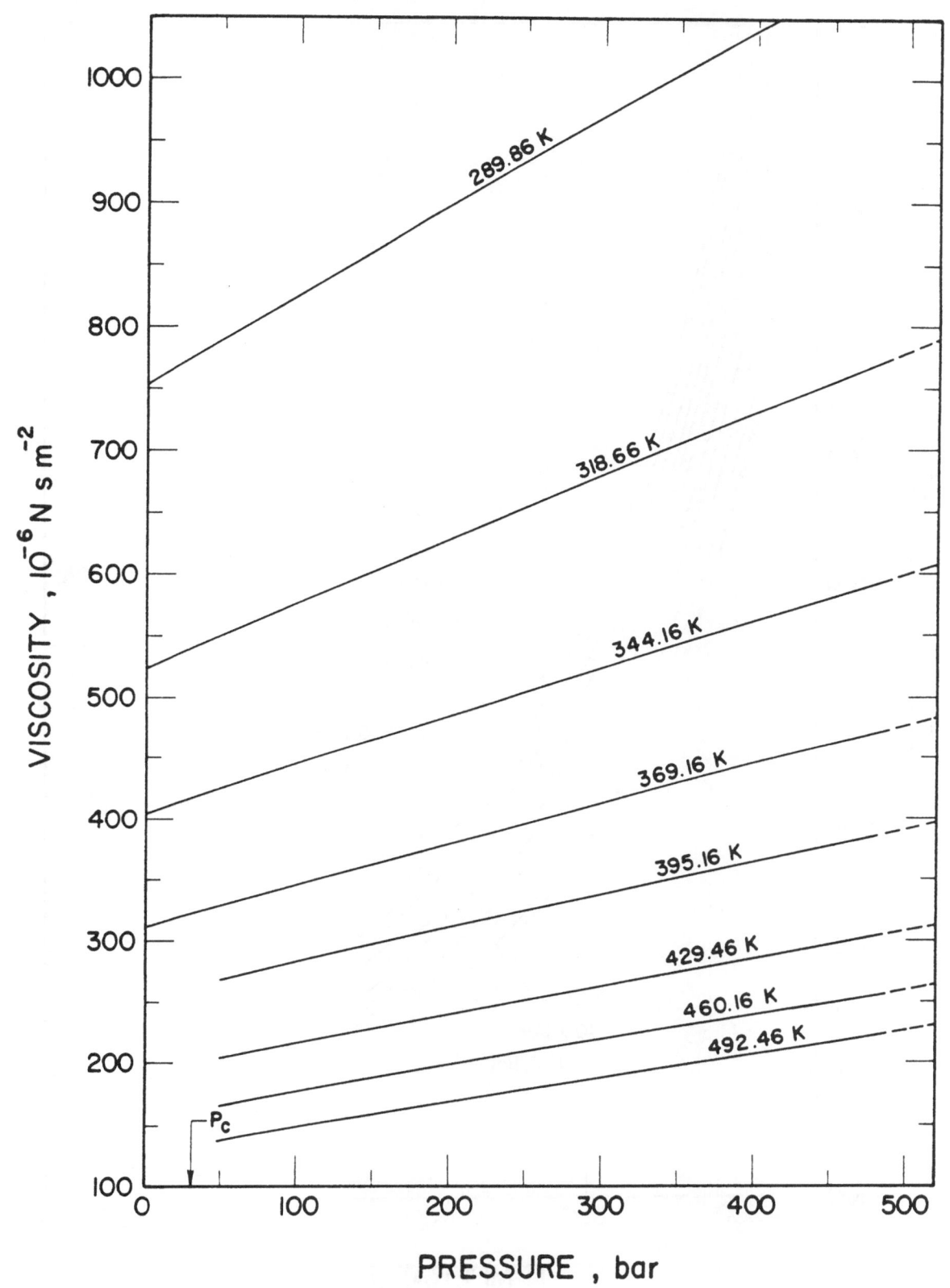

FIGURE 1. VISCOSITY OF BUTYLACETATE [70].

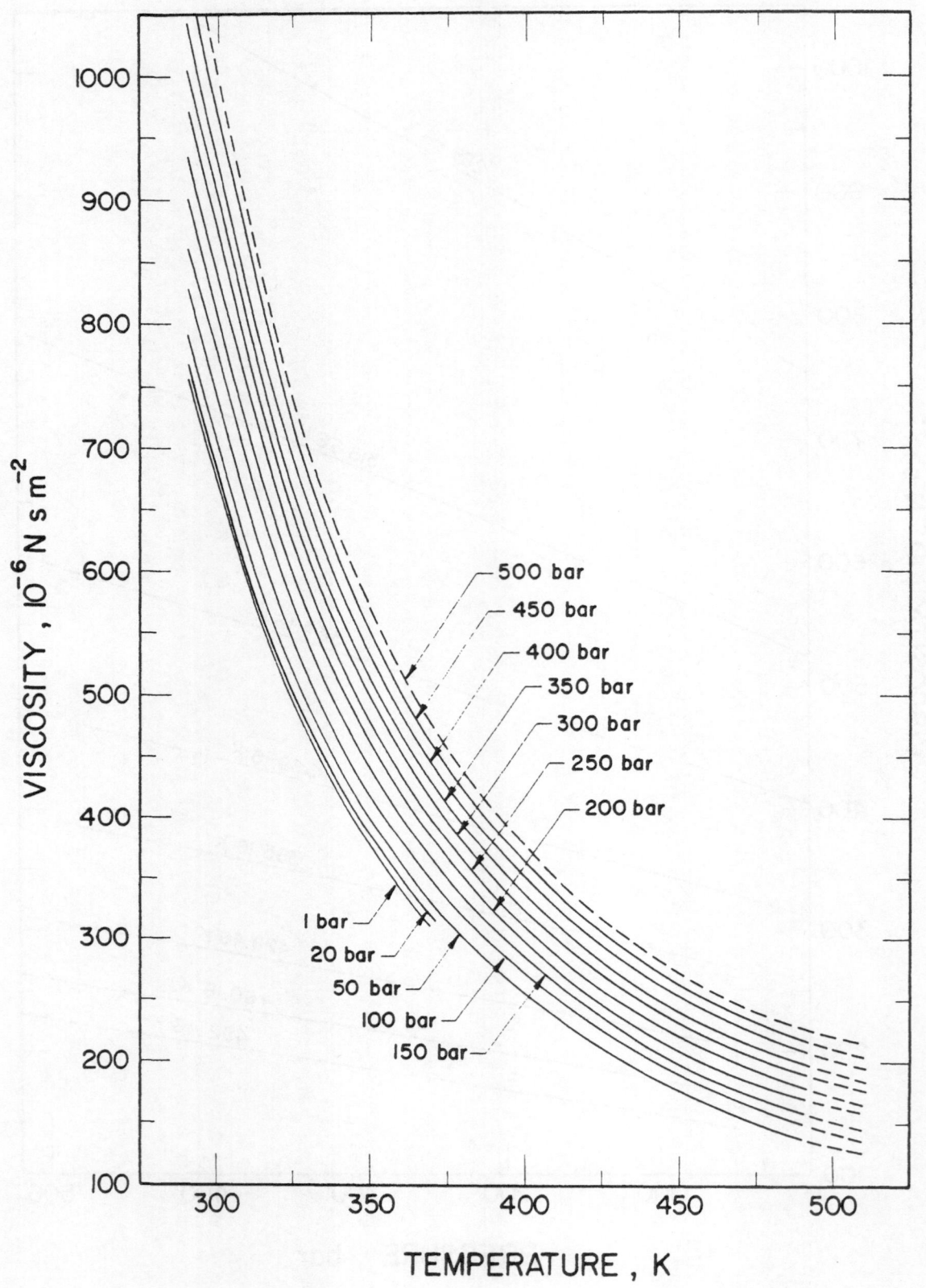

FIGURE 2. VISCOSITY OF BUTYLACETATE [70].

CARBON DIOXIDE

A considerable wealth of data exists for the pressure dependence of the viscosity of carbon dioxide.
Yet, only the dense gaseous region allows an evaluation of recommended values with some confidence.
The data of Michels et al. [146] obtained by a capillary tube viscosimeter, were used to generate the recommended values shown on the next page. From this set of data, the isotherms between T = 298.13 K and
T = 313.13 K were excluded, because they show an anomalous enhancement of viscosity in the critical
region which is too high. This has been verified by the data of Kestin et al. [111]. Using the 1 bar viscosity data of [195] and the P-V-T data of [100, 149, 150] an extrapolation from 348 K to 900 K up to a
pressure of 1000 bar was made. The measured isotherms of Michels et al. and Kestin et al. are shown in
Figure 1 in a limited pressure range. Some liquid state data at T = 298.63 K by Herreman et al. [83] are
included in Figure 1. The inconsistency with the data of Michels et al. is evident and continues to hold for
the other isotherms as well. No further data over a wide range of liquid states could be found in the literature. Therefore, the liquid state was not included in this formulation. Figure 2 shows isobars as a
function of temperature. Extrapolated values are shown as dashed lines. Saturated liquid and saturated
vapor states, which are consistent with the formulation of [195] are reproduced here.

Additional works on the pressure dependence of the viscosity of carbon dioxide are listed in the
summary table below. Reasonable agreement with the recommended values is found for the data of Kestin
et al. [109] Kurin et al. [121] Golubev et al. [57] and Haepp [75]. However, the agreement with the data
of Kiyama et al. [115] where the viscosities at high pressures appear too high, is quite unsatisfactory, as
can be seen from the departure plots in Figure 3. High deviations are also found for the data of Stakelbeck
[183] and Warburg et al. [206]. On the basis of this evaluation, the uncertainty in the recommended values is estimated to be ± 5%. This is in conformity with the data compilation in [203].

ADDITIONAL REFERENCES ON THE VISCOSITY OF CARBON DIOXIDE

Authors	Year	Ref. No.	Temperature K	Pressure bar	Method	Departure % (no. points)
Haepp	1975	75	298-475	1-475	Oscillating disc	0.2 ± 0.83 (32)
Kurin and Golubev	1974	121	293-423	98-3560	Capillary tube	1.20 ± 2.66 (15)
Herreman et al.	1971	84	283-313	density	Torsional cryst.	-
Herreman et al.	1970	83	219-303	10-192	Torsional cryst.	-
Kestin and Yata	1968	112	293-303	1-25	Oscillating disc	-
Di Pippo et al.	1967	41	293-303	1-25	Oscillating disc	-
Kestin et al.	1964	111	crit.	37-119	Oscillating disc	4.2
Kestin and Whitelaw	1963	109	295-525	1-68	Oscillating disc	1.68 ± 1.16 (20)
Kestin and Leidenfrost	1959	105	293	1-22	Oscillating disc	-
Makita	1955	138	323-573	1-101	Rolling ball	6.14 ± 6.65 (24)
Golubev and Petrov	1953	57	293-523	1-811	Capillary tube	0.74 ± 1.81 (42)
Kiyama and Makita	1952	116	323-573	1-97	Rolling ball	6.14 ± 6.65 (24)
Kiyama and Makita	1951	115	323-573	1-59	Rolling ball	6.14 ± 6.65 (24)
Comings et al.	1944	36	313	4-139	Capillary tube	6.5 (9)
Comings and Egly	1941	35	313	4-139	Capillary tube	6.5 (9)
Naldrett and Maass	1940	154	crit.	crit.	Oscillating disc	-
Schröer and Becker	1935	178	293	1-99	Falling ball	-
Stakelbeck	1933	183	258-313	5-118	Falling ball	13
Phillips	1912	161	293-313	1-122	Capillary tube	6.0 (6)
Warburg and v. Babo	1882	206	306-313	16-116	Capillary tube	18

VISCOSITY OF CARBON DIOXIDE
[μ, 10^{-6} N s m^{-2}]

T, K	1	30	40	50	60	70	73.9*	80	90	100	110	120	150	200	300	400	500	600	800	1000
310	15.4	16.1	16.5	17.1	18.2	20.2	22.0	–	–	55.4	60.0	64.5	70.3	79.6	95.6	109.4	121.5	131.9	150.0	–
315	15.7	16.3	16.7	17.3	18.1	19.6	21.0	22.6	31.4	42.0	50.8	56.0	65.1	75.6	91.5	104.8	116.6	127.1	145.2	163.2
320	15.9	16.5	16.9	17.5	18.2	19.3	20.5	21.5	25.1	32.0	42.0	48.2	60.2	71.4	87.5	100.5	111.7	122.5	140.5	158.2
325	16.1	16.7	17.1	17.7	18.3	19.2	20.1	20.9	23.4	28.0	34.2	41.3	55.0	67.1	83.5	96.3	107.4	117.9	136.0	153.6
330	16.3	17.0	17.3	17.8	18.4	19.2	20.0	20.6	22.6	25.9	29.6	35.3	50.4	62.6	79.7	92.4	103.3	113.5	131.7	149.2
335	16.6	17.2	17.5	18.1	18.6	19.2	19.8	20.5	22.2	24.4	27.4	31.0	45.8	58.0	76.0	88.8	99.4	109.6	127.7	145.0
340	16.8	17.4	17.8	18.3	18.8	19.3	19.7	20.4	21.8	23.6	25.8	28.7	41.4	54.4	72.4	85.2	95.8	105.7	124.0	140.8
345	17.0	17.7	18.0	18.4	18.9	19.4	19.8	20.4	21.6	22.9	24.7	27.0	37.2	50.4	69.0	81.9	92.3	102.1	120.4	136.8
350	17.2	17.9	18.2	18.6	19.1	19.5	19.8	20.4	21.4	22.5	24.0	26.0	33.9	47.3	65.8	78.7	89.2	98.7	116.8	133.2
360	17.7	18.4	18.6	19.0	19.4	19.8	20.0	20.5	21.3	22.0	23.3	24.9	30.5	42.5	60.1	73.0	83.4	92.8	110.4	126.2
370	18.1	18.8	19.1	19.3	19.7	20.1	20.3	20.7	21.3	22.0	23.0	24.3	28.6	38.7	55.1	67.8	78.3	87.5	104.5	120.0
380	18.6	19.3	19.5	19.7	20.1	20.5	20.6	21.0	21.4	22.0	23.0	24.0	27.6	35.8	51.1	63.2	73.7	82.8	99.1	114.0
390	19.0	19.7	19.9	20.1	20.4	20.8	20.9	21.3	21.7	22.2	23.0	23.8	26.8	33.6	47.6	59.3	69.6	78.6	94.0	108.8
400	19.4	20.0	20.2	20.4	20.7	21.1	21.2	21.6	21.9	22.4	23.2	23.8	26.4	32.1	44.8	56.0	66.0	74.8	89.8	104.0
420	20.3	20.8	21.0	21.2	21.5	21.8	21.9	22.1	22.4	22.9	23.4	24.0	26.0	30.4	40.9	50.8	60.0	68.1	82.3	95.6
440	21.2	21.7	21.8	22.0	22.3	22.5	22.6	22.8	23.1	23.4	23.8	24.4	26.0	29.6	38.3	47.1	55.2	62.8	76.1	88.6
460	22.0	22.4	22.6	22.7	23.0	23.2	23.3	23.5	23.8	24.1	24.5	24.8	26.3	29.1	36.5	44.4	51.6	58.6	71.0	82.7
480	22.7	23.3	23.4	23.5	23.8	24.0	24.1	24.2	24.5	24.7	25.0	25.3	26.6	29.0	35.5	42.5	49.1	55.3	66.8	78.0
500	23.5	24.1	24.2	24.3	24.5	24.7	24.8	24.9	25.2	25.4	25.7	25.9	27.0	29.1	34.8	41.1	47.2	52.8	63.6	74.0
550	25.4	25.9	26.0	26.1	26.3	26.5	26.6	26.7	26.8	27.0	27.2	27.4	28.3	29.9	34.1	38.9	44.1	48.8	57.8	66.5
600	27.2	27.8	27.9	28.0	28.1	28.3	28.3	28.4	28.5	28.7	28.9	29.1	29.6	30.9	34.1	38.2	42.3	46.5	54.3	61.6
650	29.0	29.4	29.5	29.6	29.7	29.9	29.9	30.0	30.1	30.3	30.4	30.6	31.0	32.0	34.7	38.1	41.6	45.3	52.2	58.5
700	30.6	31.1	31.2	31.3	31.4	31.5	31.5	31.6	31.7	31.8	31.9	32.0	32.4	33.3	35.5	38.3	41.3	44.6	50.7	56.6
800	33.9	34.1	34.2	34.3	34.4	34.5	34.5	34.6	34.7	34.8	34.9	35.0	35.2	36.0	37.6	39.5	41.9	44.5	49.3	54.4
900	36.9	37.0	37.0	37.1	37.1	37.2	37.2	37.3	37.4	37.5	37.6	37.7	38.0	38.5	39.8	41.6	43.2	45.3	49.6	53.8

Pressure, bar

*Critical pressure.

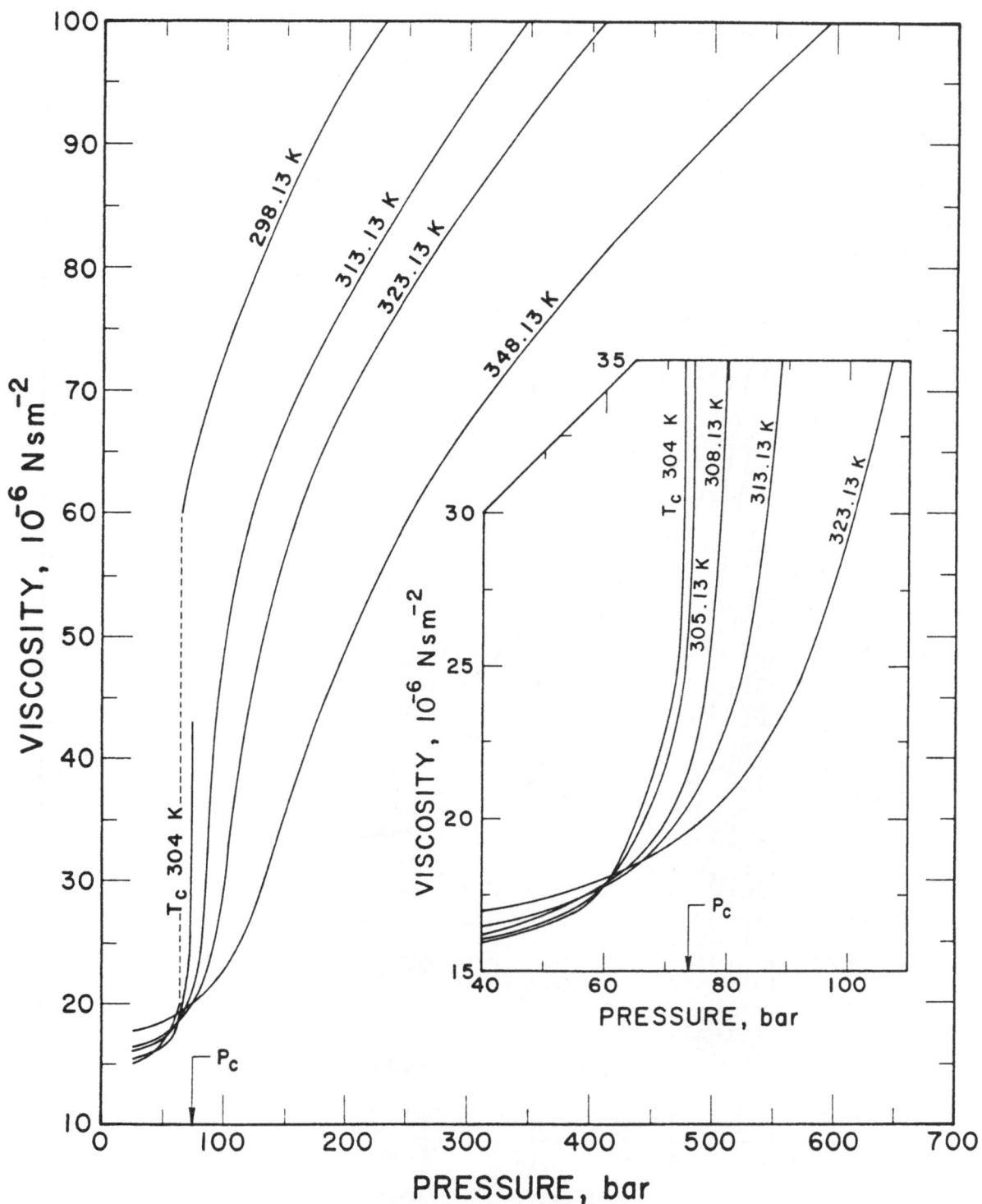

FIGURE 1. VISCOSITY OF CARBON DIOXIDE [III, 146].

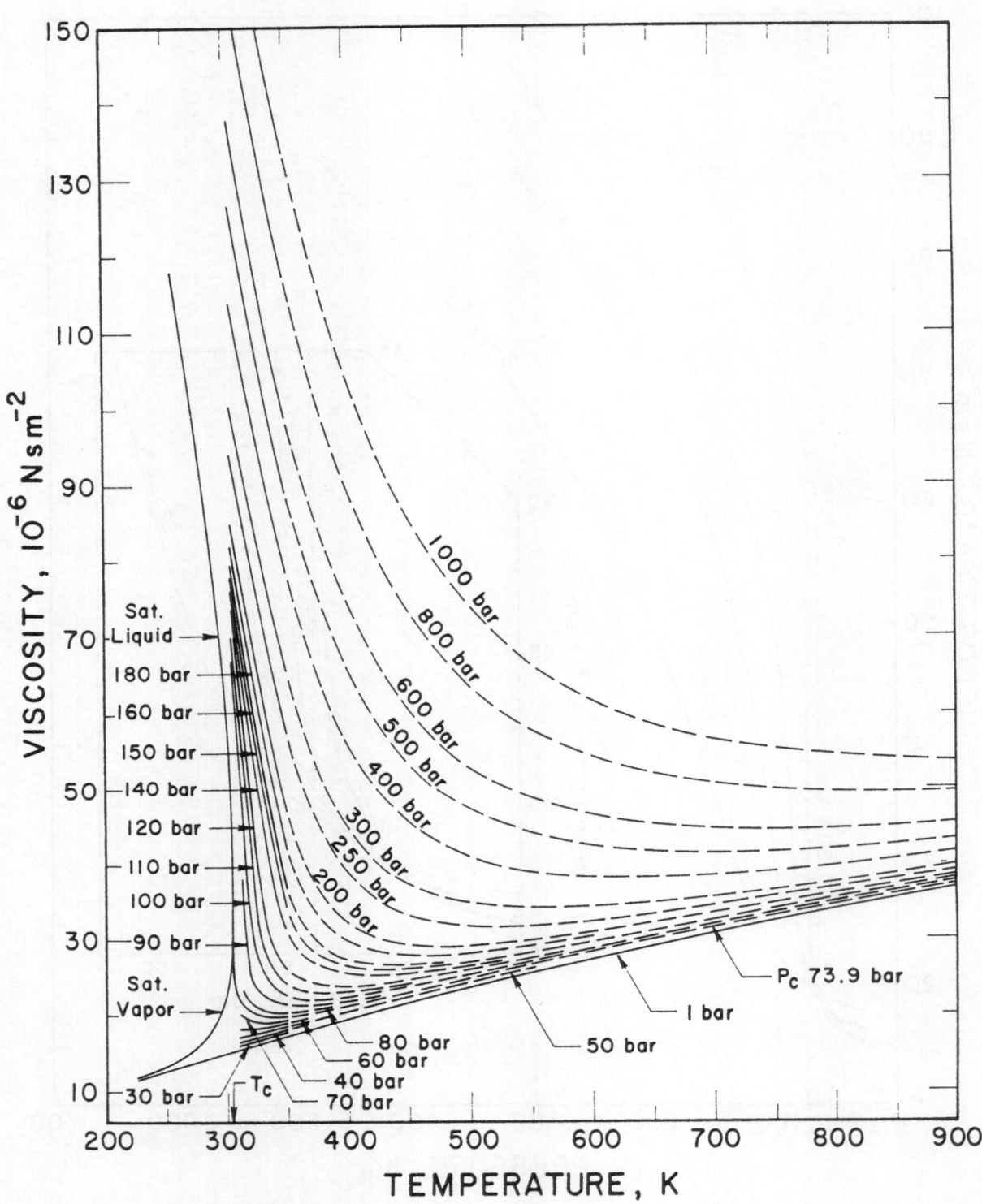

FIGURE 2. VISCOSITY OF CARBON DIOXIDE [146].

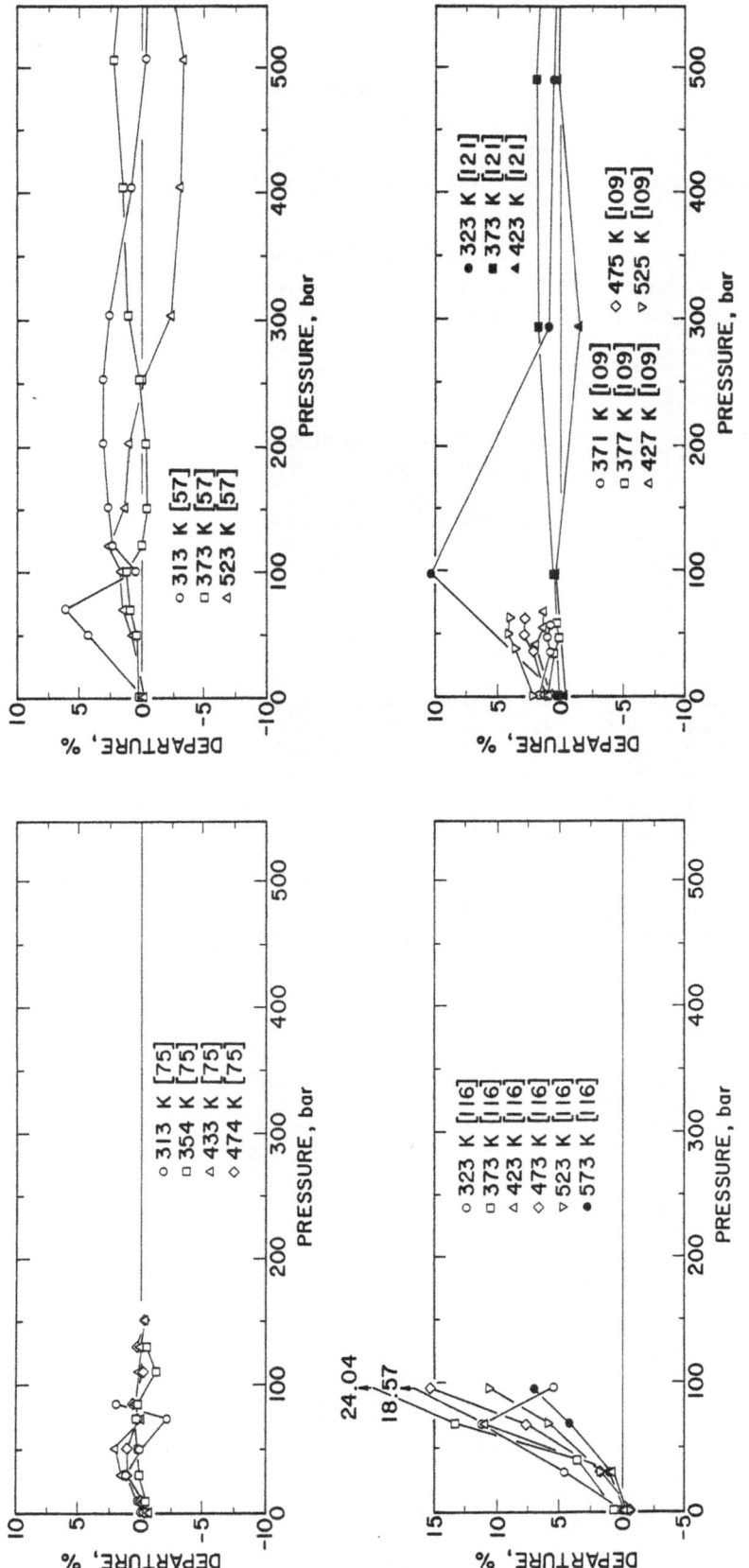

FIGURE 3. DEPARTURE PLOT ON THE VISCOSITY OF CARBON DIOXIDE.

CARBON MONOXIDE

The recommended values for the viscosity of carbon monoxide in the dense gas region were generated from references [12, 33, 57]. The data are plotted as isotherms against pressure in Figure 1. Here the isotherms from 273 K to 523 K are from Golubev [57], and those at 223 K and 248 K from Barua et al. [12]. Figure 2 shows the data plotted as isobars against temperature, where the 1 bar line was taken from [195]. The recommended values were read from this figure, and are presented below.

A comparison of the recommended values with the data of Chierici and Paratella [33] gives a mean deviation of 0.5% while the data of [12] agrees with a mean deviation of about 1% in the overlap region. The agreement is thus seen to be very satisfactory. The uncertainty in the recommended values is estimated to be ± 4%.

VISCOSITY OF CARBON MONOXIDE

$[\mu, 10^{-6} \text{ N s m}^{-2}]$

T,K	Pressure, bar													
	1	50	100	150	200	250	300	350	400	450	500	600	700	800
220	13.8	15.2	17.3											
230	14.4	15.6	17.6											
240	14.9	16.0	17.9											
250	15.4	16.4	18.2											
260	15.9	16.8	18.5											
270	16.4	17.3	18.8											
280	16.9	17.7	19.1	20.8	22.9	25.0	26.8	29.0	31.0	33.0	35.0			
290	17.4	18.1	19.4	20.9	22.9	24.8	26.6	28.6	30.6	32.4	34.4			
300	17.8	18.5	19.7	21.0	22.9	24.6	26.4	28.4	30.3	32.0	34.0	37.8	41.2	44.8
320	18.6	19.3	20.4	21.6	23.1	24.6	26.2	28.0	29.8	31.4	33.0	36.3	39.6	42.8
340	19.4	20.0	21.0	22.2	23.4	24.7	26.2	27.8	29.4	31.0	32.4	35.4	38.4	41.4
360	20.2	20.8	21.7	22.7	23.8	25.0	26.4	27.8	29.2	30.7	32.0	34.8	37.6	40.2
380	21.1	21.6	22.4	23.4	24.4	25.5	26.6	27.9	29.2	30.4	31.8	34.2	36.8	39.4
400	21.9	22.4	23.0	24.0	25.0	26.0	27.0	28.1	29.3	30.6	31.6	33.9	36.2	38.6
450	23.8	24.1	24.7	25.5	26.3	27.2	28.0	29.0	29.8	30.8	31.6	33.6	35.5	37.5
500	25.4	25.8	26.3	27.0	27.7	28.4	29.0	-	-	-	-	-	-	-

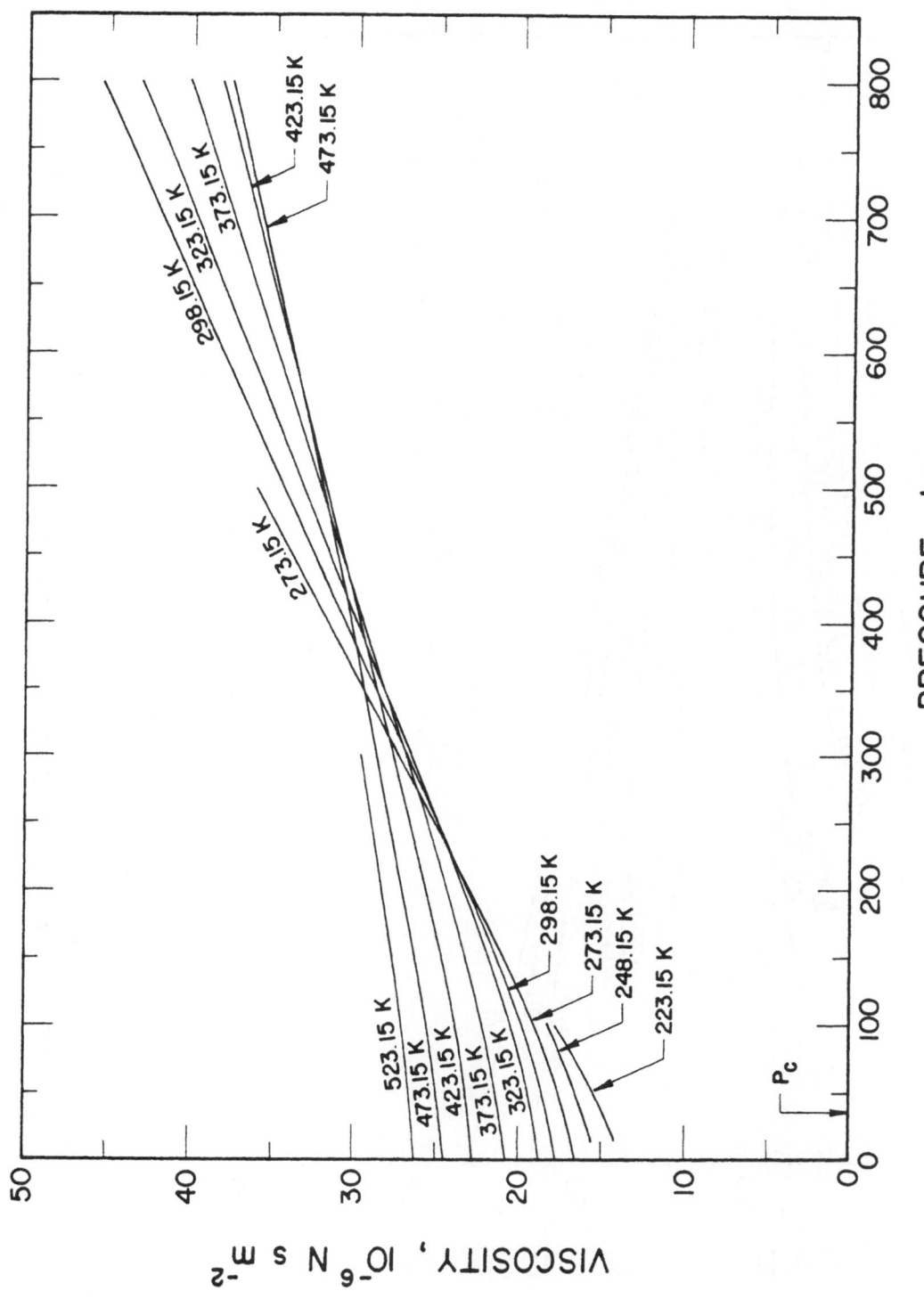

FIGURE 1. VISCOSITY OF CARBON MONOXIDE [12, 33, 57].

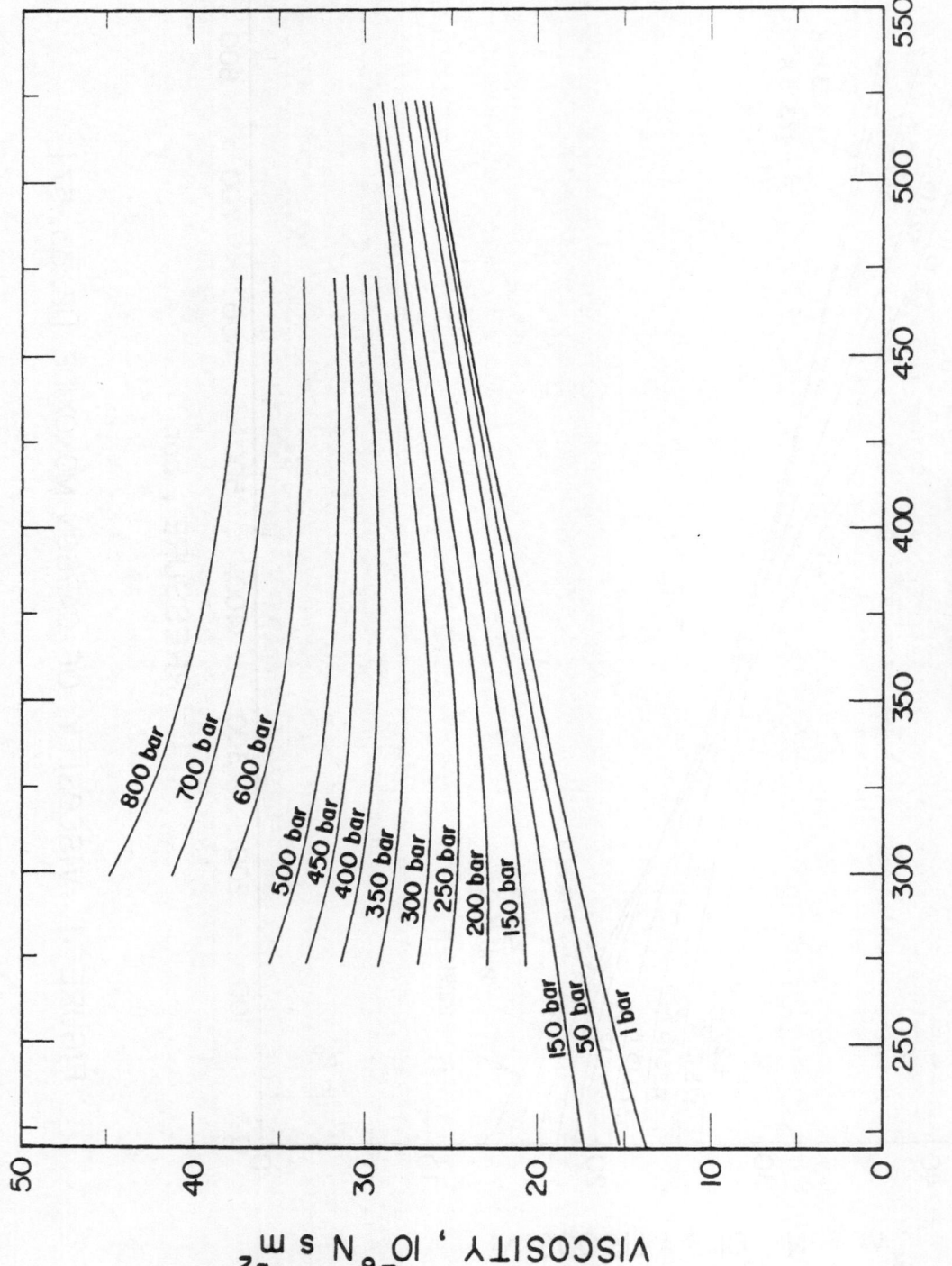

FIGURE 2. VISCOSITY OF CARBON MONOXIDE [12, 33, 57].

CHLORODIFLUOROMETHANE (R 22)

The set of data by Geller et al. [52], obtained by a capillary tube viscosimeter, was used to generate the recommended values tabulated on the next page. The data are plotted as isotherms as a function of
pressure in Figure 1, where only minor smoothing modifications were applied. Figure 2 shows isobars
plotted against temperature. The recommended values were read from this figure.

Additional works on the pressure dependence of the viscosity of chlorodifluoromethane are listed
in the summary table below. Apart from the data of Tkachev et al. [193], only small ranges of pressure,
predominantly in the gas phase, are reported such that no comparison with the recommended values was
possible. These latter authors only publish the formulae for viscosity as a function of temperature and
density as well as diagrams. They claim an experimental uncertainty of ± 1%. Comparing their formulation to the recommended values reveals deviations of a few percent. The accuracy of the recommended
values is tentatively estimated to be about ± 4%.

ADDITIONAL REFERENCES ON THE VISCOSITY OF CHLORODIFLUOROMETHANE (R 22)

Authors	Year	Ref. No.	Temperature K	Pressure bar	Method
Tkachev et al.	1972	194	253-473	1-500	Capillary tube
Kletskii	1971	117	273-473	2-20	
Tkachev and Butyrskaya	1970	198	298-473	1-500	Capillary tube
Phillips and Murphy	1970	162	201-299	0.1-20	Capillary tube
Wilbers	1961	208	235-288	1-12	Rolling ball
Tsui	1959	202	363-423	1-20	Rolling ball
Kamien and Witzell	1959	96	303-343	1-20	Rolling ball
Kamien	1956	95	303-343	1-20	Rolling ball
Makita	1955	138	298-473	1-20	Rolling ball
Makita	1954	137	298-473	1-19	Rolling ball
Benning and Markwood	1939	14	240-318	1-17	Falling ball

VISCOSITY OF CHLORODIFLUOROMETHANE, (R 22)

$[\mu,\ 10^{-6}\ \mathrm{N\ s\ m^{-2}}]$

T, K	Pressure, bar																	
	10	20	30	40	49.8*	60	80	100	120	140	160	180	200	250	300	400	500	600
250	272.5	274.5	276.3	278.4	280.5	283.1	288.0	293.1	297.3	302.3	307.0	312.0	316.5	328.5	341.0	–	–	–
260	244.5	247.0	249.5	251.5	254.0	256.5	261.5	266.5	270.8	275.5	280.1	285.0	289.5	301.0	312.5	–	–	–
270	218.7	221.5	224.3	226.5	229.2	232.0	236.7	241.5	246.0	250.5	255.0	259.6	264.0	275.0	286.0	306.5	326.0	346.0
280	195.5	198.5	201.0	204.0	206.5	209.4	214.0	219.0	223.0	227.0	232.0	236.2	240.5	251.0	261.5	281.5	300.5	320.0
290	–	177.5	180.5	183.2	186.0	188.5	193.5	198.0	202.2	206.6	211.0	215.0	219.4	229.3	239.2	258.7	277.5	295.5
300		158.5	161.5	164.2	167.3	169.6	174.4	179.0	183.2	187.5	191.8	196.0	200.1	209.7	219.6	239.3	257.0	274.0
310		141.4	144.0	146.8	149.5	152.1	156.8	161.5	165.7	170.0	174.3	178.2	182.5	192.5	202.0	222.4	238.6	255.0
320			128.0	130.6	133.2	136.0	141.0	145.5	150.0	154.4	158.7	162.7	167.0	177.5	186.6	206.8	222.5	238.0
330					118.4	121.0	126.0	131.5	136.0	140.5	145.0	149.0	153.5	164.1	173.5	192.5	208.7	223.0
340					103.5	106.9	113.0	118.6	123.5	128.5	133.0	137.2	141.6	152.0	161.6	180.0	196.0	210.0
350					88.1	93.5	101.0	107.3	112.6	117.6	122.5	126.9	131.3	141.3	150.7	168.5	184.5	198.3
360					72.1	79.5	89.5	97.0	103.0	108.2	113.0	117.6	122.0	132.0	141.3	158.5	174.0	188.1
365								92.0	98.5	104.0	108.9	113.5	117.7	127.6	136.9	154.0	169.4	183.9

*Critical pressure.

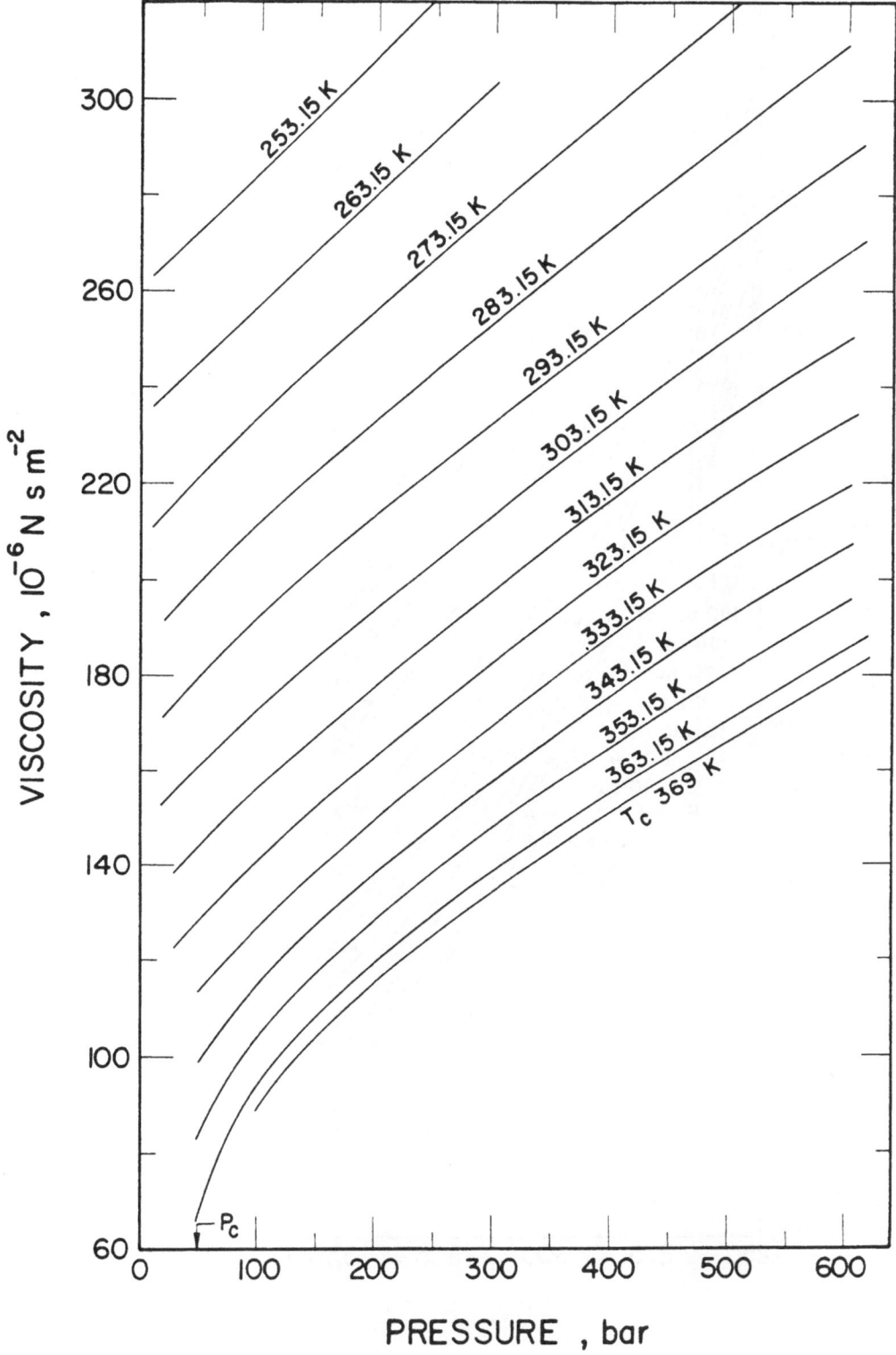

FIGURE 1. VISCOSITY OF R 22 [52].

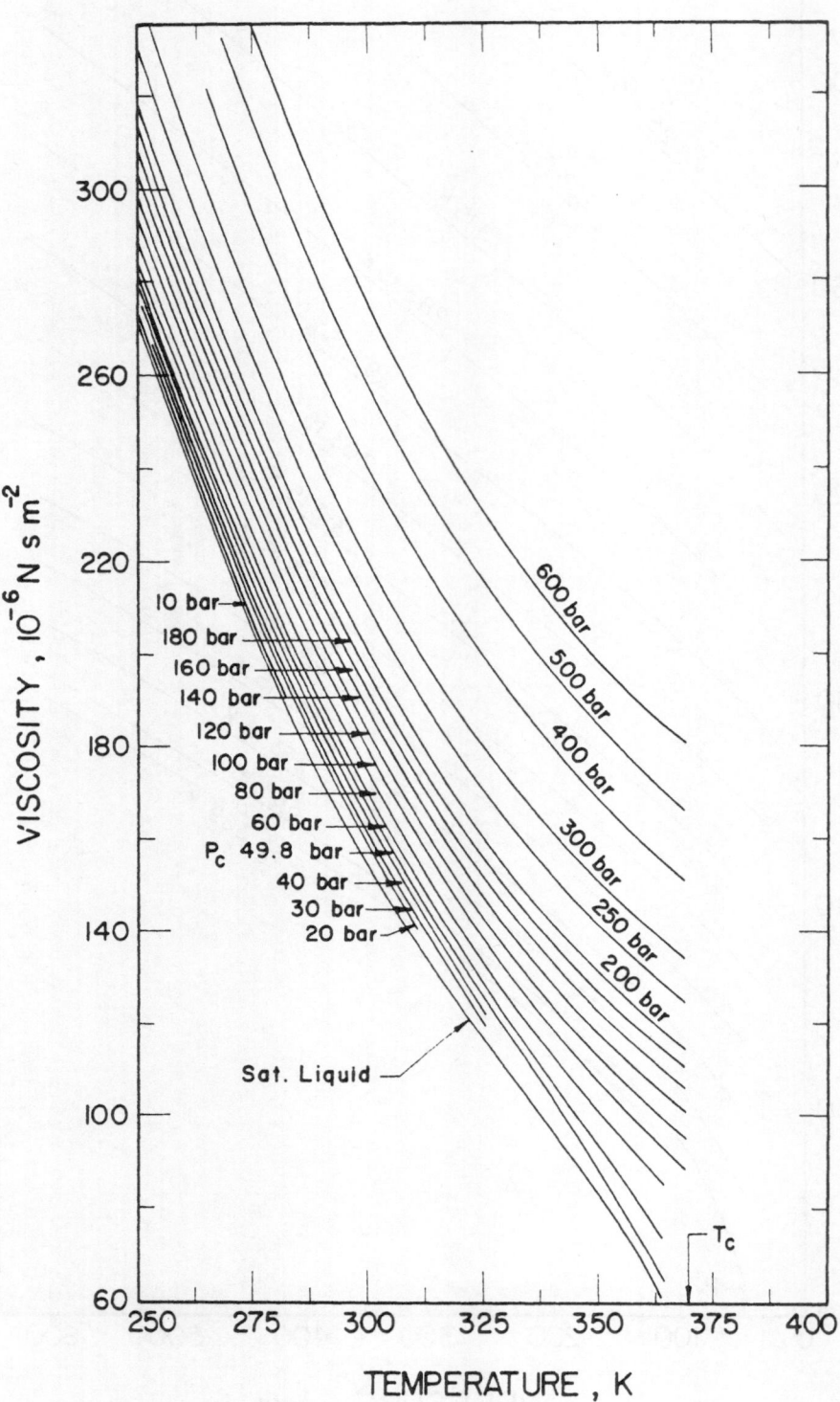

FIGURE 2. VISCOSITY OF R 22 [52].

CYCLOHEXANE

The data of Guseinov et al. [71], obtained by a capillary tube method, were used to generate the recommended values presented on the next page. The data are plotted as isotherms against pressure in Figure 1, where smoothing modifications of up to 3.3% were introduced. In Figure 2 the data are shown as isobars against temperature. The recommended values were read from this figure.

Additional works on the pressure dependence of viscosity of liquid cyclohexane are listed in the summary table below. The data of Collings et al., obtained by the torsional crystal method, reasonably agree with the recommended values, as shown in Figure 3. The same may be stated about the data of Bridgeman. The uncertainty in the recommended values is estimated to be ± 5%.

ADDITIONAL REFERENCES ON THE VISCOSITY OF CYCLOHEXANE

Authors	Year	Ref. No.	Temperature K	Pressure bar	Method	Departure % (no. points)
Collings and McLaughlin	1971	34	303–323	1–785	Torsional crystal	-1.44 ± 2.3 (15)
Kozlov et al.	1966	119	293–413	Isochores	Falling cylinder	-
Kuss	1955	123	298–353	1–1373	Falling ball	-
Bridgman	1926	21	303, 348	1–11768	Falling weight	5.7

VISCOSITY OF CYCLOHEXANE

$[\mu, 10^{-6} \text{ N s m}^{-2}]$

T, K	\multicolumn{17}{c}{Pressure, bar}
	1	10	20	40.7*	60	80	100	120	140	160	180	200	250	300	350	400	500
290	1004	1018	1034	1064	1100	1132	1164	1196	1228	1268	1292	1324	1408	1478	1555	1638	1792
300	833	847	868	896	921	940	980	1006	1028	1058	1080	1104	1169	1226	1302	1378	1505
310	716	728	744	769	788	808	847	868	886	910	930	952	1002	1050	1112	1174	1292
320	628	638	652	673	687	710	736	753	770	790	810	829	874	918	968	1020	1120
330	560	568	578	596	608	628	646	661	676	692	711	727	768	808	852	896	979
340	502	508	518	534	544	561	576	588	600	614	632	645	680	714	753	793	865
350	452	458	466	480	489	504	516	528	538	552	566	579	608	638	671	705	768
360		413	420	434	441	452	463	473	484	495	508	521	548	572	601	628	686
370		372	380	392	400	408	416	426	436	445	456	468	494	518	542	565	616
380		336	344	355	361	368	377	386	395	404	413	424	448	468	492	512	556
390		306	313	322	328	335	343	351	360	368	375	386	407	425	448	466	506
400		278	284	292	299	305	313	320	328	335	342	352	372	388	410	428	464
420			240	248	254	260	266	274	280	288	294	300	316	332	349	364	397
440			209	215	221	226	233	240	245	252	258	264	278	293	307	322	353
460				191	197	202	209	214	220	225	232	238	250	265	278	292	322
480				172	178	183	190	196	202	206	213	220	231	246	258	272	301
500				159	165	169	176	182	188	193	199	205	217	233	244	258	287
520				–	–	160	166	172	178	183	189	194	208	224	235	250	277

*Critical pressure.

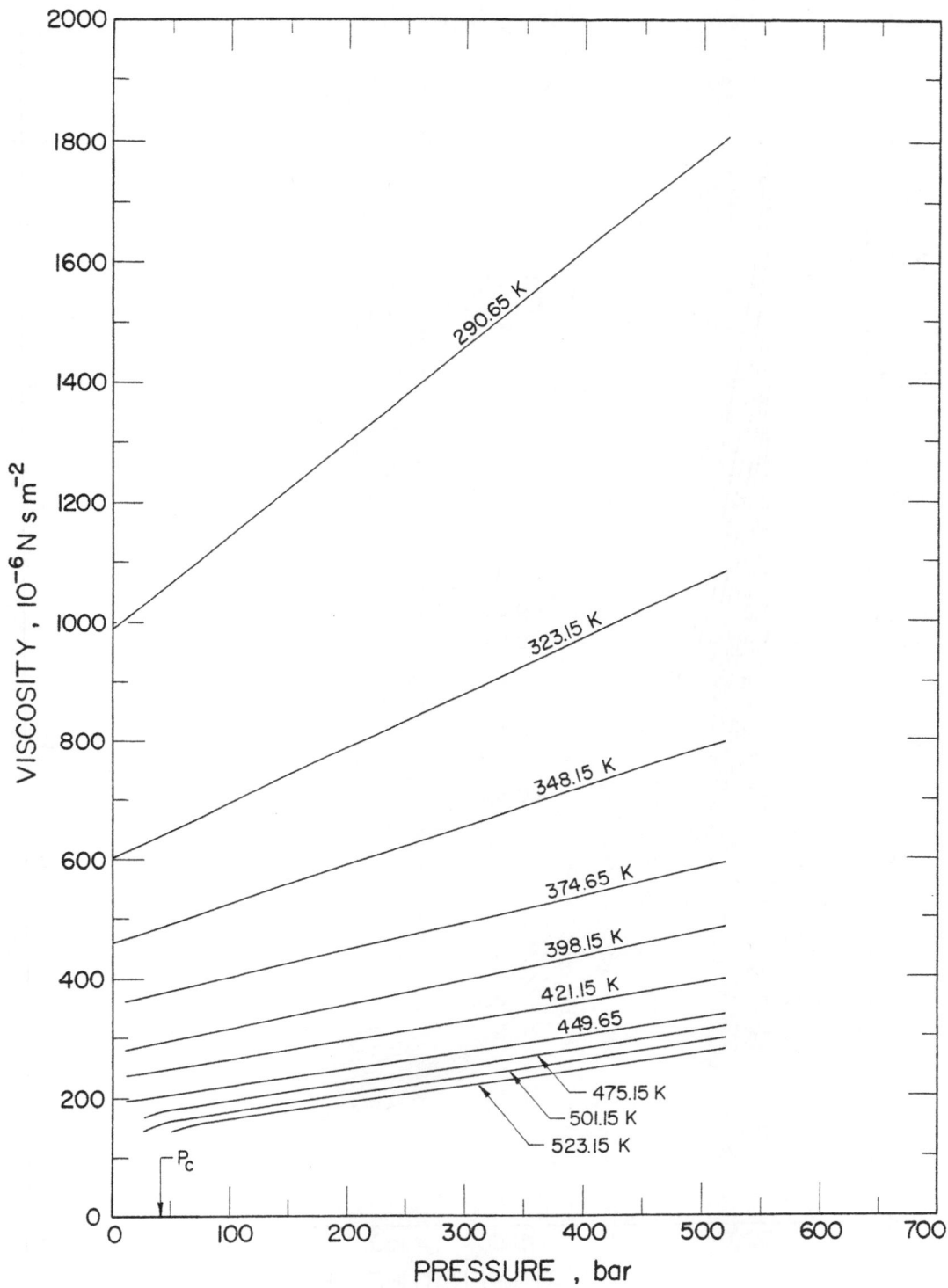

FIGURE I. VISCOSITY OF CYCLOHEXANE [71].

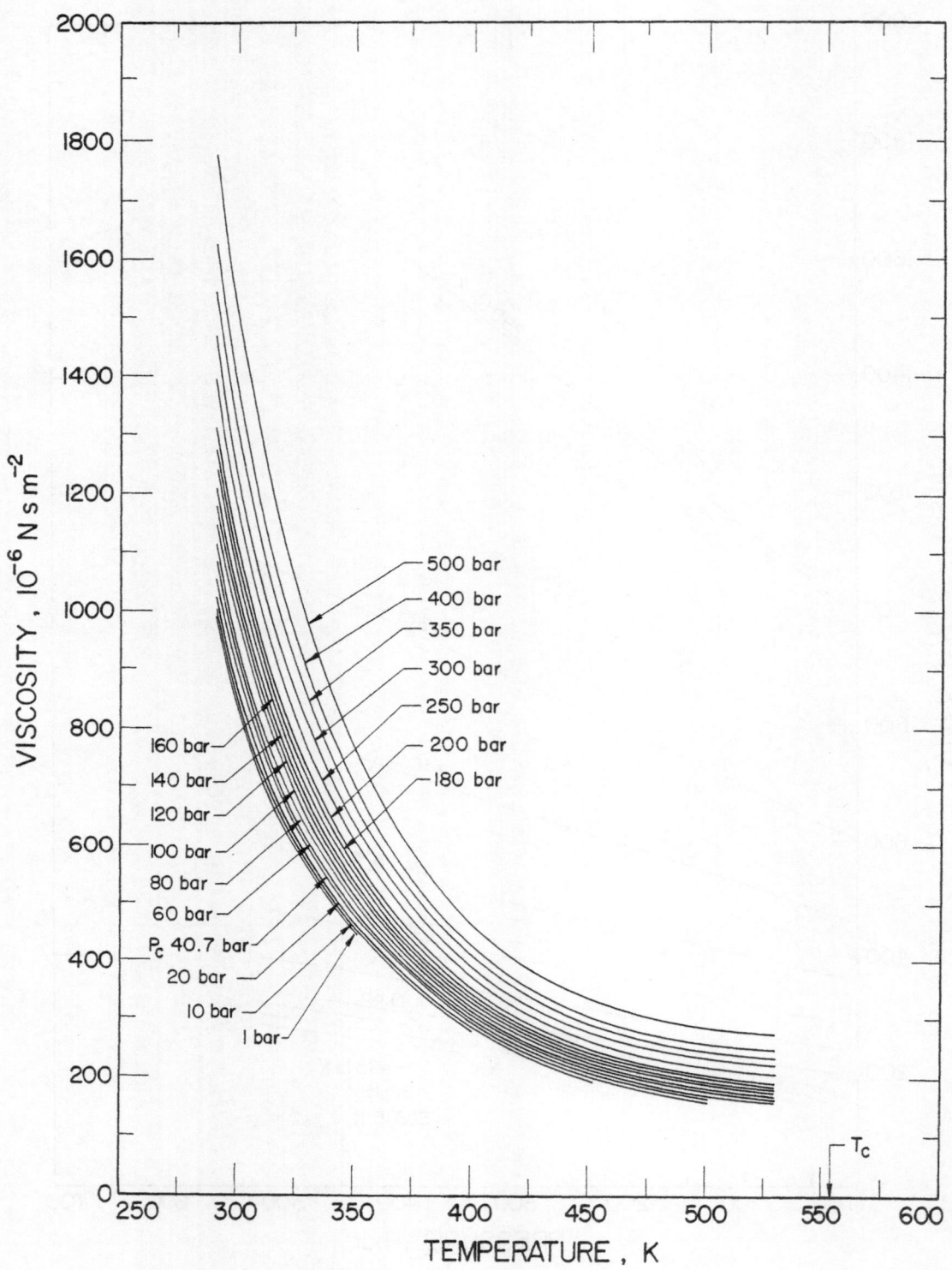

FIGURE 2. VISCOSITY OF CYCLOHEXANE [71].

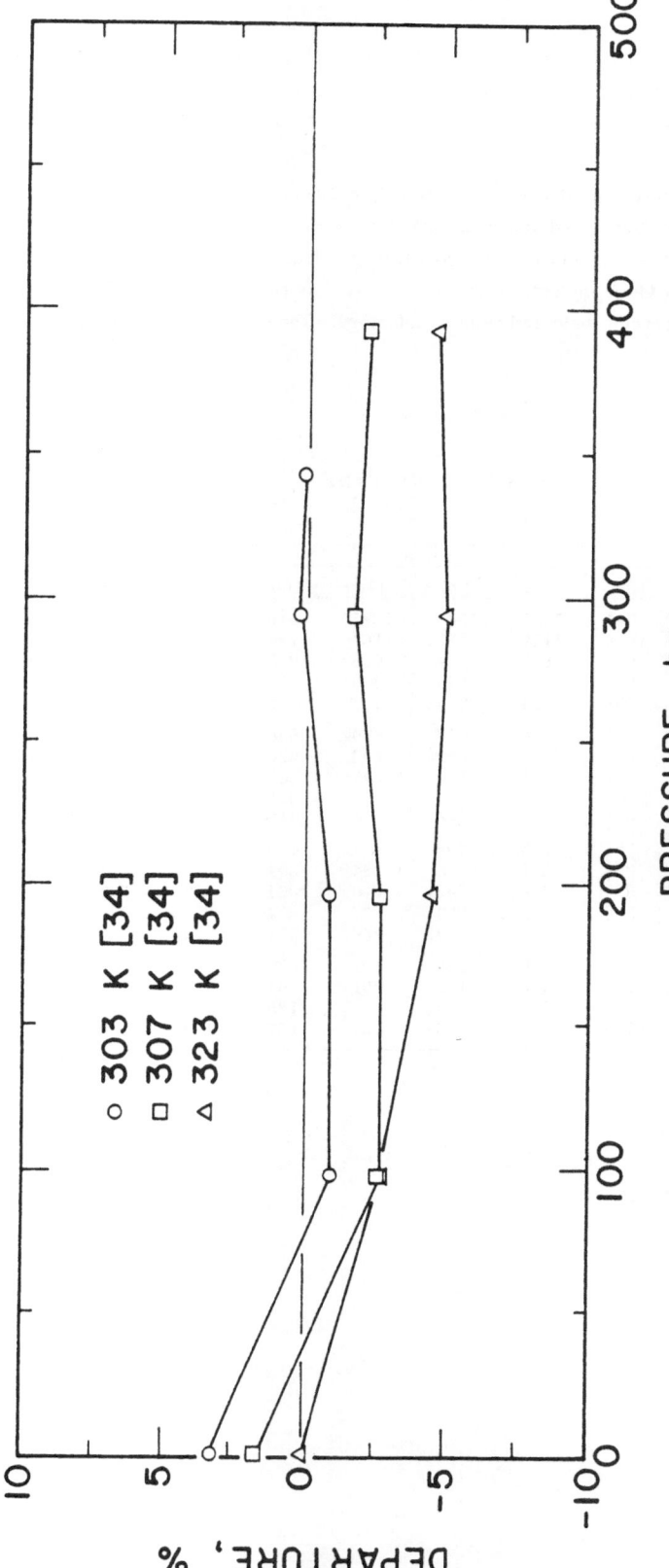

FIGURE 3. DEPARTURE PLOT ON THE VISCOSITY OF CYCLOHEXANE.

n-DECANE

The data of Carmichael et al. [26], obtained by a rotating cylinder apparatus were used to generate the recommended values presented below. The data are plotted in Figure 1 as isotherms against pressure, the smoothing modifications were less than 1%. In Figure 2, the data are plotted as isobars against temperature. From this figure, the recommended values were read.

Two additional works on the pressure dependence of the viscosity of liquid n-decane were found in the literature. Rastorguev and Keramidi [164] present formulae and diagrams covering the range from 302 K to 481 K up to 490 bar. Since no tabular data were reported no comparison was attempted. Lee and Ellington [132] present data, obtained by a capillary tube viscosimeter, in the range from 311 K to 444 K and from 14 bar to 552 bar. As seen from Figure 3, the agreement with the recommended values is within 3%, and, hence considered quite satisfactory. The uncertainty in the recommended values is estimated to be ± 3%.

VISCOSITY OF n-DECANE
$$[\mu, 10^{-5} \text{ N s m}^{-2}]$$

T, K	Pressure, bar													
	1	21.1*	40	60	80	100	120	140	160	180	200	250	300	350
280	1168	1204	1225	1254	1284	1305	1347	1372	1410	1436	1465	1544	1628	1712
290	990	1021	1040	1064	1088	1112	1135	1162	1188	1211	1240	1306	1374	1444
300	850	874	892	912	932	952	974	993	1013	1030	1060	1110	1165	1225
310	735	754	770	790	808	824	840	858	876	890	916	960	1006	1054
320	646	664	677	693	708	724	738	752	772	788	806	844	892	931
330	574	589	604	616	630	644	658	670	688	701	719	752	795	828
340	516	529	542	554	568	580	592	603	620	632	646	676	712	744
350	466	478	490	501	514	525	536	547	560	572	586	612	644	672
360	422	433	444	456	466	478	488	498	511	521	534	559	587	613
370	384	394	404	415	425	436	446	456	466	477	489	512	537	563
380	351	360	370	379	388	399	408	417	428	437	448	471	494	518
390	322	330	339	348	358	368	376	384	394	404	412	434	456	480
400	296	304	312	321	330	339	348	356	364	372	382	403	423	444
420	253	260	267	275	285	292	299	307	314	321	330	348	366	384
440	218	226	231	239	246	252	260	266	274	280	288	304	321	337
460	–	194	200	207	213	220	226	232	239	244	252	267	282	297
480	–	164	170	176	183	190	196	201	208	214	220	235	248	263
500	–	138	144	150	156	163	169	174	180	186	192	206	220	232
520	–	113	120	127	133	140	145	151	157	163	168	181	194	206

*Critical pressure.

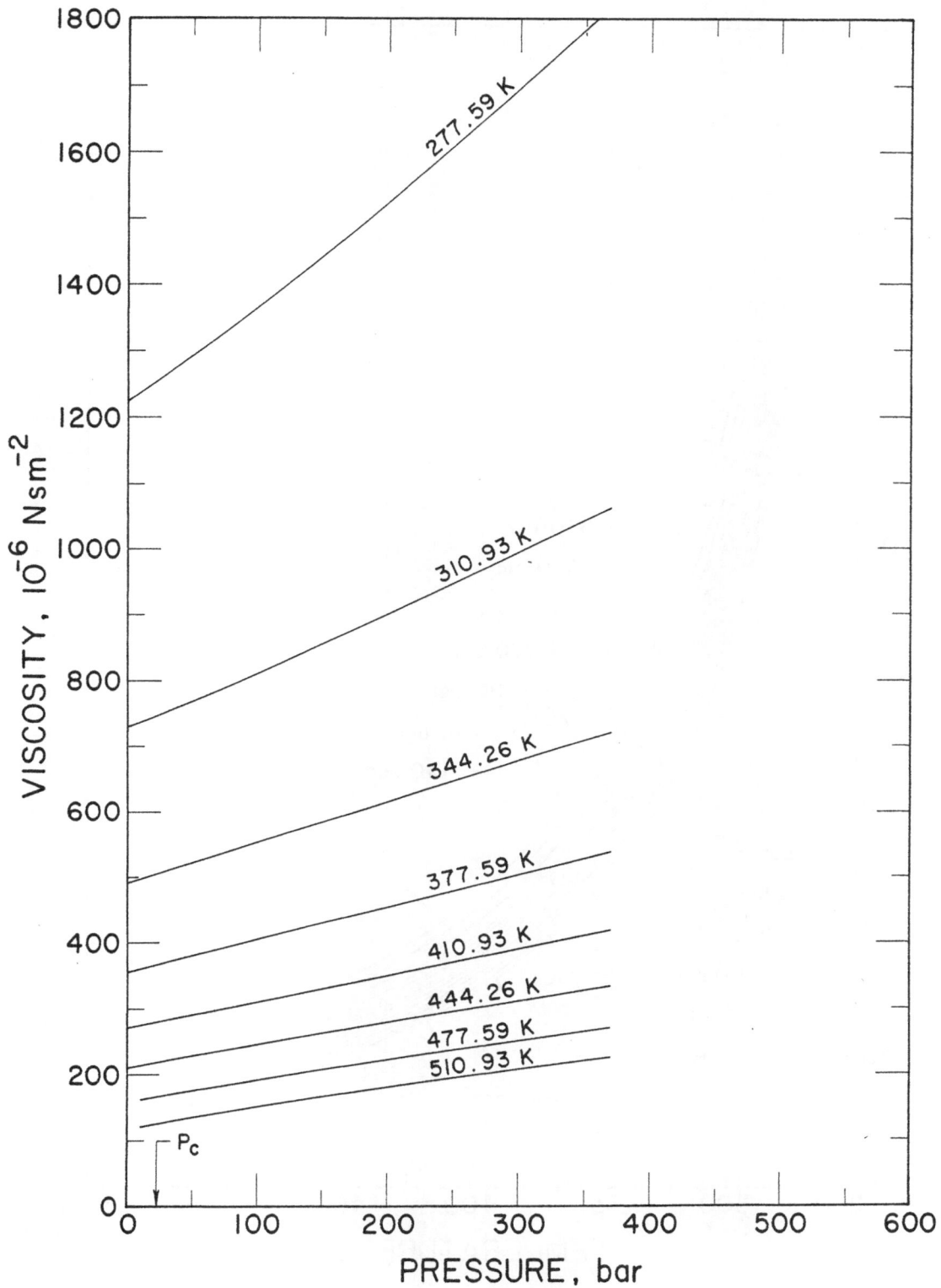

FIGURE 1. VISCOSITY OF n-DECANE [26].

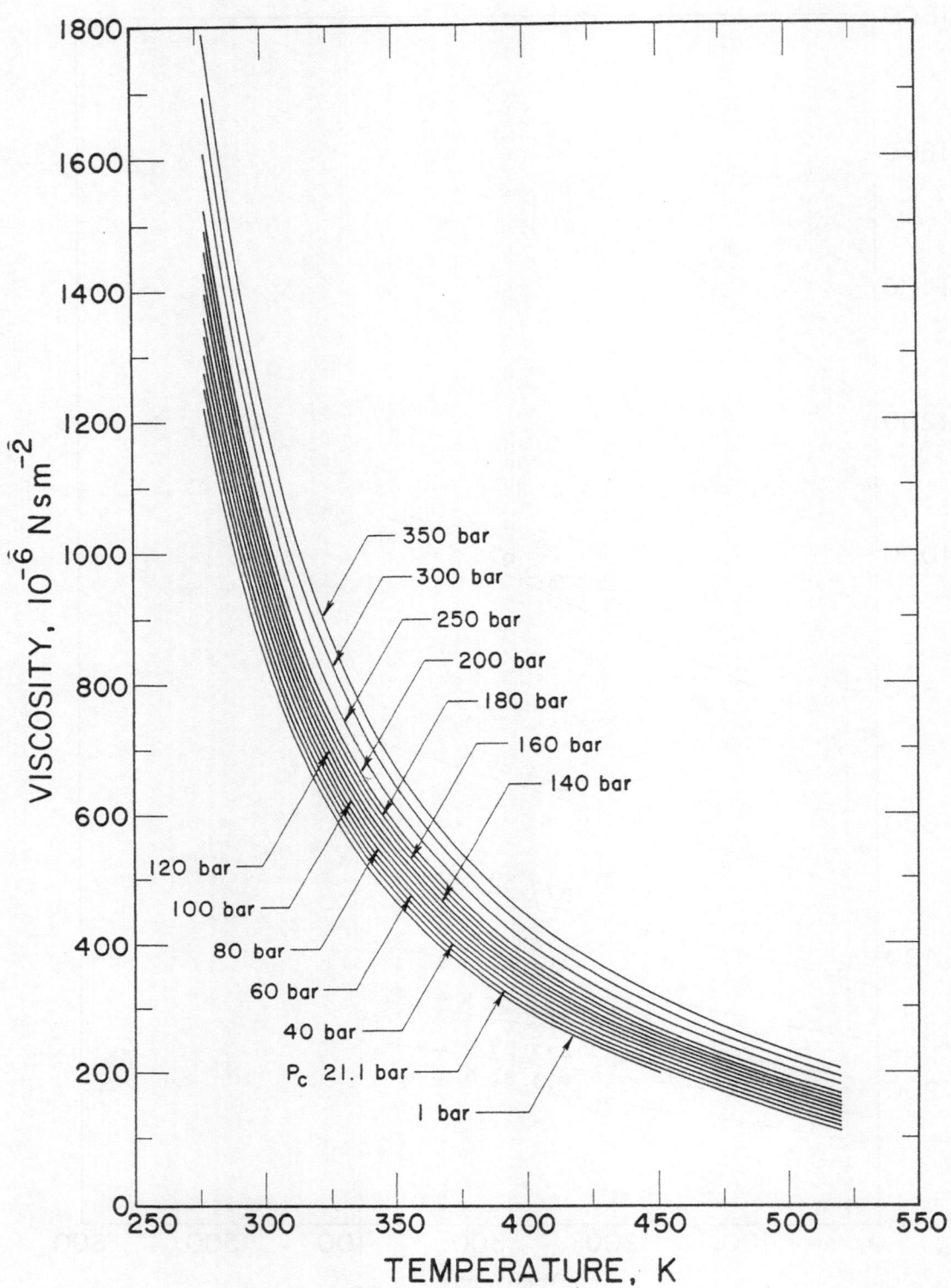

FIGURE 2. VISCOSITY OF n-DECANE [26].

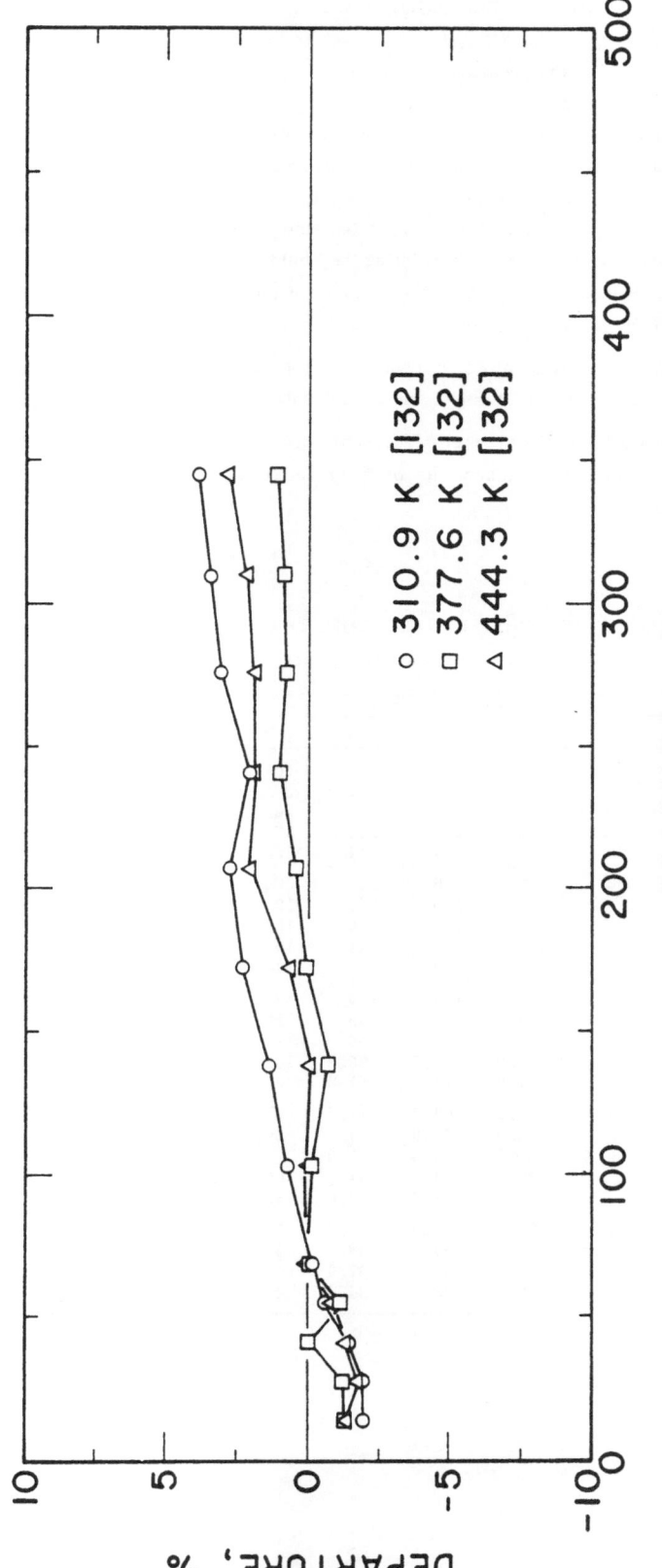

FIGURE 3. DEPARTURE PLOT ON THE VISCOSITY OF n – DECANE.

DICHLORODIFLUOROMETHANE (R 12)

A number of papers, each with a very limited pressure dependent viscosity data, have appeared in the literature [10, 45, 96, 137, 202, 207]. The only work covering a significant region of states is [32]. The data were obtained by a capillary tube viscosimeter, and the authors claim an uncertainty of = 1.5%. The given values are plotted as isotherms against pressure in Figure 1. The region of measurements is from 243.15 K to 433.15 K and from 10 bar to 600 bar. Using the dilute gas viscosities of [195] and of [51] as well as the P-V-T data of [177] it was possible to extrapolate the given data from 433.15 K to 575 K up to 200 bar using the residual viscosity concept. The extrapolated isotherms are included as dashed lines in Figure 1. The saturated liquid curve in Figure 1 was established by extrapolating the liquid isotherms to the vapour pressure from [177]. Comparing the values thus obtained to the values of [195] as well as to those of [51], large discrepancies are found. These are about 10% at low temperatures and reach as high as 40% at the critical temperature.

Figure 2 shows the measured and extrapolated values as isobars against temperature, the extrapolations being indicated by dashed lines. Recommended values were read from this plot and are tabulated below.

In view of the limited experimental data for the pressure dependence of the viscosity of R12, it is difficult to estimate the accuracy of the recommended values given here. An accuracy of 4% in the measured region and 7% in the extrapolated region appears to be a reasonable assumption.

VISCOSITY OF DICHLORODIFLUOROMETHANE (R 12)

$$[\mu, 10^{-6} \text{ N s m}^{-2}]$$

| T,K | \multicolumn{14}{c}{Pressure, bar} |
---	1	20	41.2*	50	60	80	100	120	160	200	300	400	500	600
250	10.8	327	326	329	333	340	347	354	367	382	417	438	474	–
260	11.0	287	291	297	301	307	313	321	335	348	372	402	435	468
270	11.4	256	262	268	271	278	284	290	302	315	344	372	405	434
280	12.0	233	240	245	247	254	260	265	277	290	378	345	373	402
290	12.4	211	218	222	225	232	237	244	255	267	294	319	342	369
300	12.8	190	197	202	205	272	276	223	234	246	273	297	321	343
320	13.2	153	159	163	168	174	180	188	195	205	225	256	279	299
340	14.0	123	130	133	135	144	150	156	168	178	200	223	246	266
350	14.4	106	114	118	121	130	138	144	155	165	189	210	231	252
360	14.6	93.0	100	105	108	117	126	132	144	154	178	200	218	239
370	14.8	80.0	84.0	90.0	95.0	105	114	120	132	143	165	188	206	227
380	15.2	40.0	67.0	75.0	82.5	94.0	103	109	120	132	155	716	195	215
390	15.8	25.0	32.5	53.5	65.0	81.0	92.0	100	112	122	146	167	186	204
400	16.0	21.5	27.0	38.0	50.5	68.0	81.5	90.0	103	114	138	158	177	194
410	16.5	21.0	24.5	32.0	40.5	57.0	71.5	81.0	95.5	106	130	149	167	184
420	16.7	21.0	23.5	28.5	35.0	48.0	62.0	72.5	87.5	98.5	121	140	158	175
430	17.0	20.0	23.2	27.0	32.0	42.0	54.0	64.0	80.0	91.5	113	132	149	165
440	17.5	20.5	23.0	26.5	30.0	38.5	48.0	57.0	73.0	85.0				
450	18.0	20.5	23.0	26.0	29.5	36.0	44.0	51.5	66.5	79.0				
475	19.0	21.0	23.5	25.7	28.5	34.0	38.5	44.0	56.5	67.0				
500	20.0	21.5	24.0	25.5	27.5	32.0	36.0	40.5	50.5	57.5				
525	21.0	22.0	23.5	25.5	27.5	30.5	34.5	38.0	45.5	51.5				
550	21.7	22.5	24.0	25.5	28.0	30.0	33.5	37.0	43.5	48.5				
575	22.5	23.5	24.5	26.0	28.0	30.0	33.5	36.5	42.0	47.0				

*Critical pressure.

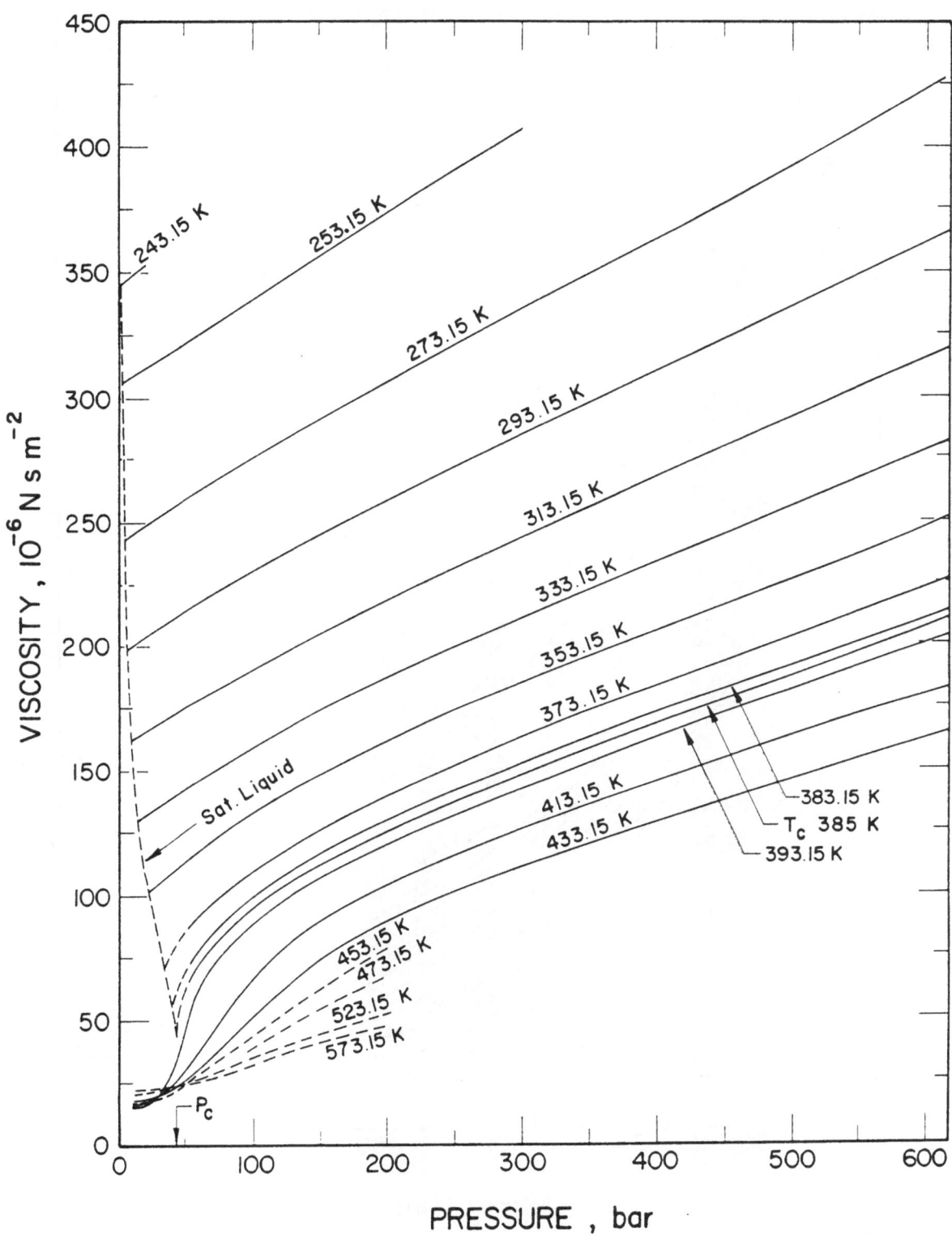

FIGURE 1. VISCOSITY OF R 12 [32].

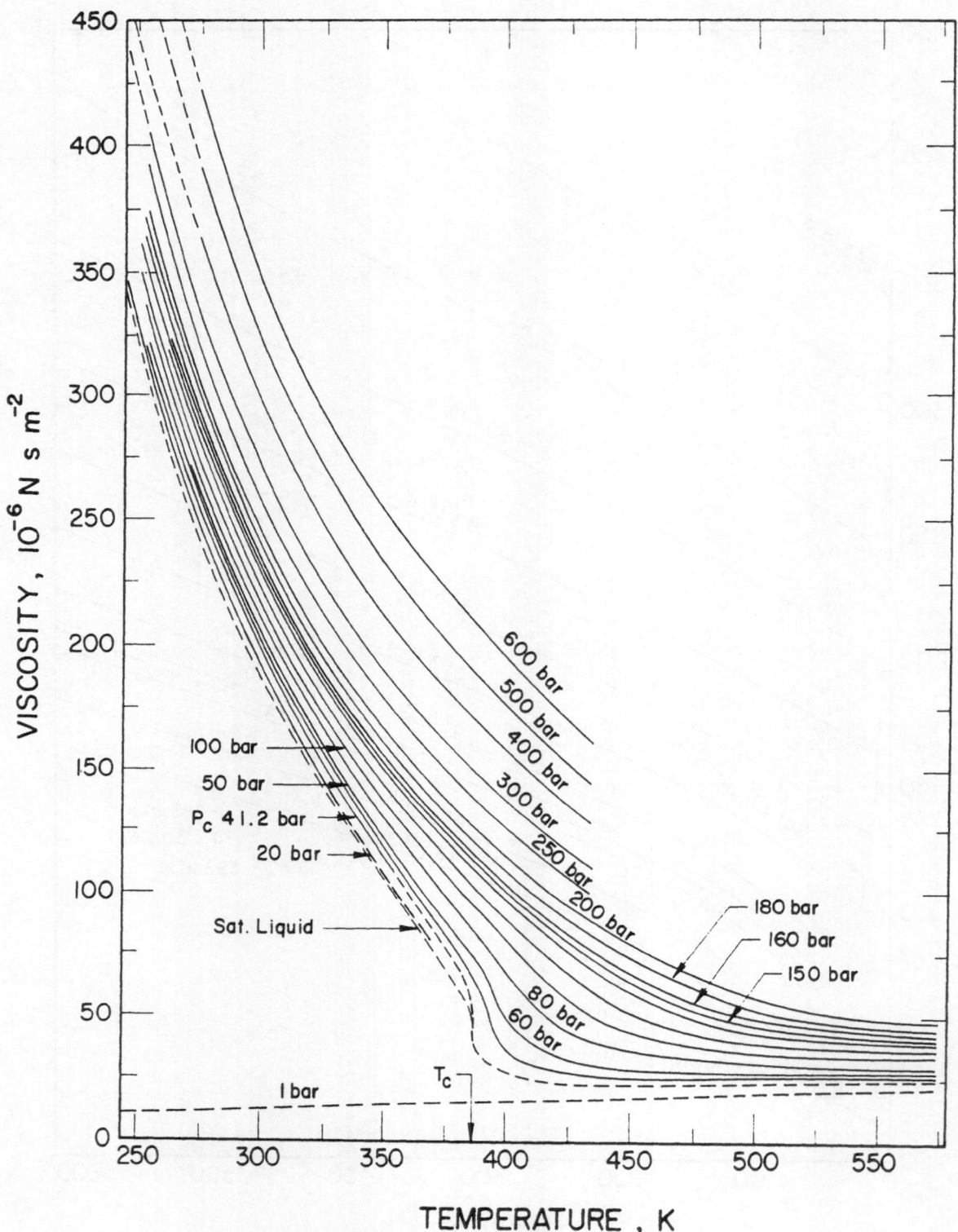

FIGURE 2. VISCOSITY OF R 12 [32].

n-DODECANE

     Only one set of data by Keramidi and Rastorguev [102] obtained by a capillary tube viscosimeter, was found in the literature for the pressure dependence of the viscosity of n-dodecane. The data are plotted as isotherms against pressure in Figure 1. Figure 2 shows isobars against temperature. The recommended values presented below were read from Figure 2. In absence of further experimental evidence, the authors' statement of uncertainty of ± 1.2% is reported here.

VISCOSITY OF n-DODECANE

$[\mu,\ 10^{-6}\ N\ s\ m^{-2}]$

T,K	Pressure, bar												
	1	18.1*	40	60	80	100	150	200	250	300	350	400	500
300	1300	1328	1372	1402	1434	1464	1548	1636	1728	1820	1912	2006	2208
310	1130	1156	1191	1220	1248	1278	1349	1424	1502	1580	1656	1734	1916
320	984	1007	1032	1060	1088	1114	1178	1244	1312	1378	1444	1509	1656
330	859	880	900	924	948	972	1028	1085	1145	1204	1260	1318	1436
340	753	772	790	811	831	852	902	952	1003	1053	1103	1153	1255
350	666	680	698	717	735	753	796	840	884	929	972	1017	1109
360	592	607	623	640	655	672	709	748	788	828	865	905	992
370	532	544	560	574	588	603	637	674	708	744	777	814	892
380	480	490	504	518	530	544	576	610	641	674	706	739	808
390	436	444	456	468	481	495	524	555	584	614	644	675	737
400	396	406	416	428	440	452	480	508	534	564	591	620	676
420	333	343	352	362	372	380	406	430	454	480	505	528	576
440	284	291	300	310	318	325	348	370	391	414	438	458	500
460	244	248	258	266	274	280	300	322	341	361	383	402	440
480	-	-	-	-	-	244	263	282	300	319	338	356	388
500	-	-	-	-	-	214	232	250	268	284	300	317	348
520	-	-	-	-	-	190	207	223	240	255	268	284	312

*Critical Pressure.

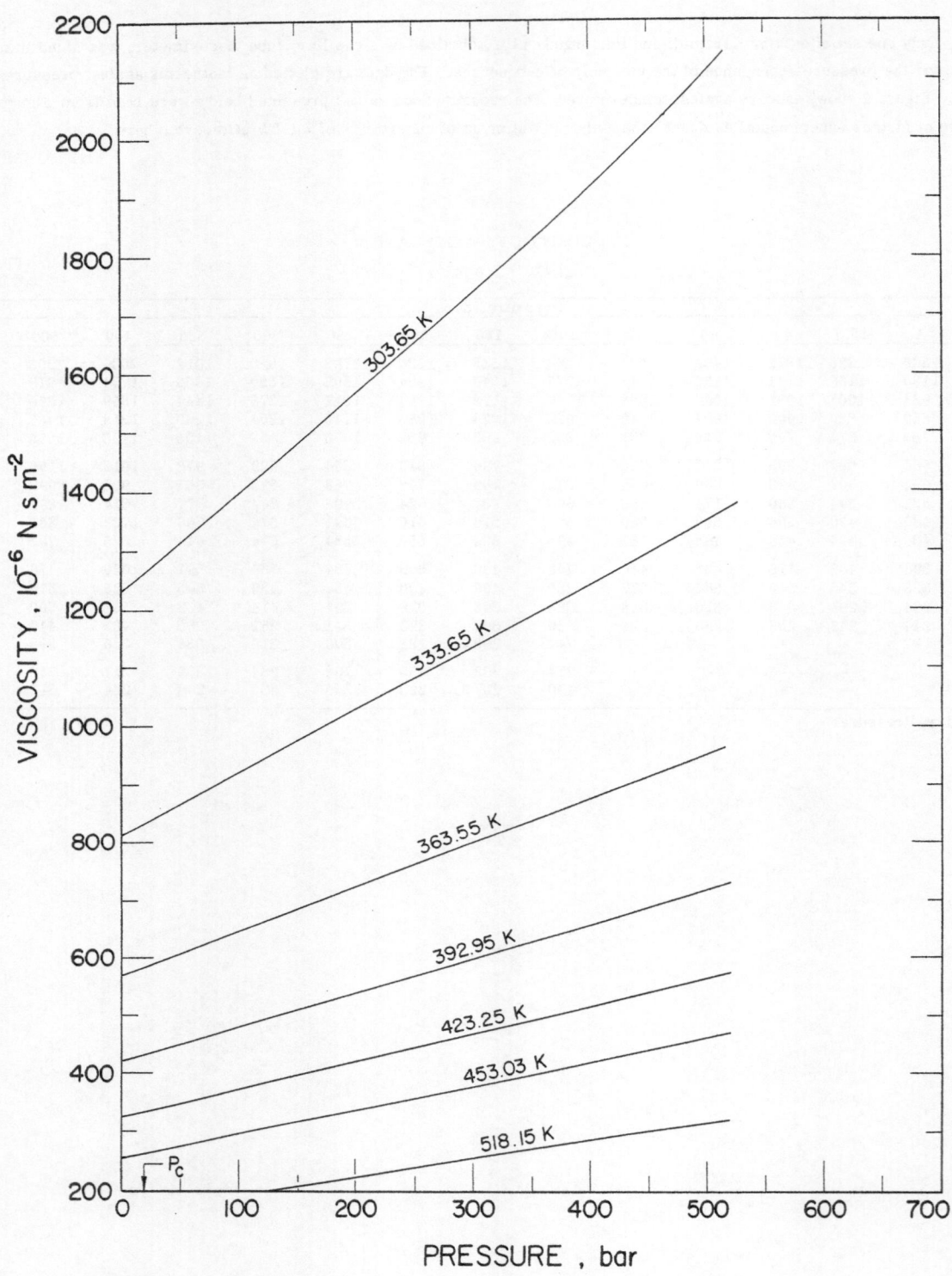

FIGURE 1. VISCOSITY OF n-DODECANE [102].

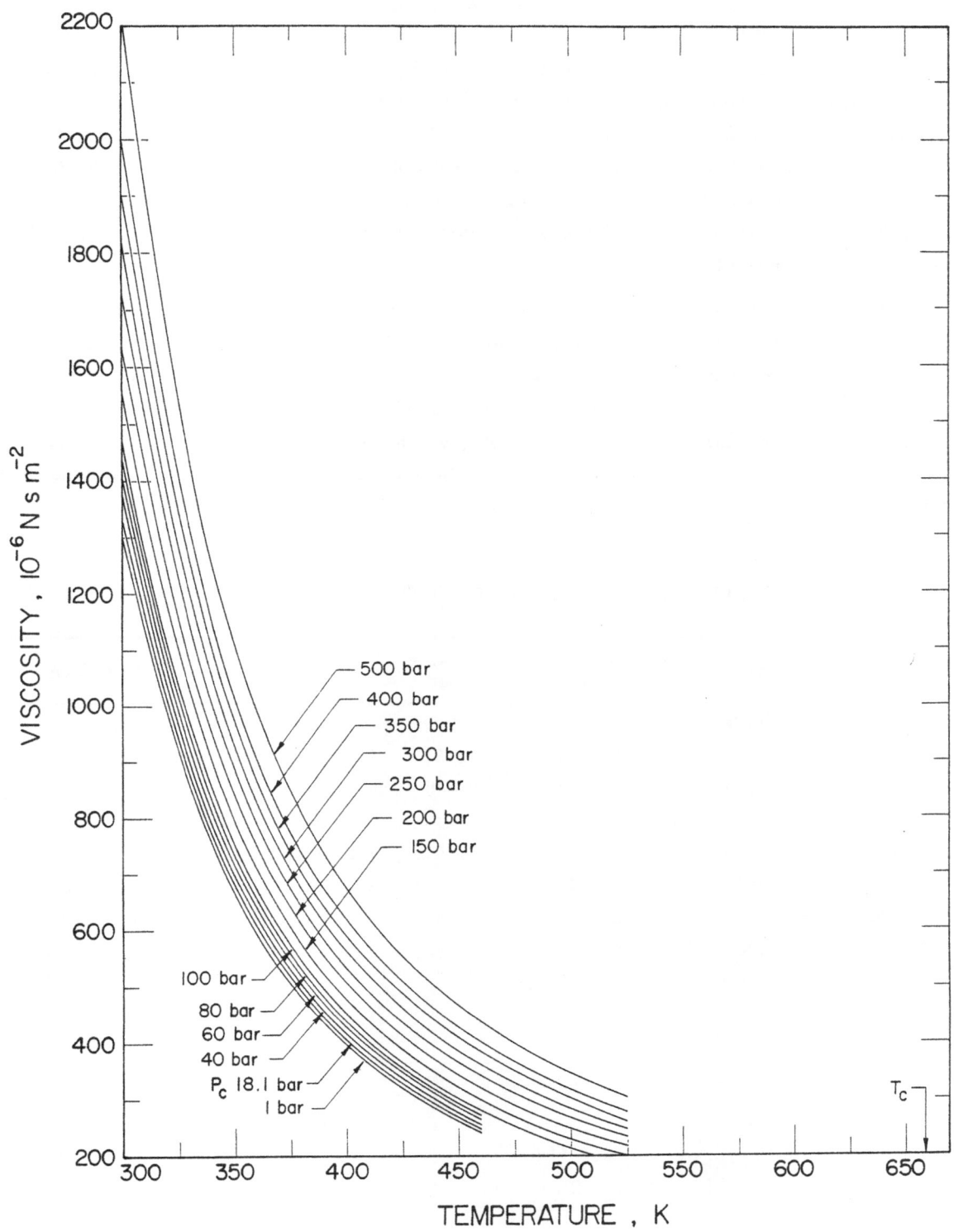

FIGURE 2. VISCOSITY OF n-DODECANE [102].

ETHANE

The data of Eakin et al. [44] obtained by a capillary tube viscosimeter, were used to generate the recommended values presented on the next page. Their data are shown as isotherms against pressure in Figure 1. The two highest isotherms, 447.59 K and 510.93 K are extrapolated values obtained by the authors [44] using the residual viscosity concept. Extrapolating the dilute gas viscosities, using the density data of [143], and the residual viscosity concept, the present authors extrapolated their data from 510 K up to 750 K and 200 bar. Figure 2 shows isobars plotted against temperature. The extrapolated values are indicated by dashed lines.

Additional works on the pressure dependence of the viscosity of ethane are listed in the summary table below. As seen from the departure plots of Figure 3, excellent agreement with the recommended values may be reported for the data of Carmichael et al. [29]. The agreement with the data of Baron et al. [11] is somewhat less satisfactory. Increasingly unsatisfactory are the comparisons with the data of Meshcheryakov et al. [57] and Smith et al. [182], which are probably of low accuracy. The uncertainty in the recommended values is estimated to be ± 3% in the measured region and ± 7% in the extrapolate region.

ADDITIONAL REFERENCES ON THE VISCOSITY OF ETHANE

Authors	Year	Ref. No.	Temperature K	Pressure bar	Method	Departure % (no. points)
Strumpf et al	1974	188	crit.	crit.	Torsional crystal	–
Carmichael and Sage	1963	29	300–478	1–345	Rotating cylinder	0.2 ± 1.0
Starling et al.	1962	184	302–311	39–60	Capillary tube	–
Swift et al.	1960	190	193–305	SL	Falling body	–
Eakin and Ellington	1959	43	298	14–552	Capillary tube	–
Baron et al.	1959	11	325–408	7–552	Capillary tube	0.02 ± 2.3 (40)
Meshcheryakov and Golubev	1954	57	258–523	1–811	Capillary tube	– 4 ± 2.9 (48)
Smith and Brown	1943	182	288–473	7–345	Rolling ball	– 7 ± 5.5 (52)

VISCOSITY OF ETHANE

[μ, 10^{-6} N s m^{-2}]

T, K	Pressure, bar																			
	1	20	30	40	45	48.8*	55	60	65	70	80	90	100	150	200	300	400	500	600	700
300	-	-	-	-	-	40.0	41.3	42.5	45.1	46.8	49.2	51.3	52.8	60.4	66.6	77.4	86.5	94.5	102.0	109.2
320	10.0	10.5	10.9	11.6	12.3	13.0	14.8	16.9	23.5	27.3	32.7	37.1	40.1	50.0	56.7	67.6	76.8	85.2	92.8	100.2
322	10.0	10.5	11.0	11.6	12.3	12.9	14.5	16.3	21.0	25.1	30.9	35.5	38.9	49.0	55.8	66.7	75.9	84.3	91.8	99.4
324	10.1	10.6	11.0	11.7	12.3	12.8	14.3	15.9	18.7	23.1	29.0	34.0	37.7	48.0	54.8	65.7	75.0	83.4	91.0	98.5
325	10.1	10.6	11.1	11.7	12.3	12.8	14.2	15.7	18.2	22.1	28.0	33.3	37.0	47.5	54.3	65.4	74.6	83.0	90.5	98.0
326	10.2	10.7	11.1	11.7	12.3	12.8	14.1	15.5	17.8	21.2	27.2	32.5	36.5	47.0	53.9	64.9	74.1	82.5	90.0	97.5
328	10.2	10.7	11.2	11.7	12.3	12.8	14.0	15.3	17.1	19.6	25.7	30.9	35.2	46.1	53.0	64.0	73.3	81.6	89.2	96.7
330	10.3	10.8	11.2	11.8	12.3	12.8	13.9	15.1	16.7	18.5	24.0	29.5	34.0	45.2	52.1	63.2	72.4	80.7	88.3	95.8
332	10.4	10.9	11.3	11.8	12.3	12.8	13.8	14.9	16.3	17.8	22.9	28.4	32.9	44.3	51.3	62.4	71.5	79.8	87.4	95.0
334	10.4	10.9	11.3	11.9	12.3	12.8	13.7	14.7	16.0	17.4	21.8	27.2	31.7	43.4	50.4	61.6	70.7	78.9	86.6	94.0
336	10.5	11.0	11.4	11.9	12.3	12.8	13.6	14.5	15.8	17.0	20.9	26.2	30.6	42.5	49.6	60.8	69.9	78.1	85.7	93.3
338	10.5	11.0	11.4	11.9	12.4	12.8	13.6	14.4	15.5	16.7	20.1	25.2	29.6	41.7	48.8	60.0	69.1	77.3	84.9	92.4
340	10.6	11.1	11.5	12.0	12.4	12.9	13.6	14.3	15.4	16.5	19.6	24.1	28.6	40.9	48.0	59.2	68.3	76.5	84.0	91.5
345	10.7	11.2	11.6	12.0	12.5	12.9	13.5	14.2	15.0	16.0	18.7	22.2	26.4	38.9	46.3	57.4	66.4	74.5	82.0	89.4
350	10.9	11.3	11.7	12.2	12.5	12.9	13.5	14.1	14.9	15.8	18.0	20.8	24.3	37.0	44.5	55.7	64.6	72.5	80.0	87.4
355	11.0	11.5	11.8	12.3	12.6	13.0	13.5	14.0	14.7	15.5	17.4	19.8	22.7	35.1	42.8	54.0	62.9	70.6	78.1	85.4
360	11.1	11.6	11.9	12.4	12.7	13.0	13.5	14.0	14.6	15.3	17.0	19.0	21.5	33.4	41.1	52.4	61.2	68.8	76.2	83.4
370	11.4	11.9	12.2	12.6	12.9	13.1	13.6	14.0	14.5	15.1	16.5	18.0	19.8	30.3	38.2	49.4	58.0	65.4	72.7	79.6
380	11.7	12.1	12.4	12.8	13.1	13.3	13.7	14.0	14.5	15.0	16.2	17.4	18.8	27.8	35.5	46.7	55.2	62.4	69.5	76.1
390	12.0	12.4	12.7	13.0	13.3	13.5	13.8	14.2	14.6	15.0	16.0	17.1	18.2	25.9	33.3	44.3	52.6	59.7	66.5	72.9
400	12.3	12.7	12.9	13.3	13.5	13.7	14.0	14.3	14.7	15.1	16.0	17.0	17.8	24.5	31.4	42.1	50.3	57.2	63.8	70.0
420	12.9	13.2	13.5	13.8	14.0	14.1	14.4	14.7	14.9	15.3	16.0	16.8	17.4	22.8	28.4	38.5	46.1	52.8	59.2	64.9
440	13.4	13.8	14.0	14.3	14.5	14.6	14.9	15.1	15.3	15.5	16.2	16.8	17.3	21.8	26.4	35.6	42.9	49.2	55.2	60.6
460	14.0	14.3	14.5	14.9	15.0	15.1	15.3	15.5	15.7	15.9	16.4	16.9	17.4	21.3	25.2	33.4	40.3	46.3	52.0	57.3
480	14.5	14.9	15.1	15.4	15.5	15.6	15.8	16.0	16.2	16.4	16.7	17.2	17.6	21.0	24.5	31.7	38.3	43.9	49.3	54.3
500	15.1	15.4	15.6	15.9	16.0	16.1	16.3	16.5	16.6	16.8	17.1	17.4	17.9	20.9	23.9	30.3	36.4	41.8	46.8	51.6
520	15.6	16.0	16.2	16.4	16.5	16.6	16.8	16.9	17.1	17.3	17.6	17.9	18.2	20.8	23.6					
540	16.1	16.5	16.7	16.9	17.0	17.1	17.3	17.4	17.5	17.7	18.0	18.2	18.5	20.9	23.4					
560	16.6	17.0	17.2	17.4	17.5	17.6	17.8	17.9	18.0	18.1	18.4	18.6	18.9	21.0	23.3					
580	17.2	17.6	17.7	17.9	18.0	18.1	18.2	18.4	18.5	18.6	18.8	19.1	19.3	21.1	23.3					
600	17.6	18.0	18.1	18.3	18.4	18.5	18.6	18.8	18.9	19.0	19.2	19.5	19.7	21.4	23.3					
620	18.1	18.5	18.6	18.8	18.9	19.0	19.1	19.2	19.3	19.5	19.7	19.9	20.1	21.6	23.4					
640	18.6	18.9	19.1	19.3	19.3	19.4	19.5	19.6	19.7	19.9	20.1	20.3	20.5	21.9	23.5					
650	18.9	19.2	19.3	19.5	19.5	19.6	19.7	19.8	19.9	20.1	20.3	20.5	20.7	22.0	23.6					
700	20.0	20.3	20.4	20.6	20.6	20.7	20.8	20.9	21.0	21.1	21.2	21.4	21.6	22.8	24.1					
750	21.0	21.3	21.4	21.6	21.6	21.7	21.8	21.9	21.9	22.0	22.1	22.3	22.4	23.6	24.7					

*Critical pressure.

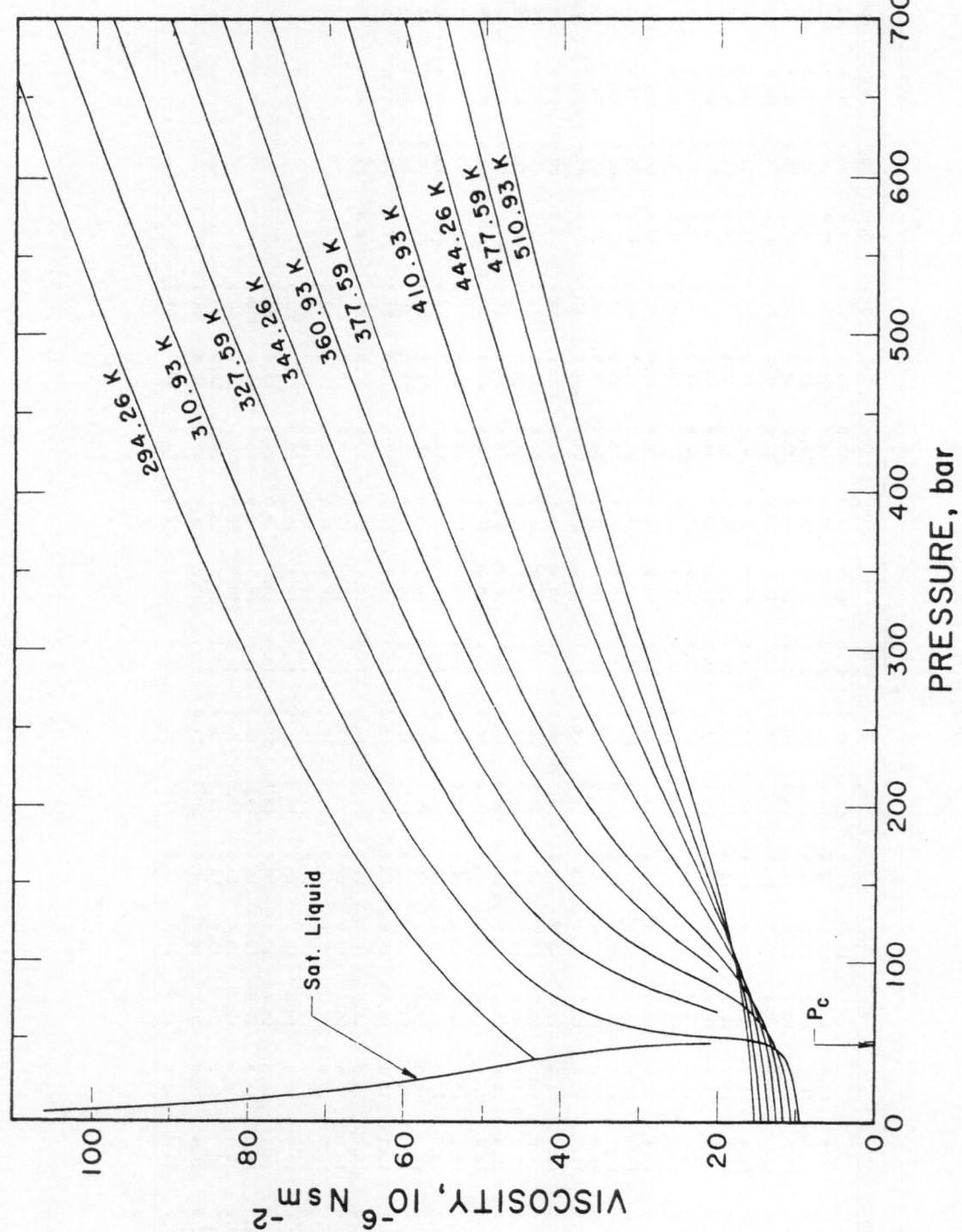

FIGURE 1. VISCOSITY OF ETHANE [44].

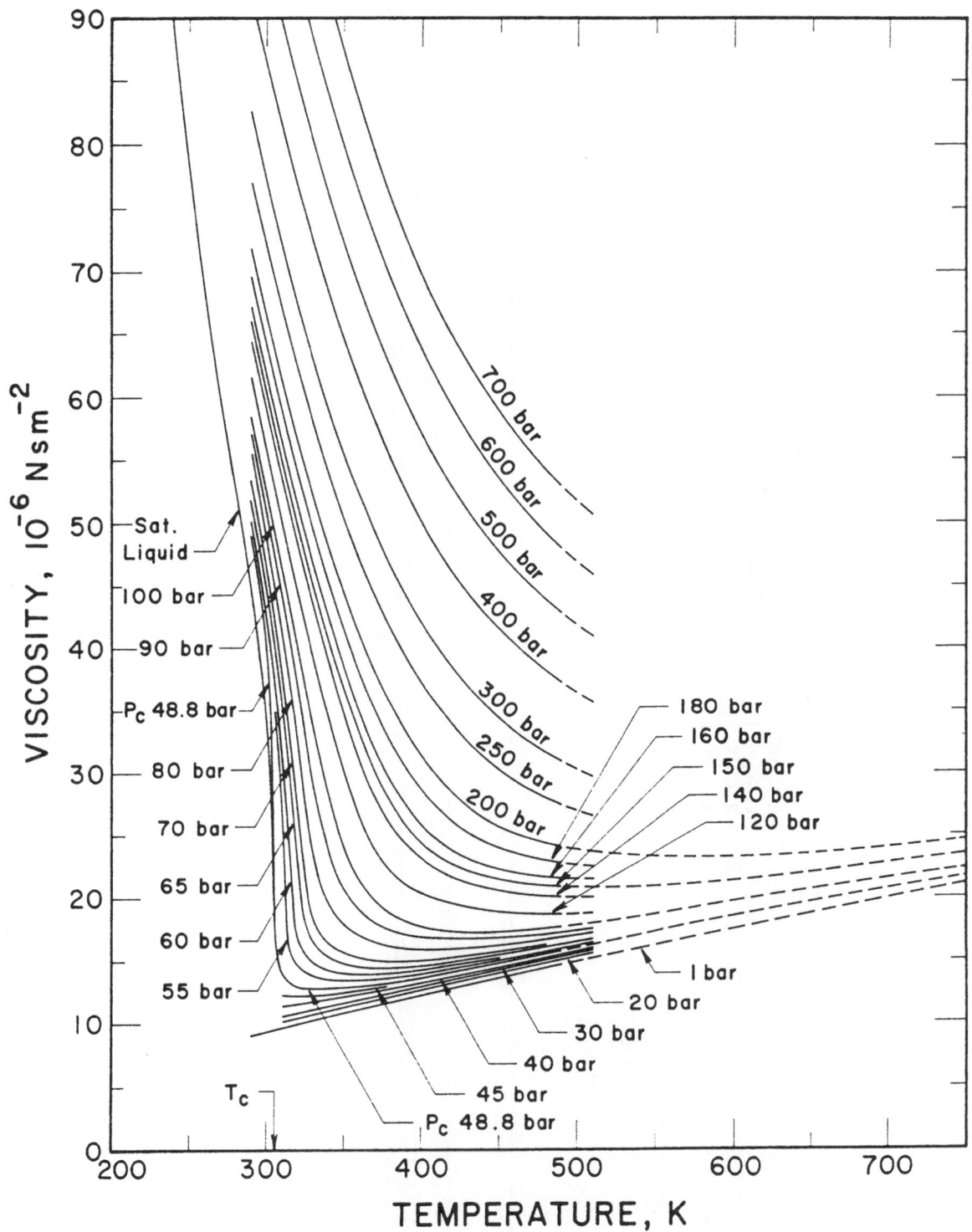

FIGURE 2. VISCOSITY OF ETHANE [44].

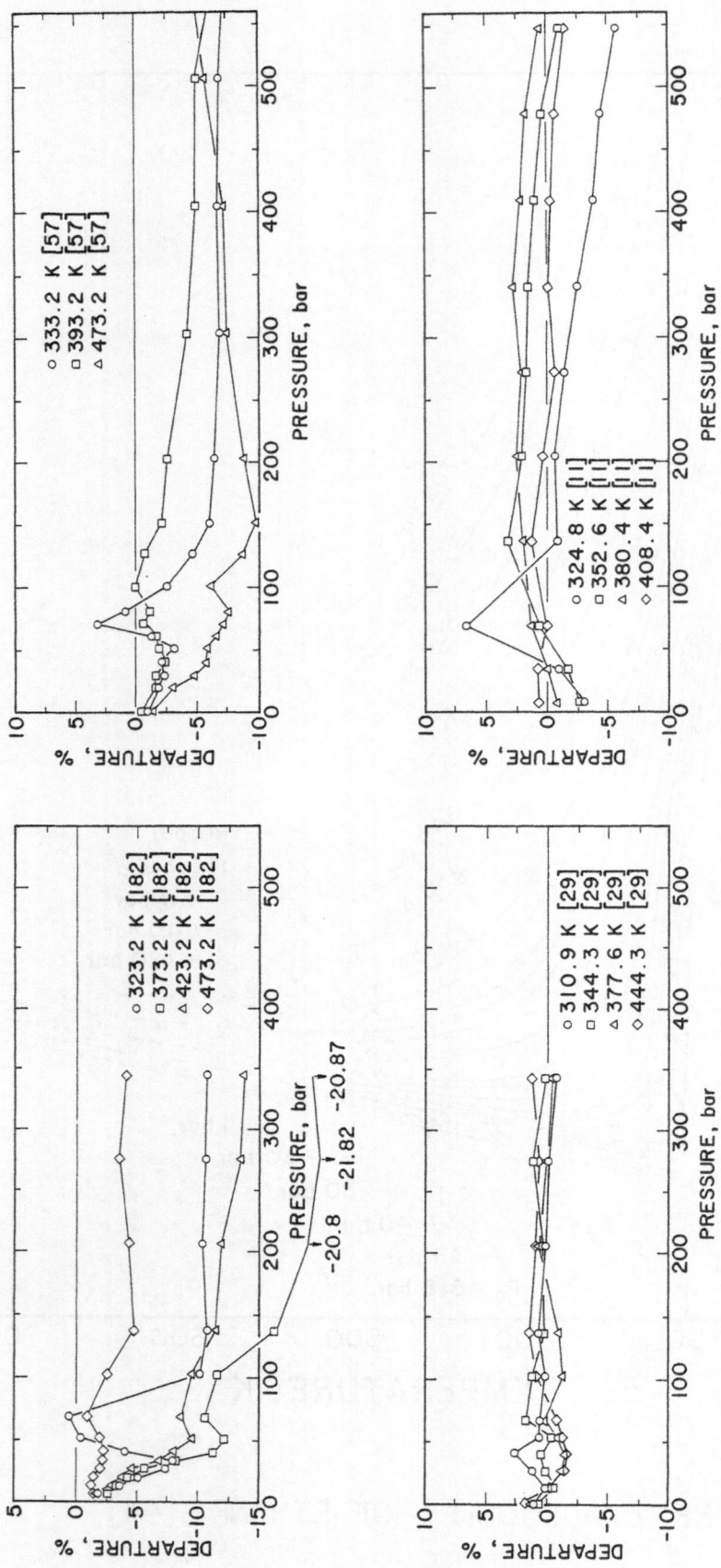

FIGURE 3. DEPARTURE PLOT ON THE VISCOSITY OF ETHANE.

ETHANOL

The data of Golubev et al. [57], obtained by the capillary tube method, were used to generate the recommended values presented below. They are plotted as isotherms against pressure in Figure 1 and as isobars against temperature in Figure 2. The recommended values were read from Figure 2. No comparison with other experimental data for ethanol at high pressures was made. The uncertainty in the recommended values is tentatively assumed to be around ± 6%.

VISCOSITY OF ETHANOL

$[\mu, 10^{-6} \text{ N s m}^{-2}]$

T, K	Pressure, bar										
	1	50	63.8*	75	100	150	200	400	600	800	1000
270	1930	1970	1980	1995	2005	2065	2100	2300	2525	2700	2950
280	1560	1590	1600	1610	1635	1670	1715	1880	2035	2230	2410
290	1265	1295	1305	1310	1340	1360	1400	1530	1665	1820	1970
300	1040	1065	1070	1075	1090	1115	1150	1255	1375	1490	1620
310	860	880	890	892	900	925	950	1045	1150	1235	1350
320	720	740	745	750	755	775	790	870	960	1040	1140
330	605	625	630	632	640	655	665	735	810	880	970
340	520	530	535	540	550	560	570	630	700	765	845
350	450	463	467	470	480	485	500	550	610	680	-
360	390	405	410	411	420	430	440	490	540	605	-
370	340	357	360	362	370	380	395	435	480	545	-
380	300	315	320	321	330	335	350	390	430	490	-
390	260	277	280	283	290	300	312	352	390	445	-
400	225	245	250	251	255	265	280	320	353	400	-
410	12.0	215	220	222	230	237	250	288	320	365	-
420	12.2	190	193	198	200	210	225	260	292	330	-
430	12.5	168	172	174	180	192	202	238	268	300	-
440	12.8	150	153	156	162	173	183	218	246	275	-
450	13.1	133	136	139	145	156	166	199	226	252	-
460	13.4	117	120	123	130	140	150	182	209	233	-
470	13.7	103	106	109	115	126	135	166	192	216	-
480	14.0	90.5	93.5	96.0	102	112	122	152	178	202	-
490	14.2	78.0	81.5	84.0	91.0	101	110	140	165	188	-
500	14.5	16.7	69.0	72.5	80.3	90.7	100	129	133	176	-
510	14.8	15.8	54.0	61.0	70.3	82.0	91.5	120	744	165	-
515	14.9	15.5	43.0	54.0	65.5	77.5	87.3	116	139	160	
520	15.1	15.4	21.0	45.0	60.5	73.5	83.5	112	136	156	-
525	15.2	15.4	19.8	30.0	56.0	70.5	80.0	109	132	151	-
530	15.3	15.5	18.8	25.0	51.5	66.0	76.0	105	128	147	-
540	15.6	15.6	18.0	20.5	42.5	59.0	69.5	98.5	121	139	-

*Critical pressure.

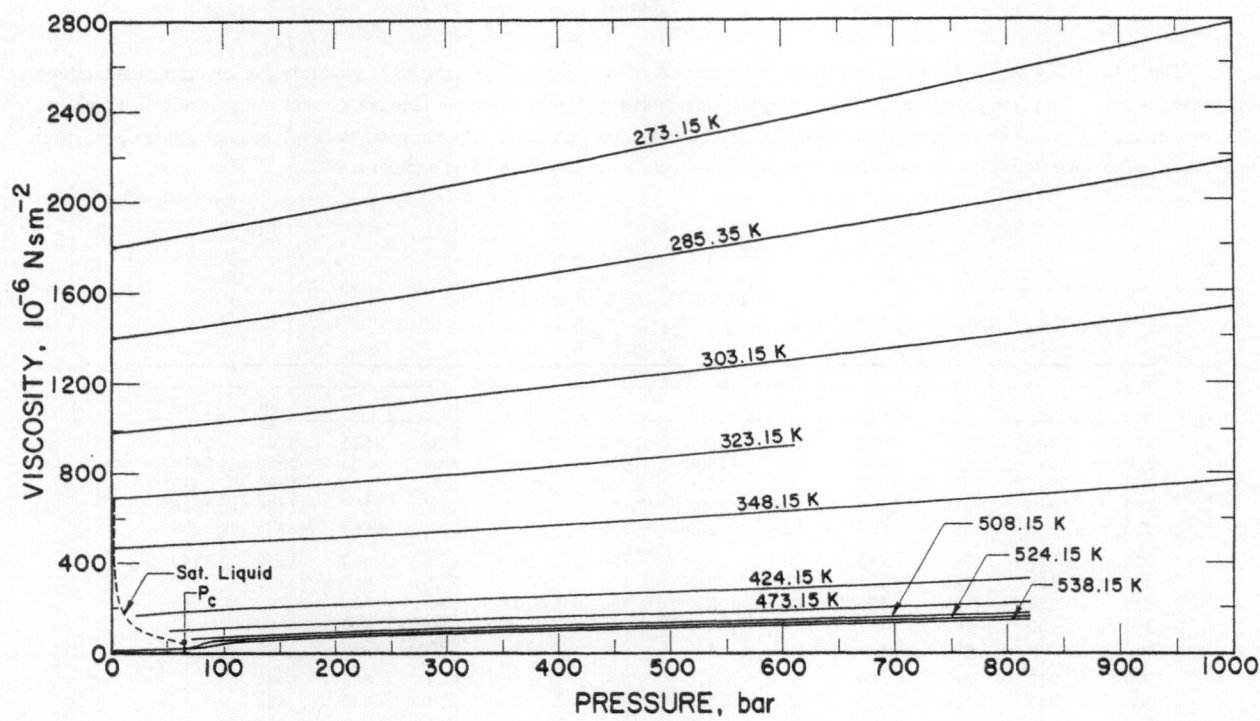

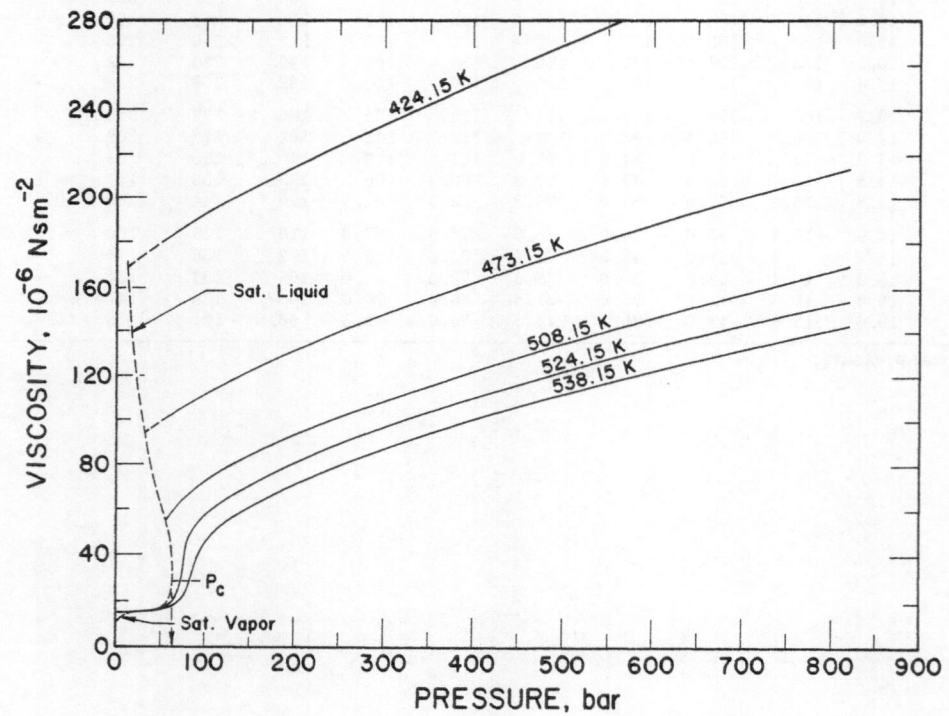

FIGURE I. VISCOSITY OF ETHANOL [57].

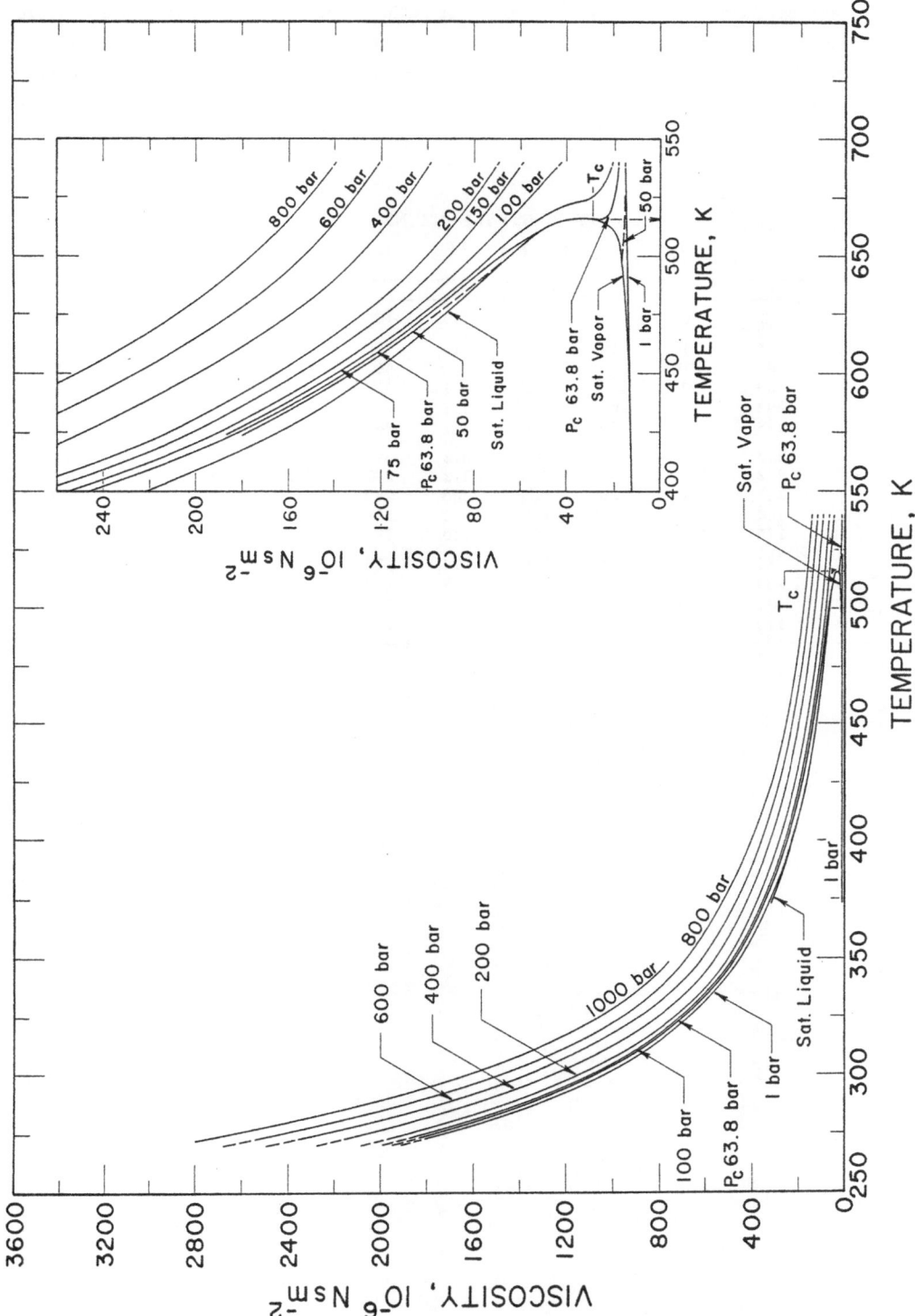

FIGURE 2. VISCOSITY OF ETHANOL [57].

ETHYLBENZENE

Only one set of data by Akhundov [6], obtained by a capillary tube viscosimeter, was found in the literature for the pressure dependence of the viscosity of ethylbenzene. The data are plotted as isotherms against pressure in Figure 1, where smoothing modifications of up to 1% were applied. Figure 2 shows isobars against temperature. The recommended values tabulated below were read from Figure 2. In the absence of additional experimental evidence the author's statement of uncertainty of ± 1.2% is reported here.

VISCOSITY OF ETHYLBENZENE

$[\mu, 10^{-6} \text{ N s m}^{-2}]$

T, K	Pressure, bar													
	1	20	36.1*	60	80	100	120	140	160	180	200	250	300	400
300	658	672	683	701	709	719	731	741	750	760	771	800	823	877
320	535	545	554	563	570	579	587	596	604	615	625	645	667	710
340	443	451	458	465	471	480	486	494	502	512	520	537	557	594
360	372	380	386	392	398	405	411	417	425	432	439	456	474	507
380	318	324	330	338	343	350	355	361	368	373	379	394	408	440
400	273	278	284	292	297	302	307	312	318	324	328	343	354	383
420	–	241	247	252	258	262	268	272	277	281	286	298	309	334
440	–	212	216	221	226	230	235	238	243	247	251	262	273	295
460	–	187	191	196	200	204	208	212	216	220	223	234	244	264
480	–	166	169	175	179	183	188	190	194	198	201	210	221	238
500	–	146	150	156	160	164	168	172	176	179	182	192	202	218
520	–	129	134	139	143	147	151	155	159	162	166	175	184	200
540	–	114	118	124	128	132	136	140	144	148	152	161	169	184
560	–	100	104	110	114	118	122	126	130	134	139	147	155	170

*Critical pressure.

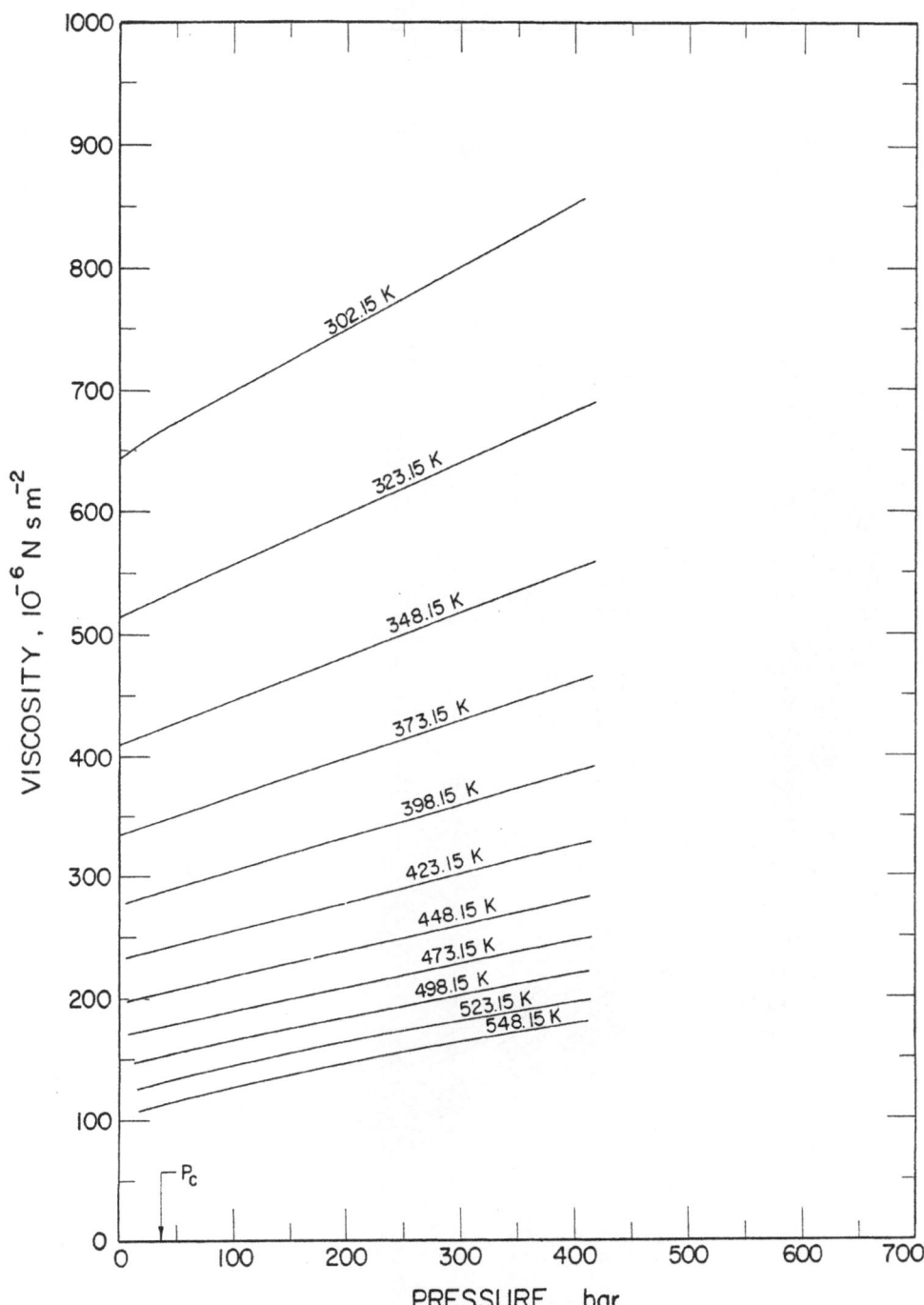

FIGURE I. VISCOSITY OF ETHYLBENZENE [6].

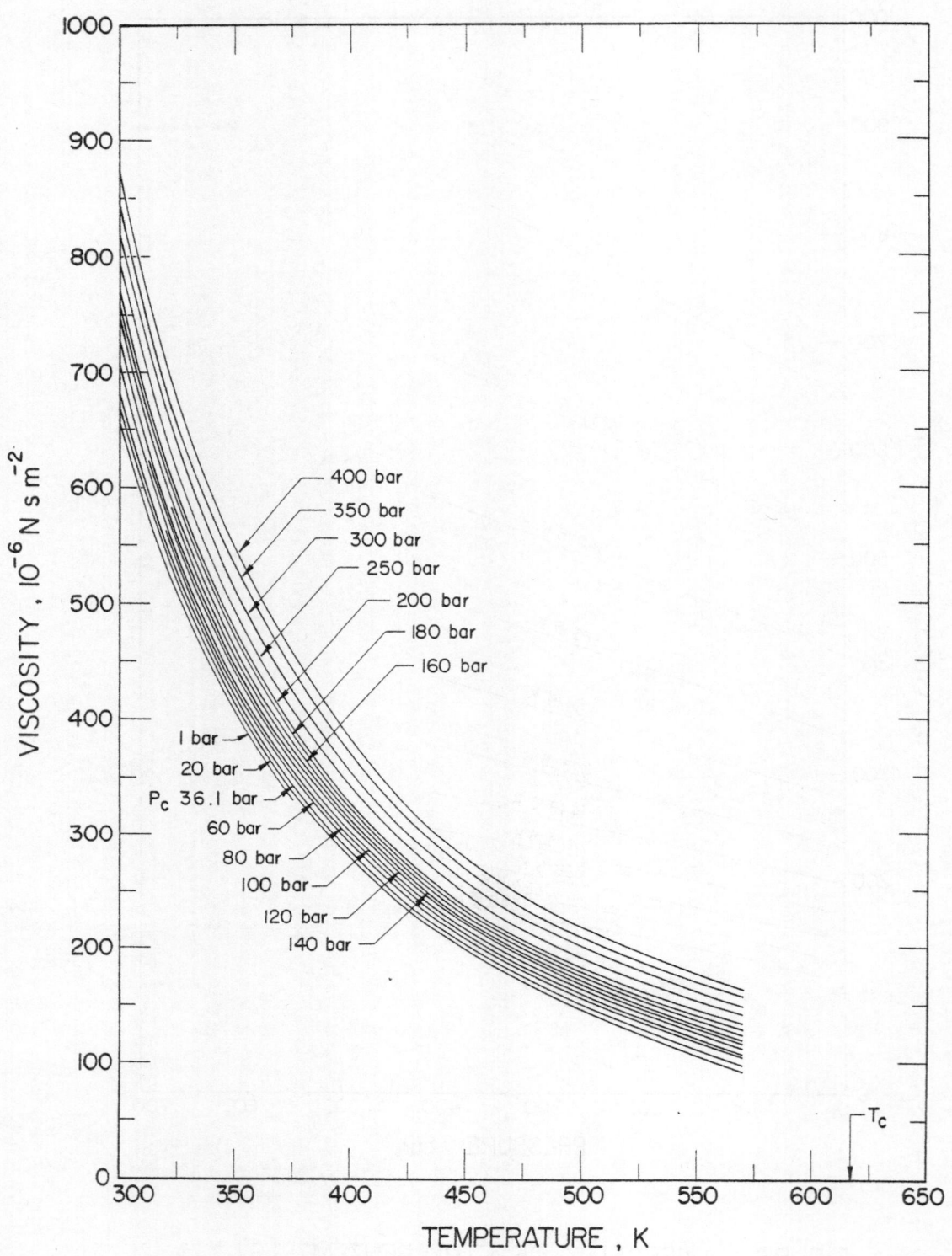

FIGURE 2. VISCOSITY OF ETHYLBENZENE [6].

ETHYLCYCLOHEXANE

The only available set of data that was found in the literature for the pressure dependence of the viscosity of liquid ethylcyclohexane is that of Guseinov et al. [73]. No statements about the experimental procedure nor the precision of the data are made by the authors. The data are shown as isotherms against pressure in Figure 1 and as isobars against temperature in Figure 2. The recommended values are tabulated below.

VISCOSITY OF ETHYLCYCLOHEXANE

$[\mu,\ 10^{-6}\ N\ s\ m^{-2}]$

T, K	Pressure, bar																			
	10	15	20	25	30	35	40	50	60	70	80	90	100	150	200	250	300	350	400	500
290	905	967	910	912	915	917	920	925	932	939	945	955	965	1020	1077	1140	1192	1250	1310	1435
300	825	827	830	833	836	839	842	847	854	861	870	877	885	935	987	1030	1082	1137	1190	1300
320	645	648	651	654	657	660	663	670	674	678	682	691	700	745	782	817	850	887	920	1040
340	507	510	512	515	517	520	522	527	537	536	540	547	555	580	607	635	670	712	750	825
360	415	417	420	422	425	427	430	435	438	442	445	454	452	475	500	525	547	580	607	660
380	347	350	352	354	356	358	360	365	368	372	375	378	382	402	425	447	467	492	512	560
400	292	295	207	299	301	303	305	310	313	137	320	322	325	345	365	385	402	425	442	480
420	247	249	251	253	255	257	258	262	266	207	275	278	282	300	317	335	352	370	387	420
440	210	212	213	215	217	219	221	225	229	233	237	341	245	260	275	192	310	325	340	372
460	185	186	188	190	192	194	196	200	204	208	212	215	217	232	247	262	275	290	307	337
480	–	–	–	–	–	–	–	180	183	187	190	192	195	210	225	240	252	267	280	307
500	–	–	–	–	–	–	–	160	163	167	170	173	177	192	207	220	235	247	260	285
520	–	–	–	–	–	–	–	140	145	150	155	158	162	177	192	205	220	232	245	267
530	–	–	–	–	–	–	–	132	137	142	147	150	157	172	185	200	212	227	237	262

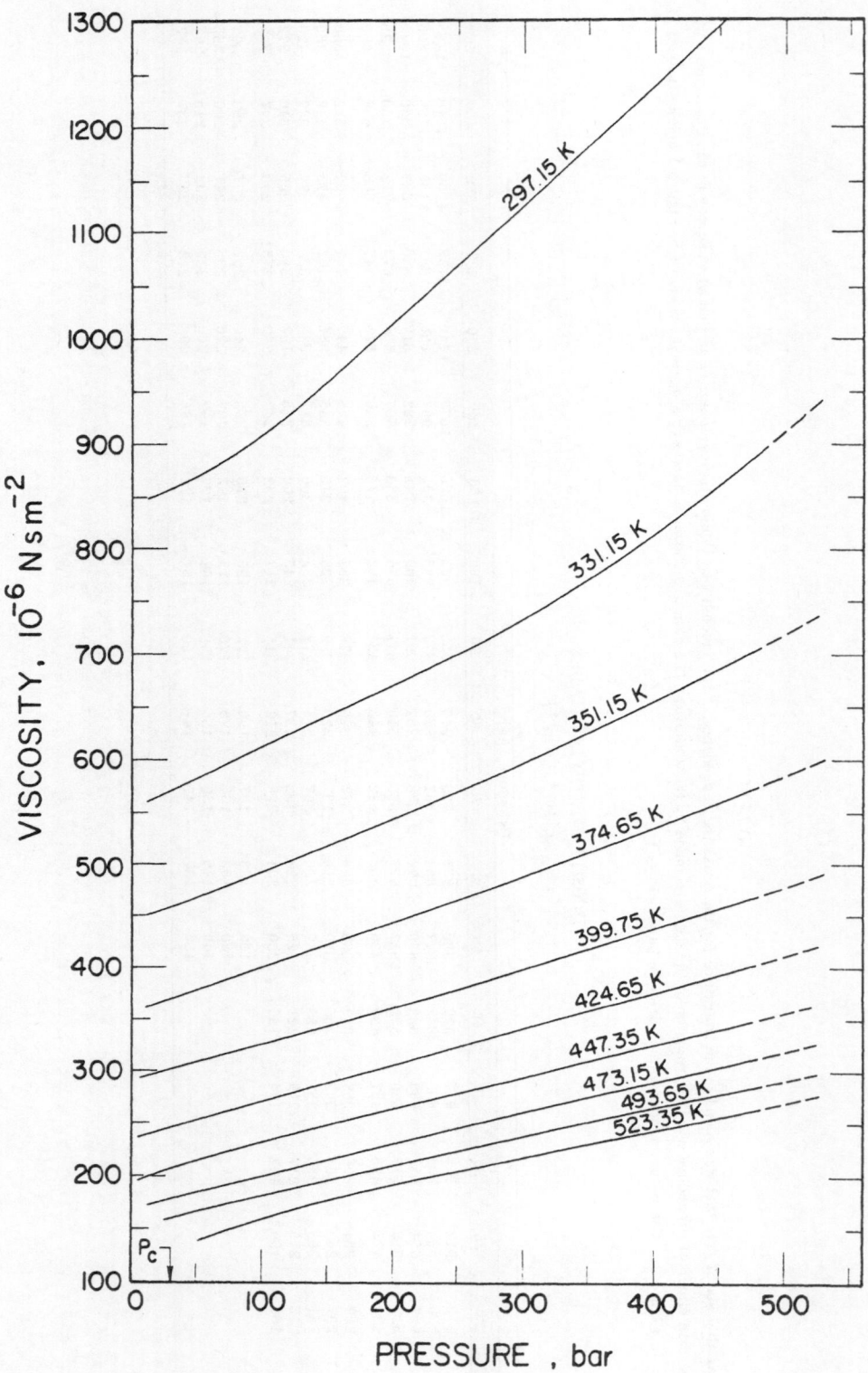

FIGURE 1. VISCOSITY OF ETHYLCYCLOHEXANE [73].

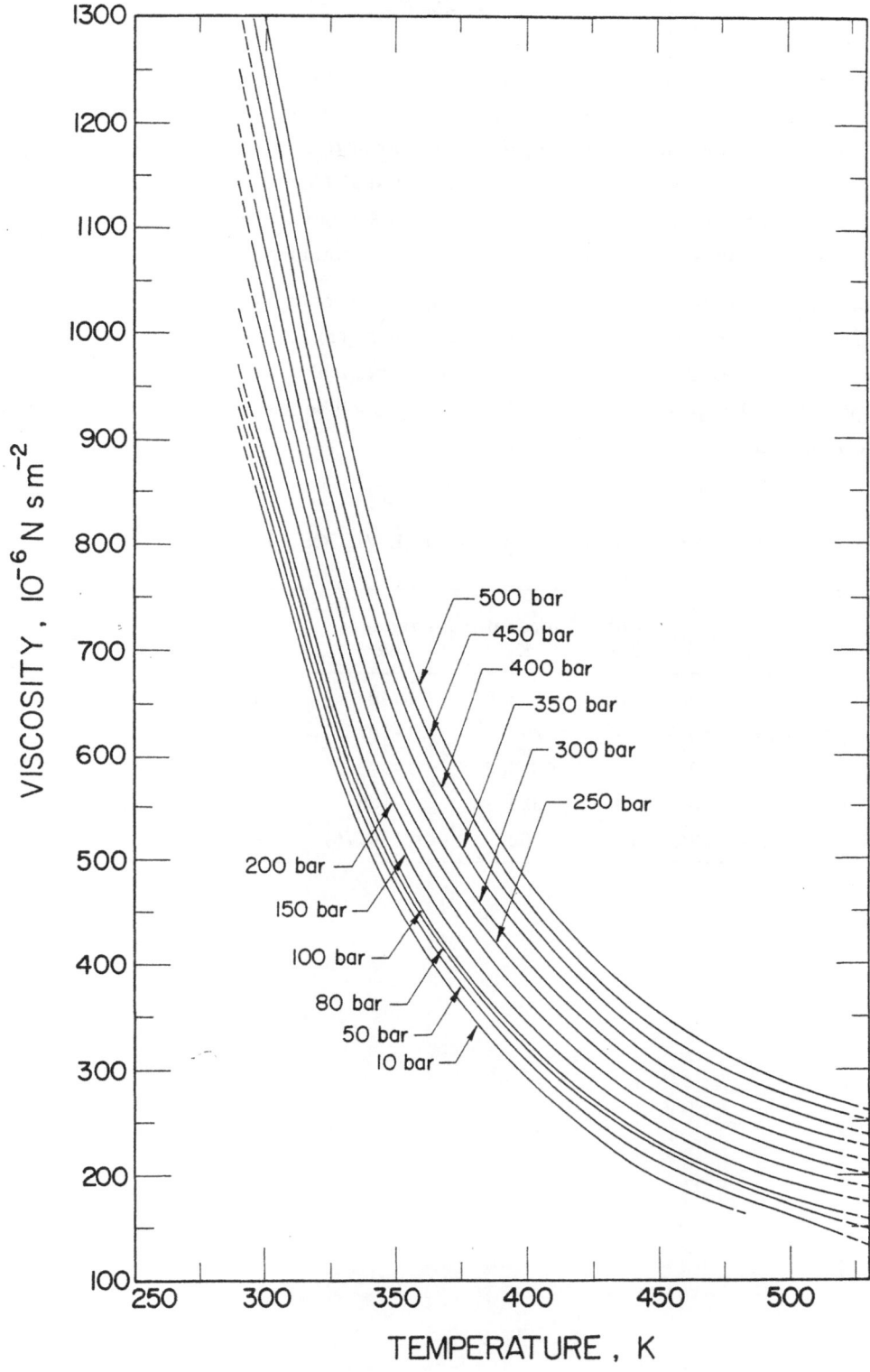

FIGURE 2. VISCOSITY OF ETHYLCYCLOHEXANE [73].

ETHYLENE

The data of Golubev and Petrov [57], obtained by a capillary tube viscosimeter were used to generate the recommended values tabulated on the next page. They are plotted as isotherms against pressure in Figure 1 where the maximum smoothing modifications amount to 2%. From Figure 1, the isobars in Figure 2 were read. Using the P-V-T data of [143], the dilute gas viscosities of [195] and the residual viscosity concept, an extrapolation from 423 K to 700 K up to 800 bar was possible. These extrapolated values are shown as dashed lines in Figure 2, from which the recommended values were read.

Additional works on the pressure dependence of the viscosity of ethylene are listed in the summary table below. The data of Comings et al. [35] reasonably confirm the recommended values, while the data of Mason et al. [142] and Neduzhii et al. [158] cover the critical region and the liquid region, respectively, at comparatively low pressures only. The data of Gonikberg et al. [61] show deviations of up to 9% from the recommended values, (see Figure 3).

ADDITIONAL REFERENCES ON THE VISCOSITY OF ETHYLENE

Authors	Year	Ref. No.	Temperature K	Pressure bar	Method	Departure % (no. points)
Neduzhii and Khmara	1968	158	190-290	1-40	Capillary tube	–
Gonikberg and Vereshchagin	1947	61	297	101-1013	Oscillating disc	9.0
Comings et al.	1944	36	303-368	1-173	Capillary tube	-0.06 ± 1.06
Comings and Egly	1941	35	313	4-139	Capillary tube	3.7
Mason and Maass	1940	142	283	pcr (50.76)	Oscillating disc	–

VISCOSITY OF ETHYLENE

$[\mu,\ 10^{-6}\ N\ s\ m^{-2}]$

T, K	1	20	30	40	45	50.6*	55	60	70	80	100	140	180	200	300	400	500	600	700	800
300	10.4	10.9	11.4	12.2	13.0	13.8	14.8	16.8	22.2	30.0	37.0	45.2	51.0	53.5	63.3	71.4	78.9	86.0	92.2	97.2
305	10.5	11.0	11.5	12.2	12.8	13.5	14.2	15.7	18.5	24.6	32.6	42.3	48.3	50.9	60.8	68.8	76.2	83.2	89.2	94.3
310	10.7	11.2	11.6	12.2	12.8	13.3	13.8	14.9	17.0	21.3	30.2	39.8	46.0	48.6	58.4	66.4	73.8	80.6	86.6	91.8
315	10.8	11.3	11.7	12.3	12.7	13.2	13.6	14.4	16.2	19.2	27.3	37.5	43.9	46.6	56.4	64.5	71.7	78.4	84.3	89.4
320	11.0	11.5	11.8	12.4	12.7	13.1	13.5	14.1	15.6	17.9	24.7	35.2	41.9	44.6	54.4	62.5	69.7	76.4	82.2	87.4
330	11.3	11.8	12.1	12.5	12.8	13.1	13.4	13.8	15.1	16.8	20.5	31.2	38.4	41.0	51.0	59.0	66.1	72.6	78.4	83.6
340	11.6	12.1	12.3	12.7	12.9	13.2	13.5	13.8	14.9	16.1	18.8	27.8	35.2	37.9	48.2	56.1	63.0	69.4	75.2	80.4
350	11.9	12.4	12.6	12.9	13.1	13.4	13.6	13.9	14.8	15.8	18.0	25.4	32.4	35.1	45.6	53.4	60.2	66.5	72.2	77.5
360	12.2	12.6	12.8	13.2	13.3	13.6	13.8	14.0	14.8	15.6	17.5	23.7	30.0	32.6	43.3	51.0	57.8	64.0	69.6	74.9
370	12.5	12.9	13.1	13.4	13.6	13.8	14.0	14.2	14.9	15.6	17.4	22.4	28.0	30.5	41.2	48.9	55.6	61.6	67.3	72.6
380	12.8	13.2	13.5	13.8	14.0	14.2	14.4	14.6	15.1	15.8	17.3	21.7	26.6	29.1	39.4	47.1	53.6	59.6	65.2	70.5
390	13.1	13.5	13.8	14.1	14.3	14.5	14.7	14.9	15.4	16.0	17.3	21.3	25.7	28.0	37.8	45.4	51.9	57.8	63.4	68.6
400	13.4	13.8	14.1	14.4	14.6	14.8	15.0	15.2	15.7	16.3	17.5	21.0	25.1	27.2	36.5	44.0	50.4	56.2	61.6	66.8
420	13.9	14.4	14.8	15.0	15.2	15.5	15.6	15.8	16.2	16.7	17.8	20.6	24.1	25.9	34.2	41.4	47.7	53.3	58.6	63.6
440	14.5	15.0	15.4	15.7	15.8	16.0	16.1	16.3	16.7	17.1	18.0	20.4	23.4	24.9	32.5	39.2	45.3	50.9	55.8	60.6
450	14.8	15.2	15.6	15.9	16.1	16.3	16.4	16.5	16.9	17.3	18.2	20.3	23.1	24.5	31.7	38.3	44.2	49.7	54.4	59.2
500	16.2	16.6	16.9	17.1	17.3	17.5	17.6	17.7	18.0	18.3	19.0	20.5	22.5	23.5	29.1	34.5	39.7	44.6	48.8	53.0
550	17.5	18.0	18.2	18.3	18.4	18.6	18.7	18.8	19.1	19.4	19.9	21.1	22.6	23.4	27.9	32.3	36.6	40.7	44.6	48.3
600	18.8	19.2	19.4	19.5	19.6	19.8	19.9	19.9	20.1	20.4	20.9	21.8	23.1	23.8	27.2	31.1	34.8	38.1	41.8	45.0
650	20.0	20.3	20.5	20.6	20.7	20.9	21.0	21.0	21.2	21.5	21.9	22.7	23.8	24.3	27.1	30.5	33.7	36.7	40.0	42.7
700	21.1	21.4	21.6	21.8	21.9	22.0	22.1	22.1	22.3	22.5	22.8	23.7	24.4	24.9	27.4	30.3	33.2	36.1	38.9	41.0

Pressure, bar

*Critical pressure.

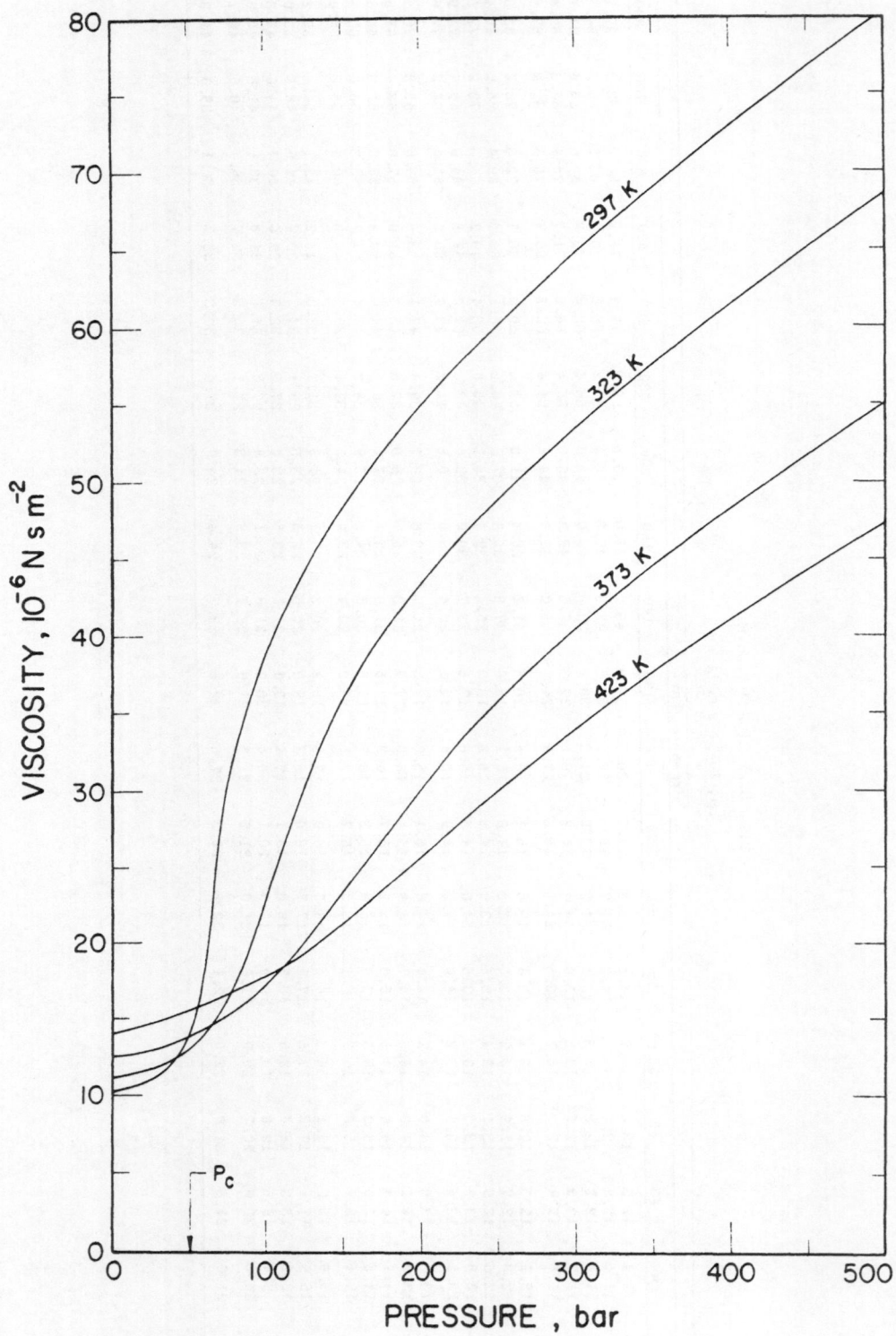

FIGURE I A. VISCOSITY OF ETHYLENE [57].

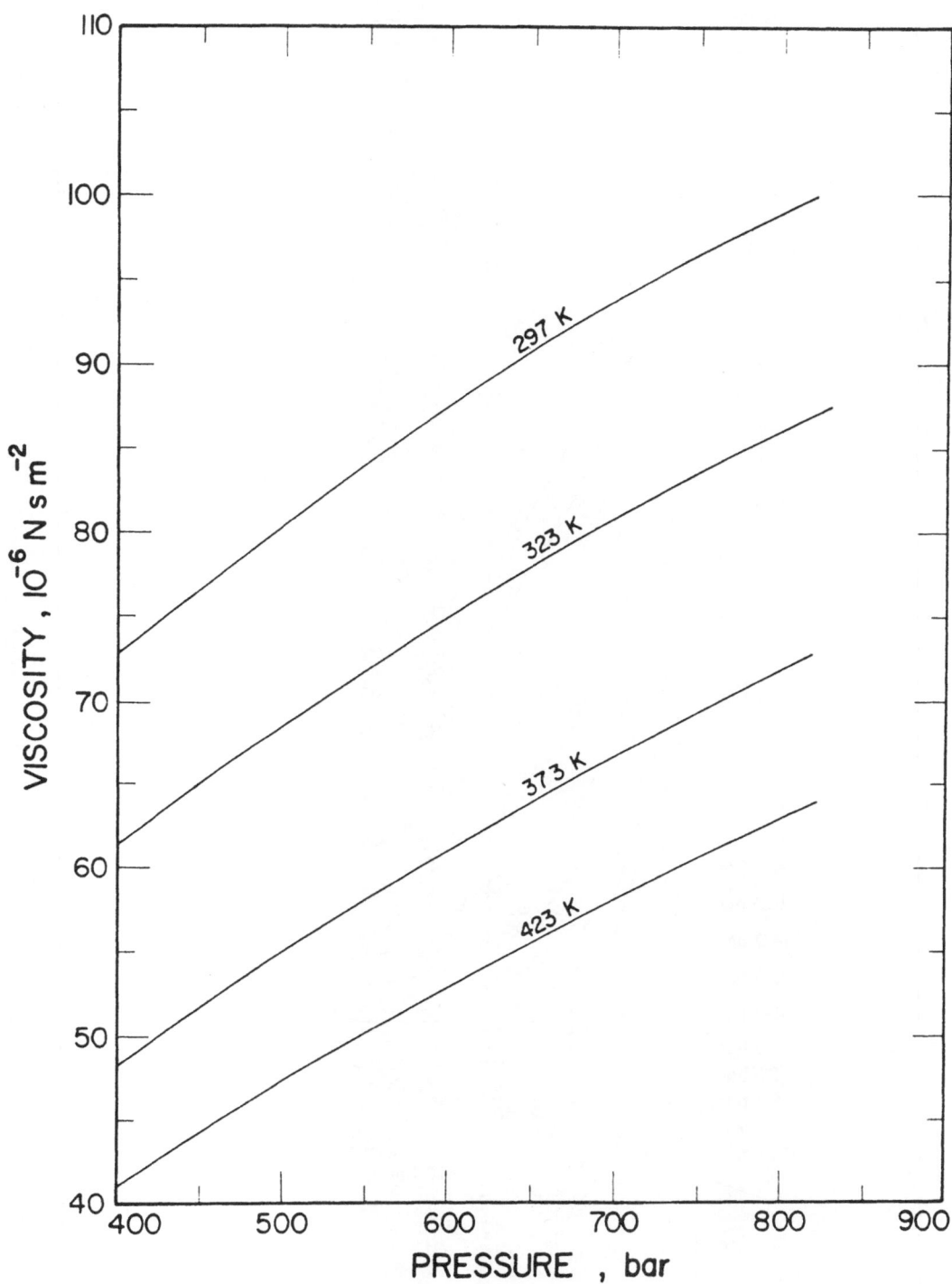

FIGURE 1B. VISCOSITY OF ETHYLENE [57].

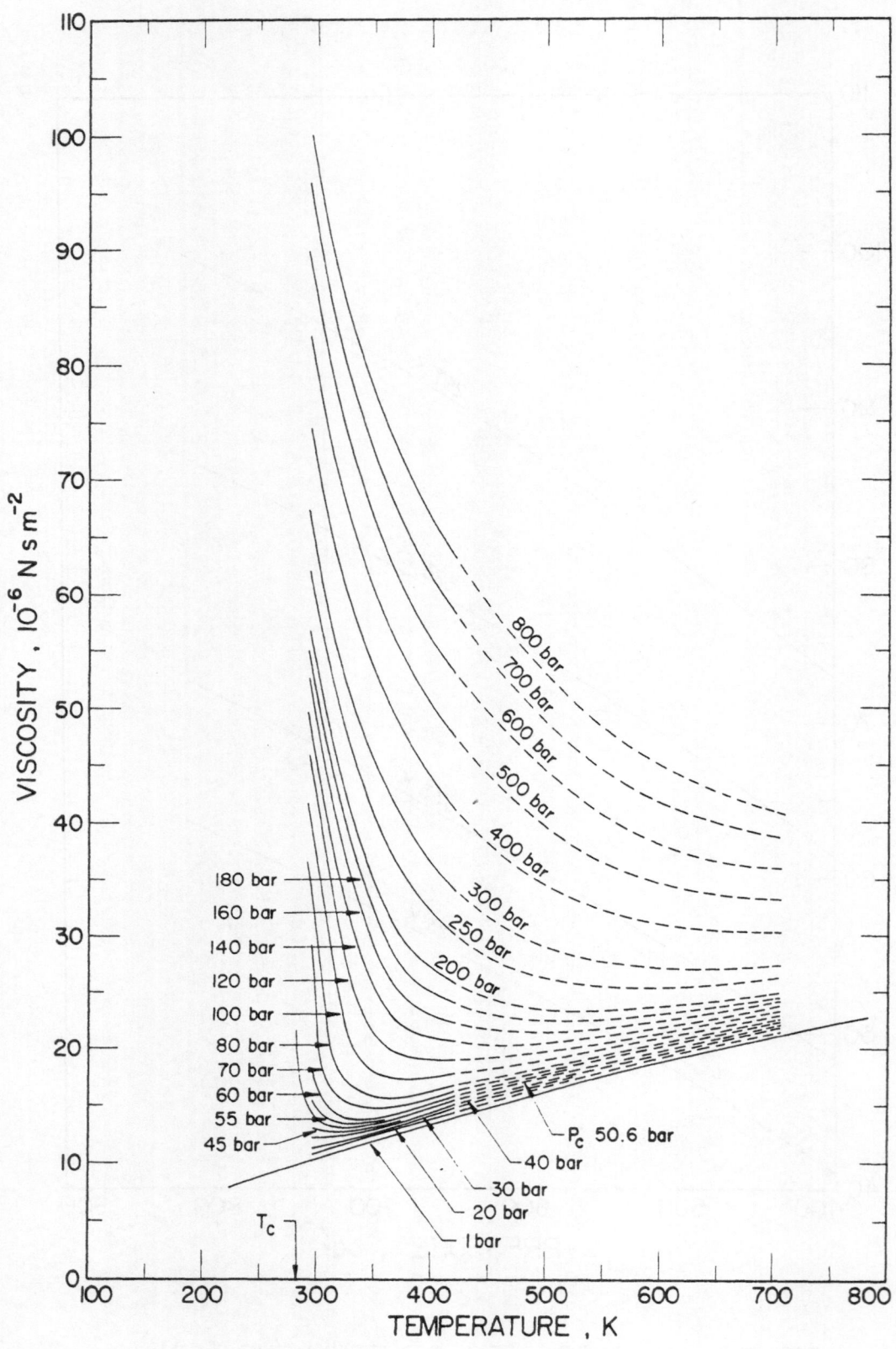

FIGURE 2. VISCOSITY OF ETHYLENE [57].

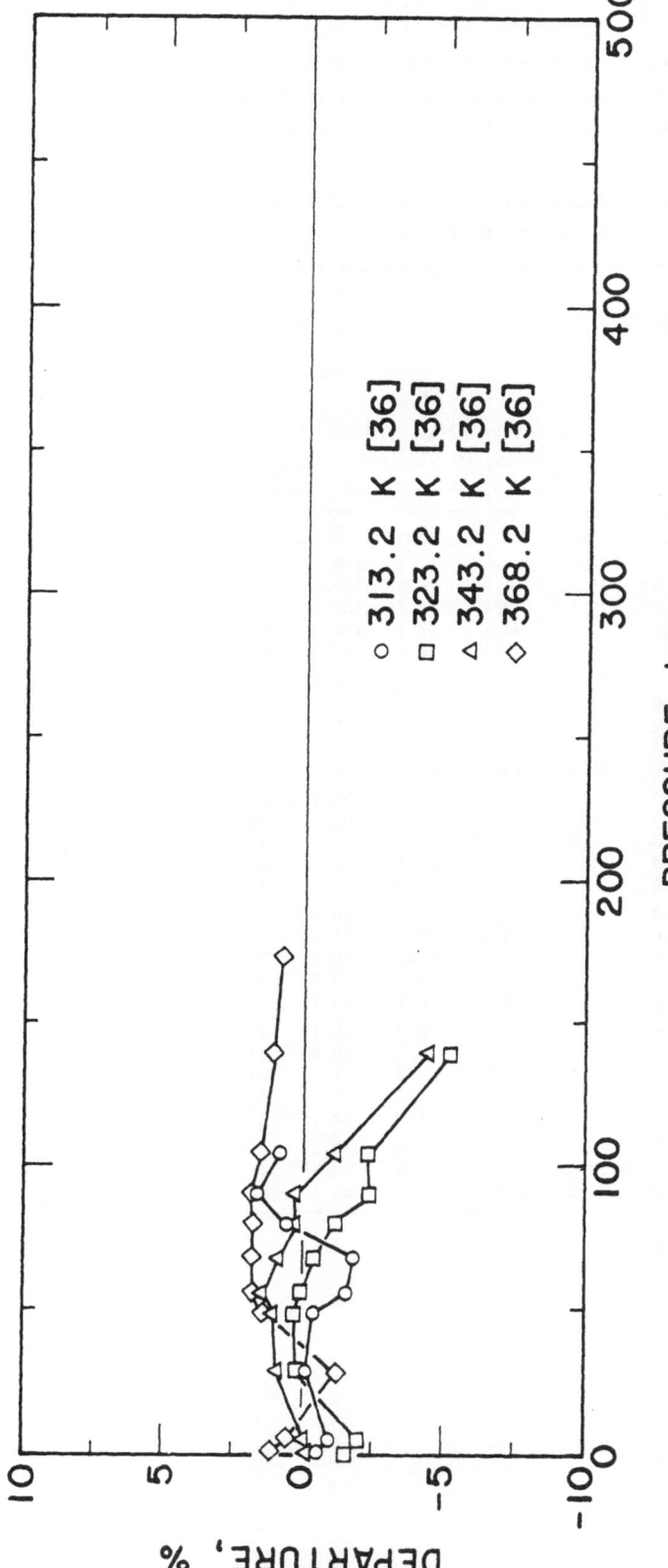

FIGURE 3. DEPARTURE PLOT ON THE VISCOSITY OF ETHYLENE.

FLUORINE

The data of Haynes [78], obtained by the torsional crystal method, were used to generate the recommended values tabulated below. The data are plotted as isotherms against pressure in Figures 1A and 1B, where smoothing modifications up to 1% were applied. Figure 2 shows isobars plotted against temperature. The recommended values were read from the latter figure.

No additional works on the pressure dependence of the viscosity of fluorine were found in the literature. The data of Elverum and Doescher [46] are limited to the saturated liquid, and, according to [78], show systematic deviations from the recommended values of 3-5 %. In the absence of additional experimental evidence, the author's estimate of accuracy 0.5% is given here.

VISCOSITY OF FLUORINE
$[\mu, 10^{-6} \text{ N s m}^{-2}]$

| T, K | \multicolumn{14}{c}{Pressure, bar} |
	1	20	30	40	52.2*	60	70	80	90	100	120	140	160	200
90	7.51	206.8	209.2	211.5	214.1	215.6	217.5	219.4	221.3	223.1	226.6	230.2	233.6	240.4
100	8.30	156.7	158.9	161.0	163.5	164.9	166.8	168.6	170.4	172.1	175.4	178.6	181.7	187.7
110	9.09	123.3	125.3	127.2	129.5	130.8	132.6	134.3	136.0	137.6	140.7	143.7	146.7	152.4
120	9.90	96.9	98.8	100.7	103.0	104.4	106.1	107.8	109.4	111.1	114.2	117.3	120.3	126.0
130	10.71	–	–	–	80.0	82.3	84.9	87.2	89.1	90.8	94.1	97.3	100.4	106.4
140	11.51	–	–	–	56.8	60.4	63.8	66.5	68.8	70.9	75.0	78.6	82.0	88.4
155	12.68	13.5	14.1	15.0	16.7	18.6	23.2	31.6	38.4	43.4	50.7	56.2	60.7	68.2
156	12.76	13.6	14.2	15.0	16.7	18.4	22.4	29.9	36.4	41.7	49.5	55.0	59.4	67.0
157	12.84	13.7	14.3	15.1	16.7	18.3	21.7	28.3	34.3	40.1	48.1	53.7	58.2	65.8
158	12.92	13.7	14.3	15.1	16.6	18.2	21.1	27.1	32.6	33.6	46.8	52.4	57.0	64.7
159	12.99	13.8	14.4	15.2	16.6	18.1	20.7	26.1	31.1	33.2	45.5	51.1	55.8	63.6
160	13.07	13.9	14.5	15.3	16.6	18.0	20.4	25.3	29.9	35.8	44.2	49.9	54.6	62.4
165	13.45	14.2	14.8	15.5	16.7	17.7	19.4	22.3	25.6	29.7	38.0	44.0	48.8	57.0
170	13.83	14.6	15.1	15.8	16.8	17.7	19.1	20.9	23.1	26.1	32.7	38.6	43.7	52.2
175	14.20	14.9	15.4	16.1	17.0	17.7	18.9	20.2	21.9	24.2	29.4	34.7	39.4	48.0
180	14.58	15.3	15.7	16.3	17.2	17.8	18.8	20.0	21.3	23.2	27.4	32.0	36.4	44.6
185	14.96	15.7	16.1	16.6	17.4	18.0	18.9	19.9	21.1	22.5	26.0	29.9	33.8	41.4
190	15.33	16.0	16.4	16.9	17.7	18.2	19.0	19.9	21.0	22.2	25.1	28.4	32.0	39.0
195	15.71	16.4	16.7	17.2	17.9	18.4	19.2	20.0	21.0	22.0	24.4	27.3	30.5	36.9
200	16.08	16.7	17.0	17.5	18.2	18.6	19.4	20.1	21.0	21.9	24.0	26.5	29.3	35.2
210	16.82	17.4	17.7	18.1	18.7	19.1	19.8	20.4	21.1	22.0	23.6	25.6	27.8	32.7
220	17.54	18.0	18.3	18.7	19.2	19.6	20.2	20.8	21.4	22.1	23.6	25.2	27.0	31.0
230	18.25	18.7	19.0	19.3	19.8	20.2	20.7	21.2	21.8	22.4	23.8	25.2	26.8	29.9
240	18.92	19.4	19.7	20.0	20.5	20.8	21.3	21.7	22.3	22.8	24.1	25.3	26.6	29.4
250	19.57	20.0	20.3	20.6	21.1	21.4	21.8	22.2	22.7	23.2	24.3	25.4	26.6	29.1
260	20.20	20.7	21.0	21.2	21.7	22.0	22.4	22.8	23.2	23.7	24.7	25.7	26.7	28.9
280	21.45	21.9	22.3	22.5	22.9	23.2	23.6	23.9	24.3	24.7	25.5	26.3	27.2	29.0
300	22.71	23.2	23.5	23.8	24.2	24.4	24.8	25.1	25.4	25.8	26.5	27.1	27.9	29.5

*Critical pressure.

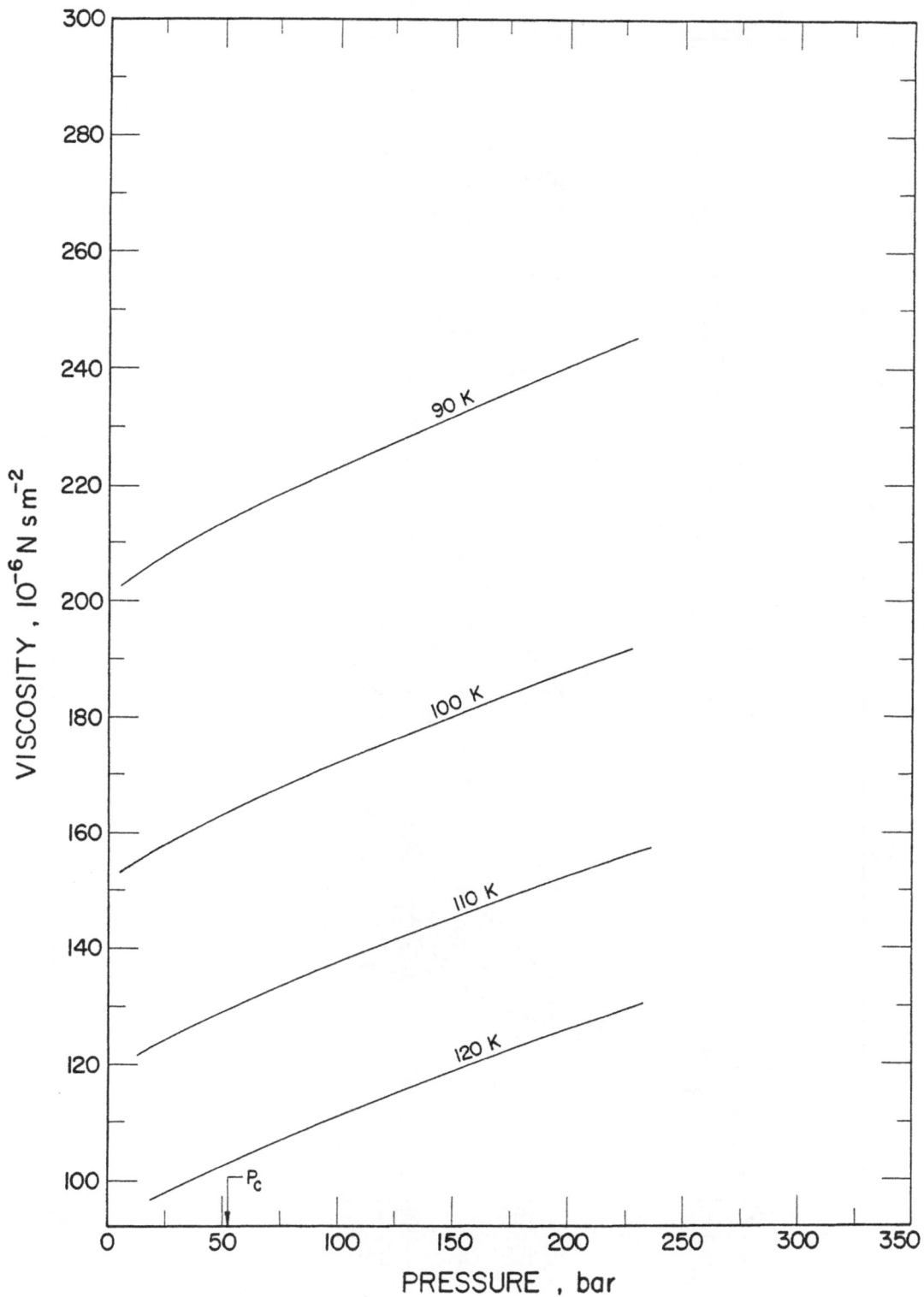

FIGURE IA. VISCOSITY OF FLUORINE [78].

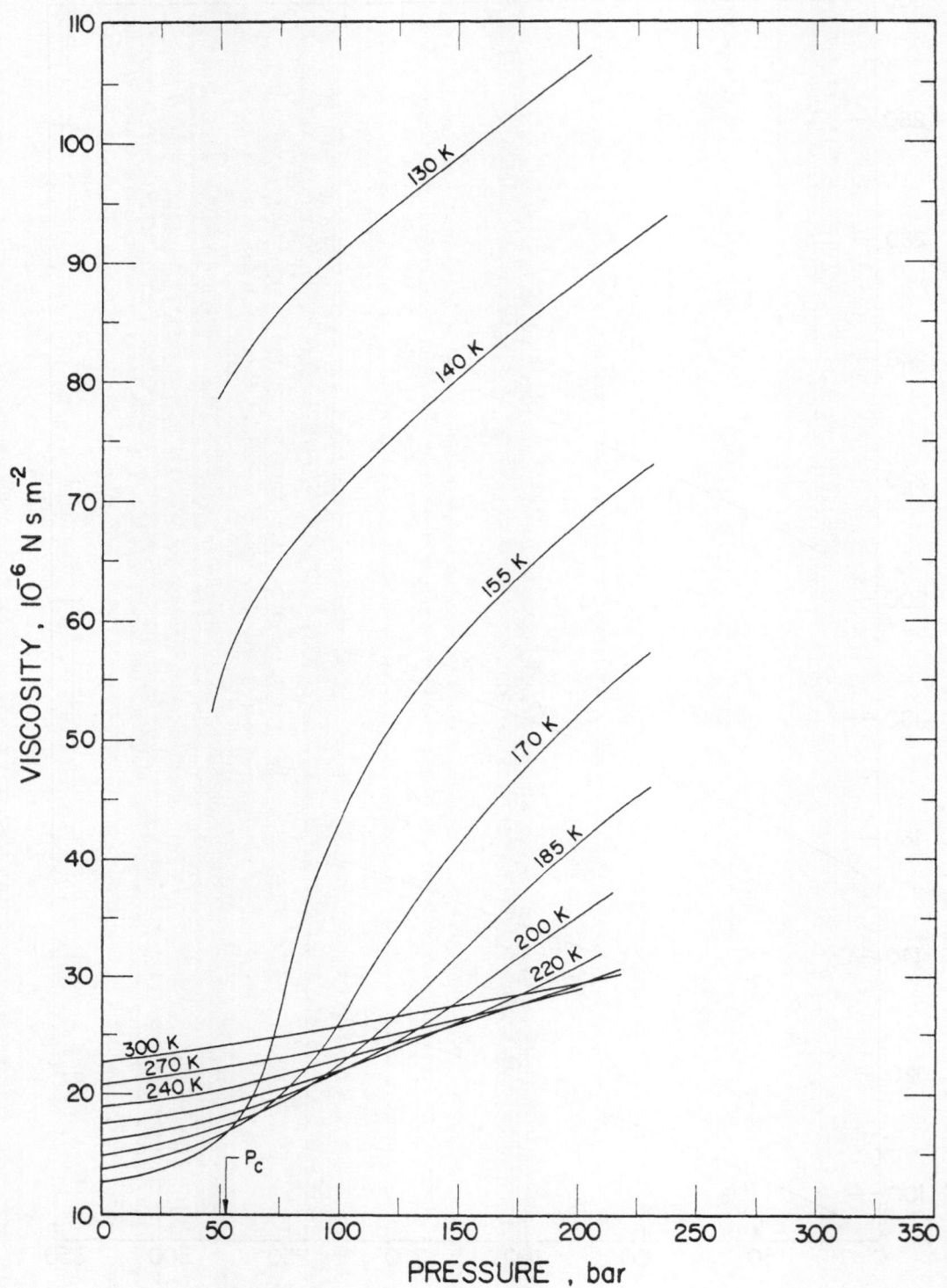

FIGURE IB. VISCOSITY OF FLUORINE [78].

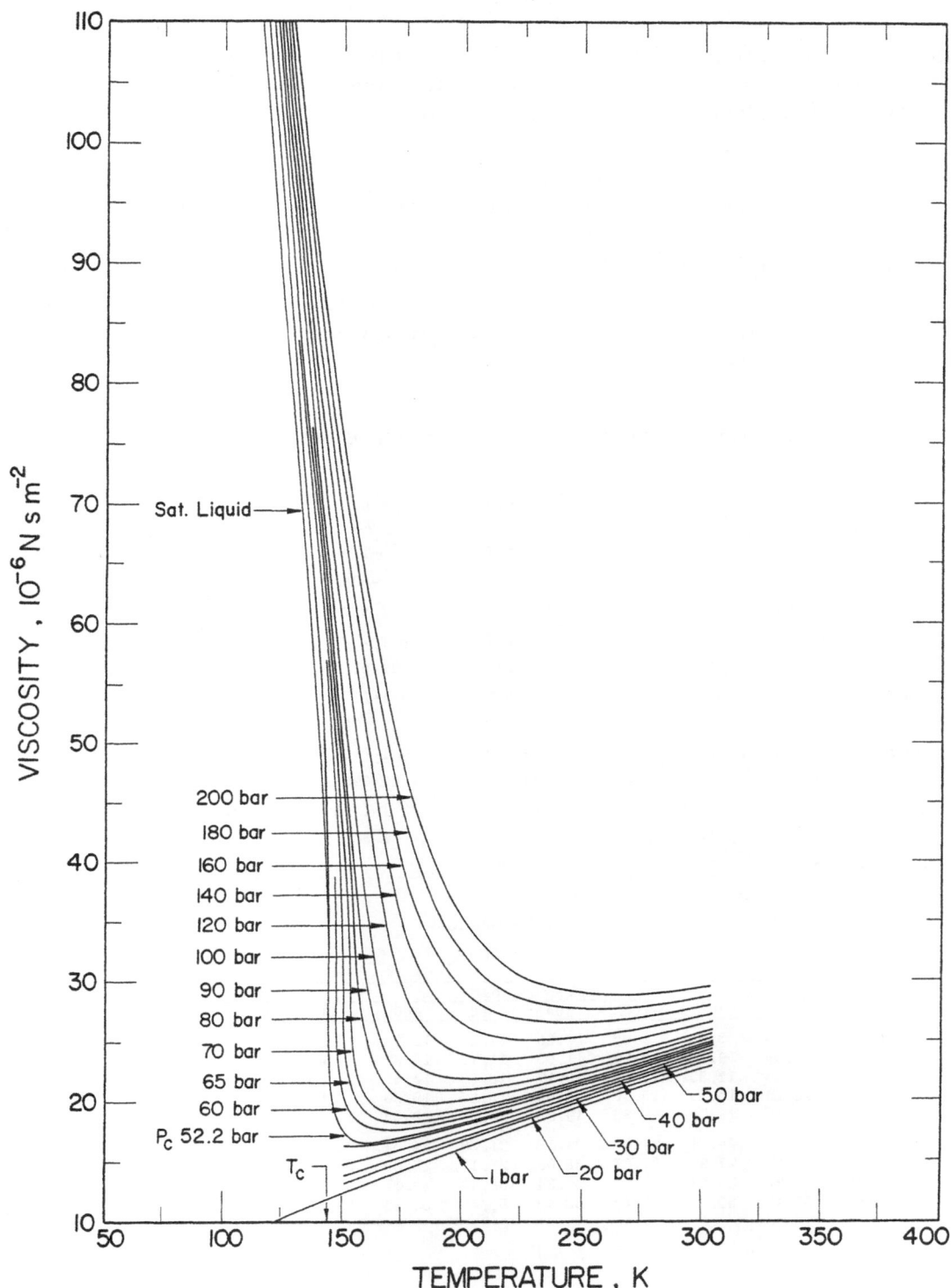

FIGURE 2. VISCOSITY OF FLUORINE [78].

HELIUM

The values of [203] were used to generate a set of recommended values in the dense gas region. They are plotted in Figure 1 as isotherms against pressure. Only at the lower temperatures, an appreciable pressure dependence is noted, the curves being shaped somewhat irregularly. Figure 2 shows a plot of viscosity against temperature for even values of pressure. This plot was used to read the recommended values presented below.

A number of additional works on the viscosity of helium at high pressures are listed in the summary table below. With the exception of the data of Tsederberg et al. [201] and of Robinson [170] who only publishes a small diagram, no significant pressure effects are shown in all these works although the pressures used reach several hundred bars. Therefore, it was not considered worthwhile to compare these data with the recommended values.

The recommended values were compared with the data of Tsederberg et al. [201] which were the only second set of data available with a significant pressure dependence. As seen from Figure 3, the agreement was found to be very good. The estimated uncertainty in the recommended values is ± 2%.

ADDITIONAL REFERENCES ON THE VISCOSITY OF HELIUM

Authors	Year	Ref. No.	Temperature K	Pressure bar	Method
Gracki et al.	1969	65	183–298	5–170	Capillary tube
Kao and Kobayashi	1967	97	183–323	1–500	Capillary tube
Reynes and Thodos	1966	168	373–473	77–830	Capillary tube
Golubev and Gnezdilov	1965	58	273–523	1–785	Capillary tube
Popov and Tsederberg	1963	136	283–918	41–523	Capillary tube
Kestin and Whitelaw	1963	109	295–520	1–117	Oscillating disk
Flynn et al.	1963	49	223–373	22–170	Capillary tube
Kestin and Leidenfrost	1959	104	298–510	1–140	Oscillating disk
Ross and Brown	1957	171	223–298	69–690	Capillary tube
Robinson	1955	170	16–90	1–1750	Falling body
Tjerkstra	1952	192	2–4	5–50	Capillary tube
Tsederberg et al.	1974	201	80–273	1–400	Capillary tube

VISCOSITY OF HELIUM
$[\mu,\ 10^{-6}\ N\ s\ m^{-2}]$

T, K	Pressure, bar								
	1	50	100	200	300	400	500	600	800
80	8.36	8.55	9.00	10.20	11.25	12.98	14.92		
100	9.59	9.77	10.16	11.00	11.73	13.32	14.67		
150	12.66	12.89	13.05	13.17	13.50	14.57	15.28		
200	15.29	15.34	15.40	15.60	15.87	16.30	17.04		
300	19.76	19.79	19.83	19.93	20.22	20.43	20.80	21.08	21.62
400	23.98	24.01	24.05	24.18	24.39	24.54	24.70	24.84	25.03
500	27.88	27.92	27.97	28.03	28.17	28.33	28.38	28.52	28.64
800	38.25	38.27	38.29	38.34	38.40	38.47	38.53		
1300	53.42	53.42	53.42	53.43	53.45	53.48	53.50		

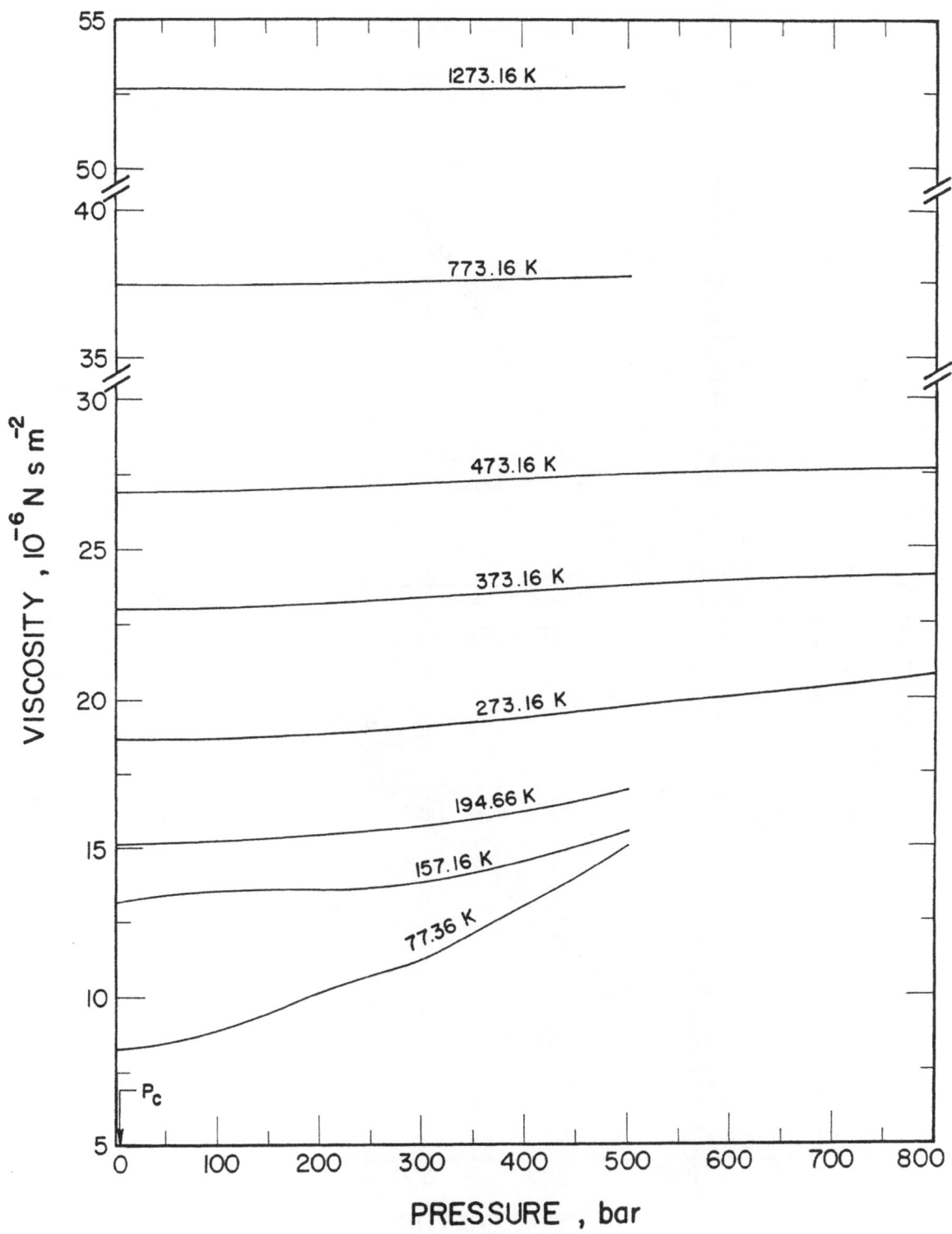

FIGURE 1. VISCOSITY OF HELIUM [203].

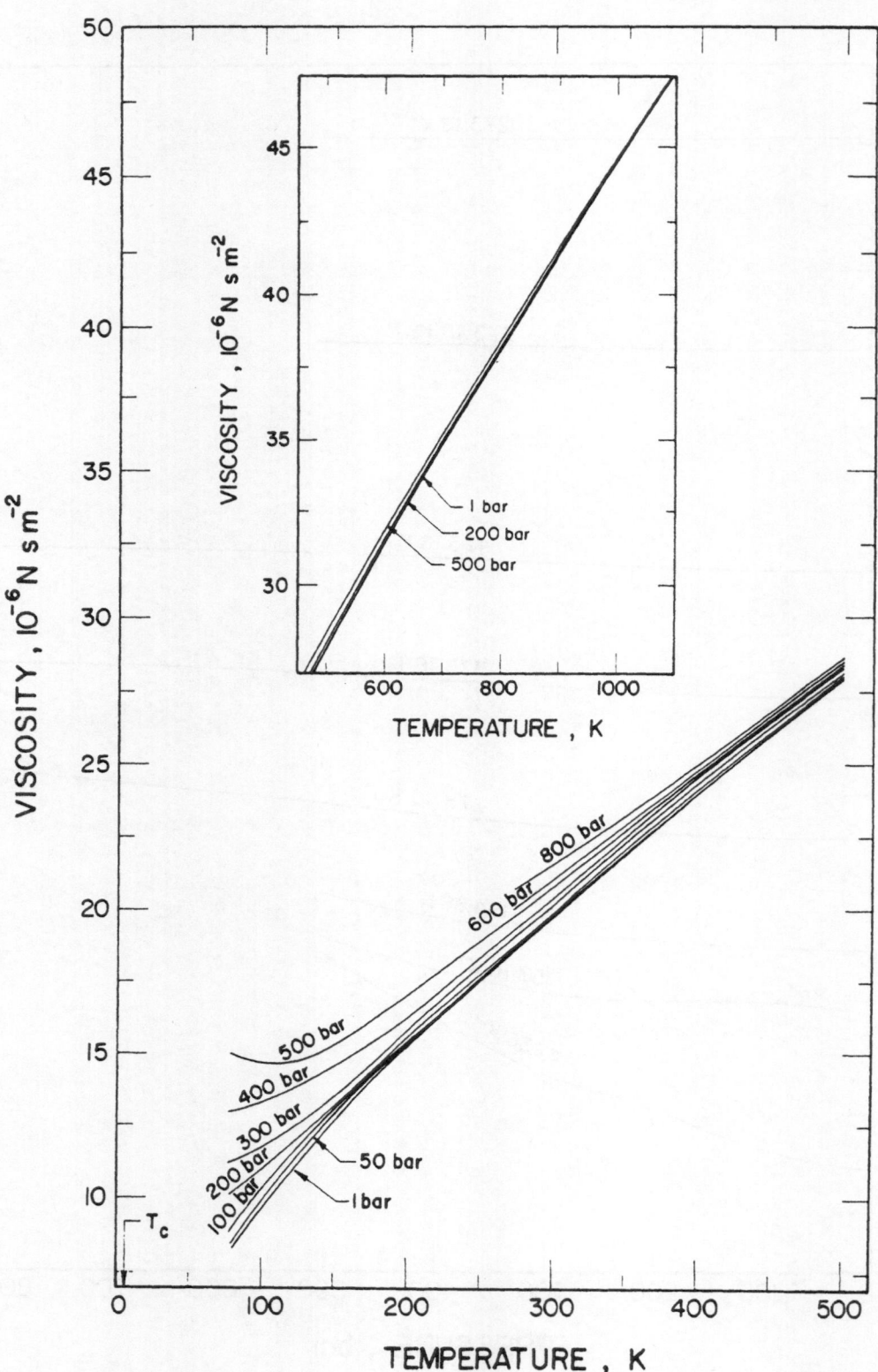

FIGURE 2. VISCOSITY OF HELIUM [203].

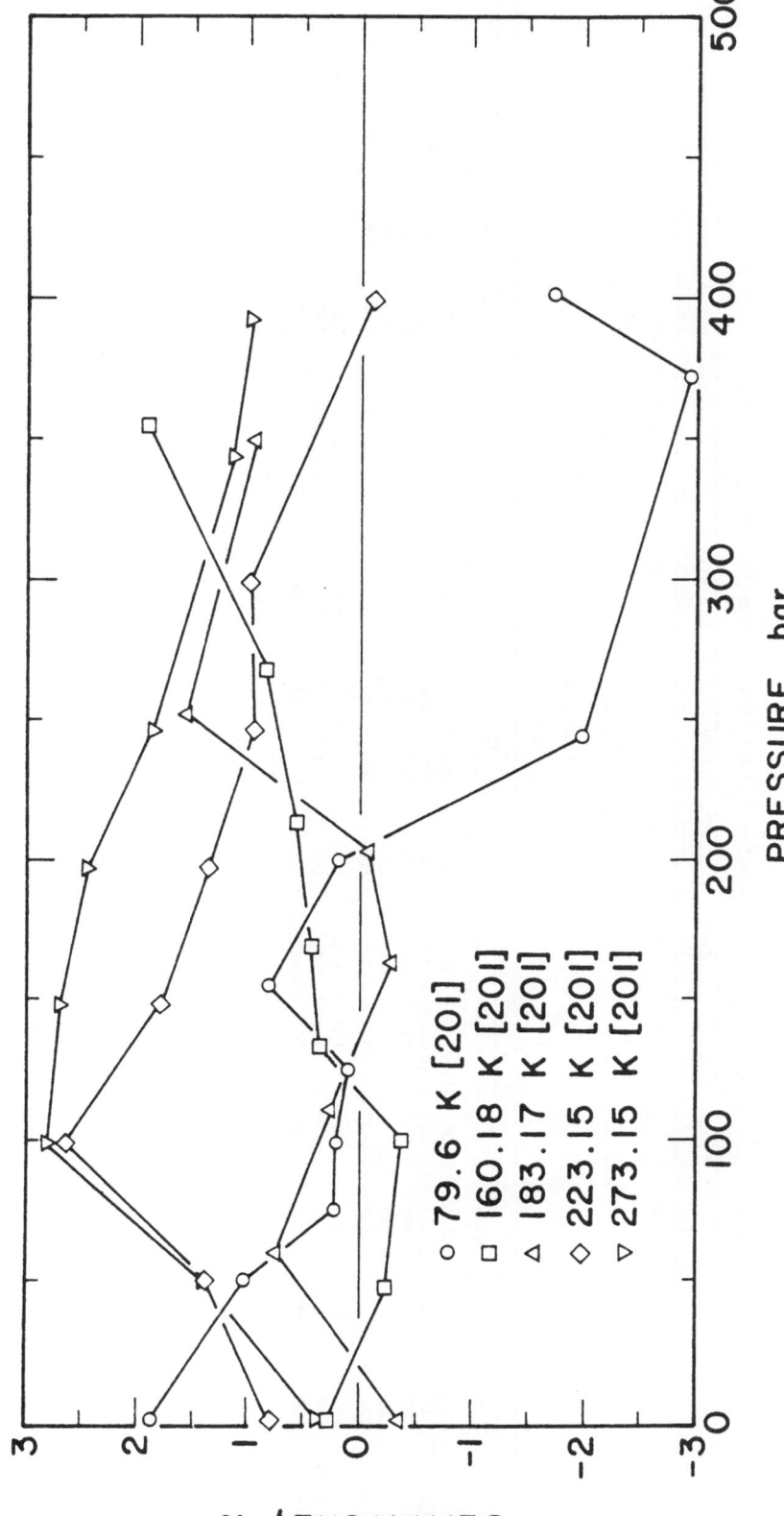

FIGURE 3. DEPARTURE PLOT ON THE VISCOSITY OF HELIUM.

n – HEPTANE

The data of Agaev and Golubev [3], obtained by a capillary tube viscosimeter, were used to generate the recommended values presented below. The data are plotted as iso-therms as a function of pressure in Figures 1A and 1B, where the smoothing modifications did not surpass 1%. Using the P–V–T data of [128], the dilute gas data of [195] with slight extrapolation and the residual viscosity concept, it was possible to extrapolate from 548 K to 620 K up to 500 bar. Figure 2 shows isobars plotted against temperature, the extrapolated values being indicated by dashed lines.

Additional works on the pressure dependence of the viscosity of n-heptane are listed in the summary table below. The data are those of Kuss et al. [124] and Khalilov [113] for the saturated liquid. While good agreement is stated with Khalilov in [113], deviations of the recommended values from Kuss et al. [124] are about 5%. The uncertainty in the recommended values is estimated to be ± 5% in the region of experiments and ± 7% in the region of extrapolation.

ADDITIONAL REFERENCES ON THE VISCOSITY OF n-HEPTANE

	Year	Ref. No.	Temperature K	Pressure bar	Method
Kuss and Pollmann	1969	124	313	1 - 1471	Capillary Tube
Khalilov	1939	113	293 - 538	SL	Capillary Tube

VISCOSITY OF n-HEPTANE

$[\mu, \ 10^{-6} \ \text{N s m}^{-2}]$

T, K	1	10	20	27.4*	30	35	40	50	60	80	100	150	200	250	300	400	500
300	381	386	390	393	394	396	398	403	407	415	424	470	470	495	519	570	626
320	313	317	321	324	325	326	328	332	336	343	351	371	391	411	431	474	517
340	260	263	266	269	270	271	273	276	280	286	293	310	327	344	361	396	431
360	218	221	224	226	227	228	229	233	236	242	248	263	278	293	308	339	368
380	7.55	187	190	192	193	194	196	199	201	207	213	227	240	254	268	294	320
400	7.93	160	163	164	165	167	168	171	173	179	184	197	210	223	235	259	282
420	8.31	137	139	141	142	143	144	147	150	155	160	173	185	197	208	231	251
440	8.68	116	119	121	122	123	124	126	129	134	139	151	163	174	185	206	226
460	9.06	97.7	101	103	104	105	107	110	112	118	123	135	146	157	167	186	205
480	9.44	10.7	85.5	88.1	89.0	90.5	92.0	95.2	98.0	103	109	120	131	141	151	170	188
500	9.81	10.9	70.5	74.0	75.0	77.0	78.8	82.0	84.9	90.4	95.7	107	118	128	137	155	173
520	10.2	11.2	14.6	58.8	60.5	63.3	65.8	70.0	73.2	79.0	84.4	96.0	106	116	125	143	159
530	10.4	11.4	14.3	50.6	52.7	56.2	59.2	64.0	67.6	73.8	79.4	91.0	101	111	120	137	153
535	10.5	11.4	14.1	45.9	48.2	52.6	56.0	61.1	64.9	71.4	76.9	88.6	98.6	108	117	134	150
550	10.7	11.7	13.8	18.1	27.0	40.0	45.8	52.6	57.2	64.3	69.8	81.7	91.8	101	109	126	142
555	10.8	11.8	13.7	17.6	20.0	34.4	41.6	49.7	54.6	62.0	67.6	79.5	89.6	98.6	107	123	139
560	10.9	11.9	13.7	17.1	18.9	28.9	36.0	47.0	52.1	59.8	65.4	77.4	87.4	96.4	109	121	136
565	11.0	12.0	13.7	16.7	18.4	24.6	30.8	44.2	49.6	57.6	63.2	75.3	85.2	94.2	102	118	134
570	11.1	12.1	13.7	16.5	17.9	21.7	27.4	41.4	47.2	55.6	61.2	73.2	83.2	92.1	100	116	131
580	11.3	12.3	13.7	16.1	17.3	19.7	24.1	36.0	42.6	51.6	57.2	69.4	79.3	88.2	96.0	111	127
590	11.4	12.4	13.8	16.0	16.8	18.9	22.2	31.4	38.4	48.0	53.8	66.0	75.8	84.4	92.0	107	122
600	11.6	12.6	13.9	15.8	16.6	18.4	20.9	28.0	35.0	44.8	50.8	62.9	72.5	81.0	88.8	103	118
610	11.8	12.8	14.1	15.9	16.5	18.0	20.1	25.8	32.2	41.8	48.2	60.2	69.4	77.7	85.6	99.4	114
620	12.0	13.0	14.4	16.0	16.6	17.8	19.6	24.0	29.6	39.0	45.8	57.6	66.5	74.6	82.8	96.5	111

*Critical pressure

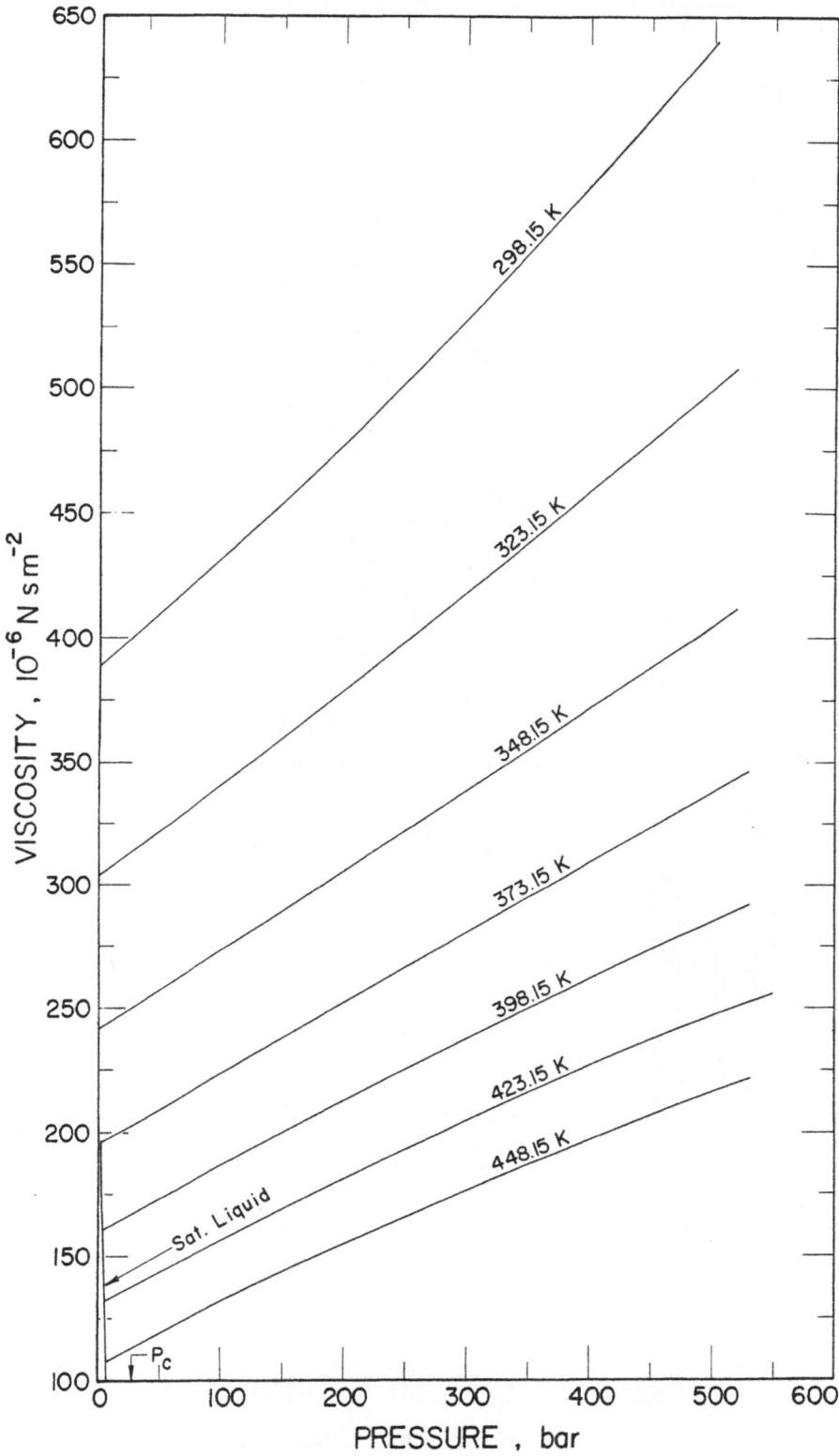

FIGURE IA. VISCOSITY OF n-HEPTANE [3].

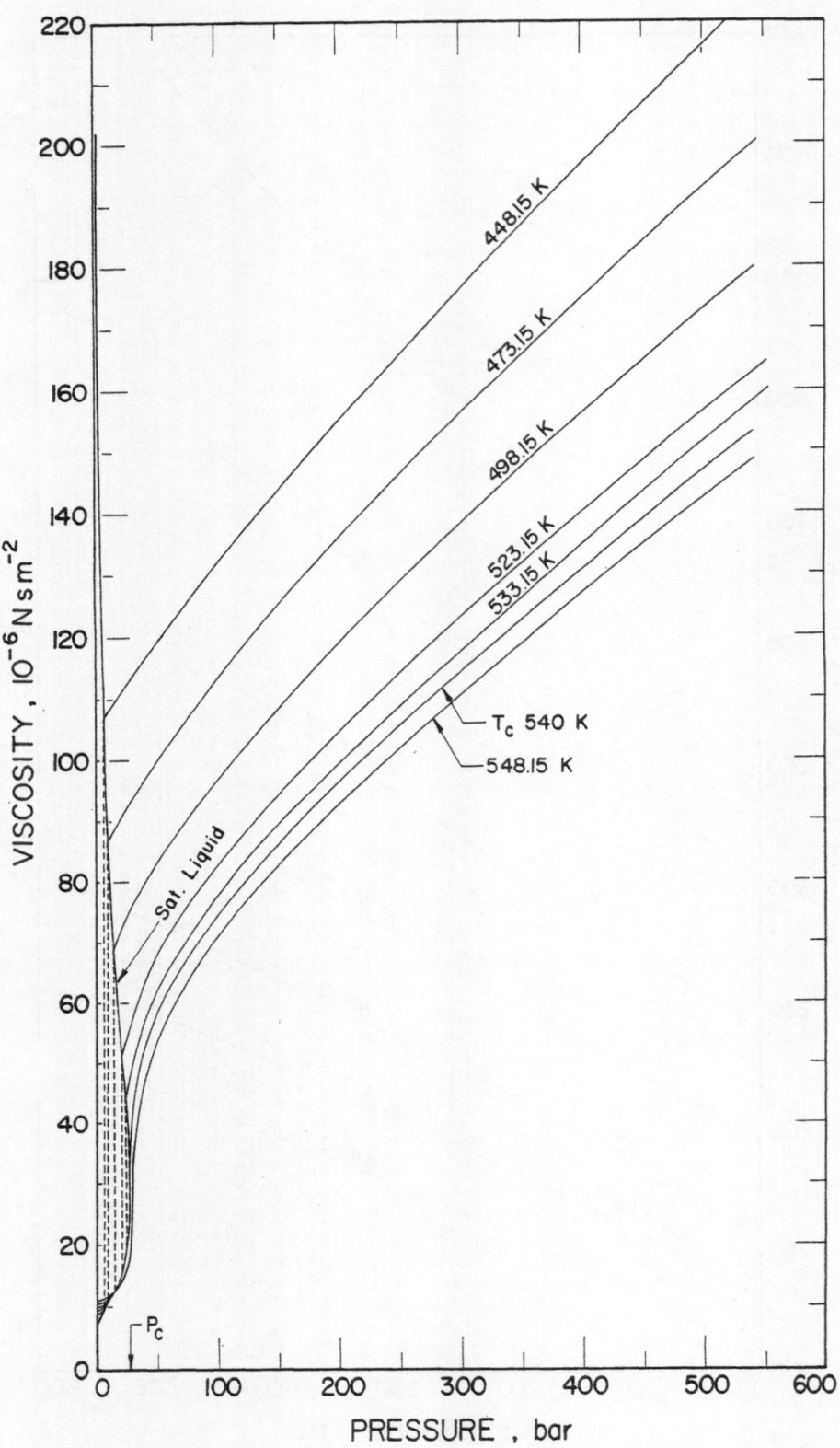

FIGURE IB. VISCOSITY OF n-HEPTANE [3].

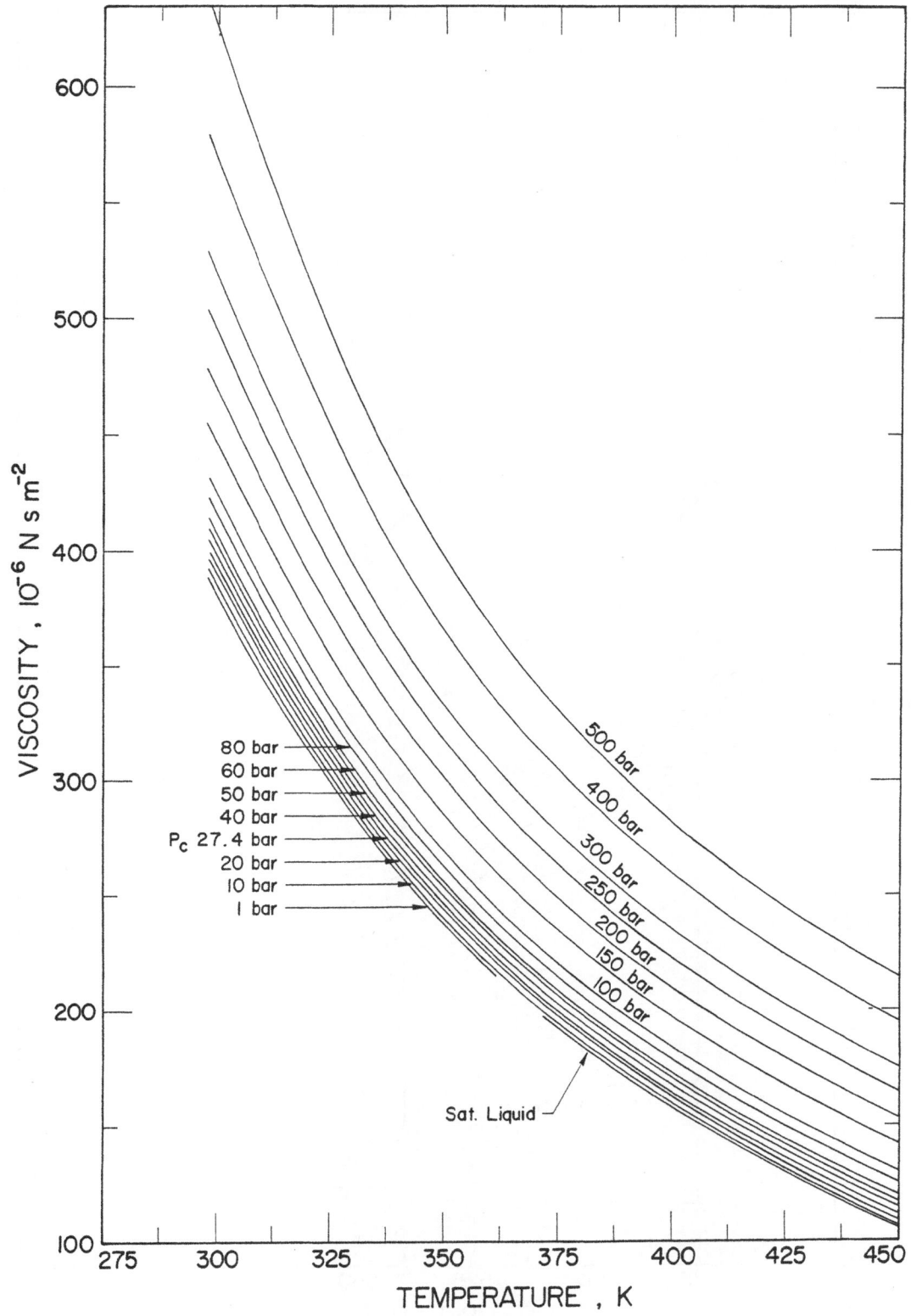

FIGURE 2A. VISCOSITY OF n-HEPTANE [3].

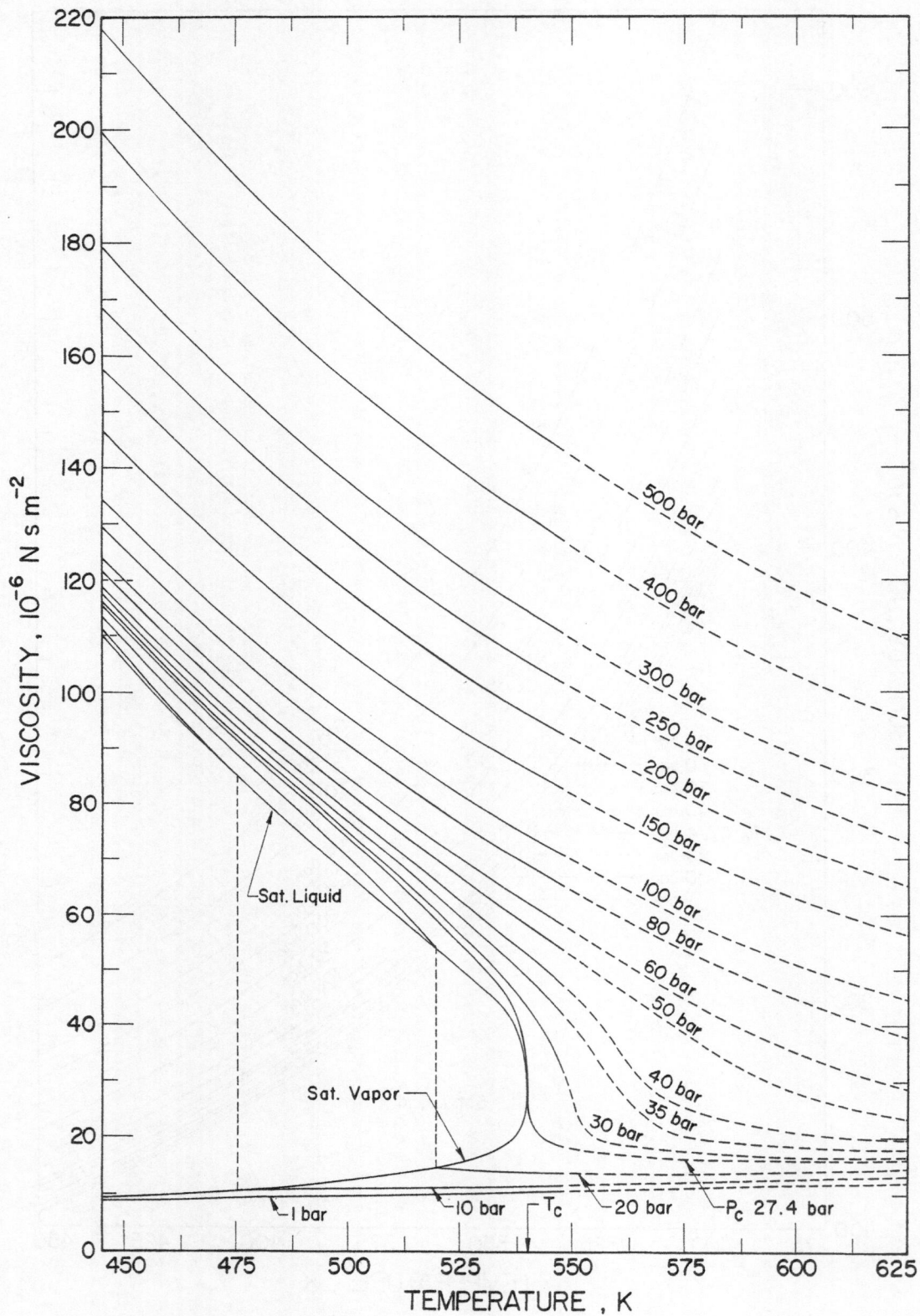

FIGURE 2B. VISCOSITY OF n-HEPTANE [3].

n-HEPTENE

The data of [157], obtained by a falling body method, were used to generate the recommended values presented below. They are plotted as isotherms against pressure in Figure 1, where some smoothing modifications were applied. The measured isotherms were extrapolated to the vapour pressure of n-heptene which was estimated from the Harlacher-Braun-constants as described in [166]. From Figure 1 even isobars were plotted against temperature in Figure 2, including slight extrapolations, from where the recommended values were read. No comparison with other works was possible. The accuracy of the recommended values is estimated to be ± 5%.

VISCOSITY OF n-HEPTENE
$[\mu, \ 10^{-6} \ \text{N s m}^{-2}]$

T, K	Pressure, bar												
	1	30	50	70	100	150	200	250	300	350	400	450	500
300	322	335	344	354	366	390	405	424	444	464	482	503	524
305	308	320	330	338	348	370	385	403	422	440	456	476	495
310	294	307	315	322	332	352	367	385	404	419	437	455	473
320	270	282	290	296	304	322	337	354	371	386	402	419	435
330	249	259	267	272	281	297	312	327	344	357	372	388	402
340	230	240	247	252	260	276	290	304	319	332	346	360	374
350	212	221	228	233	241	256	270	283	296	309	322	335	347
360	196	204	210	216	224	237	250	263	275	288	299	310	322
370	–	188	194	200	207	220	233	244	255	268	278	288	300
380	–	174	130	185	193	204	217	228	238	250	259	269	279
400	–	150	155	160	168	180	190	201	210	220	228	237	246
420	–	128	133	138	145	156	167	177	186	196	204	212	221
440	–	110	114	120	126	137	148	156	166	174	182	190	199
460	–	94	99	104	110	121	131	139	148	156	163	170	130
480	–	83	88	91	98	108	118	127	134	141	149	157	167
490	–	81	85	88	96	105	116	124	131	138	145	154	164

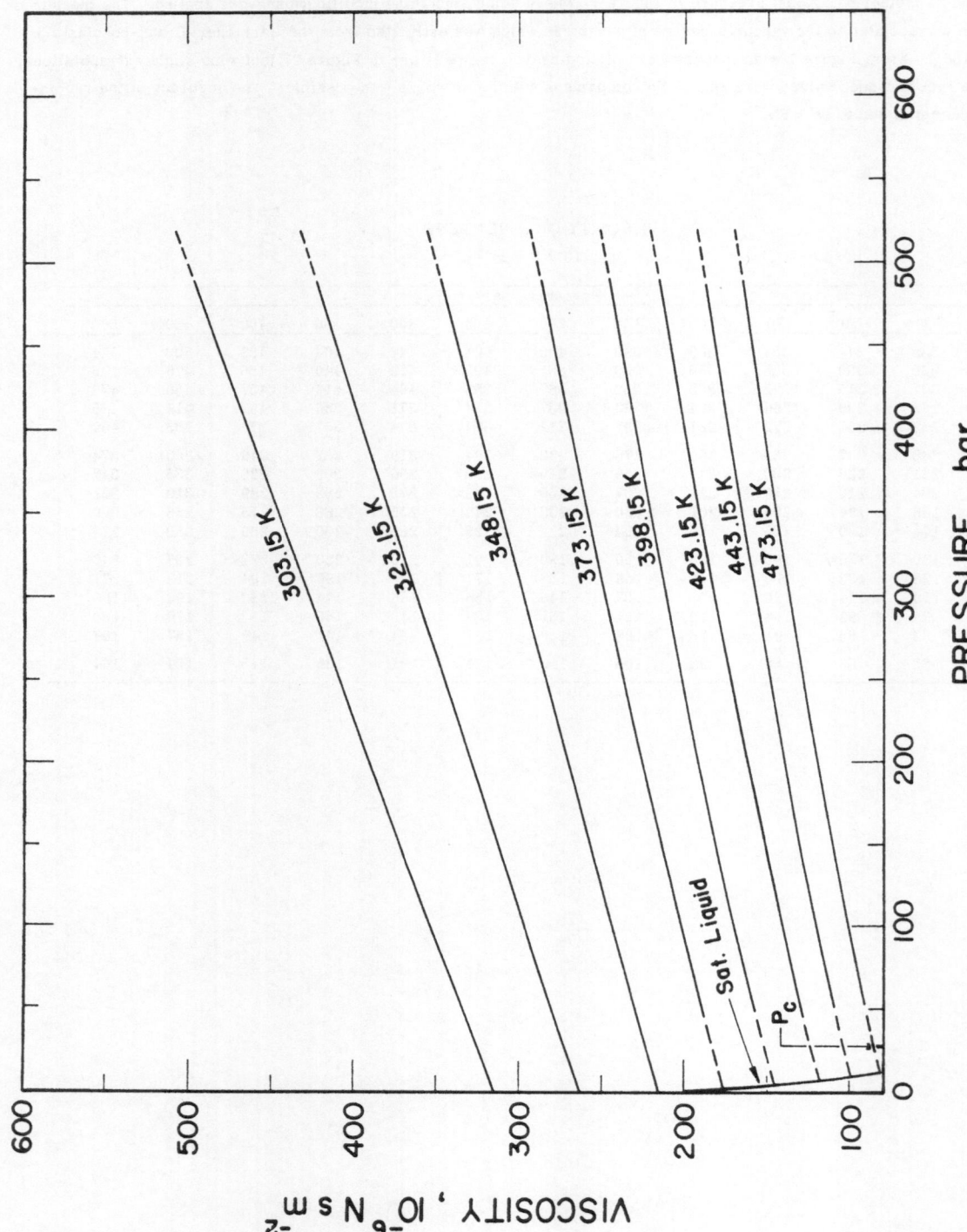

FIGURE I. VISCOSITY OF *n*-HEPTENE [157].

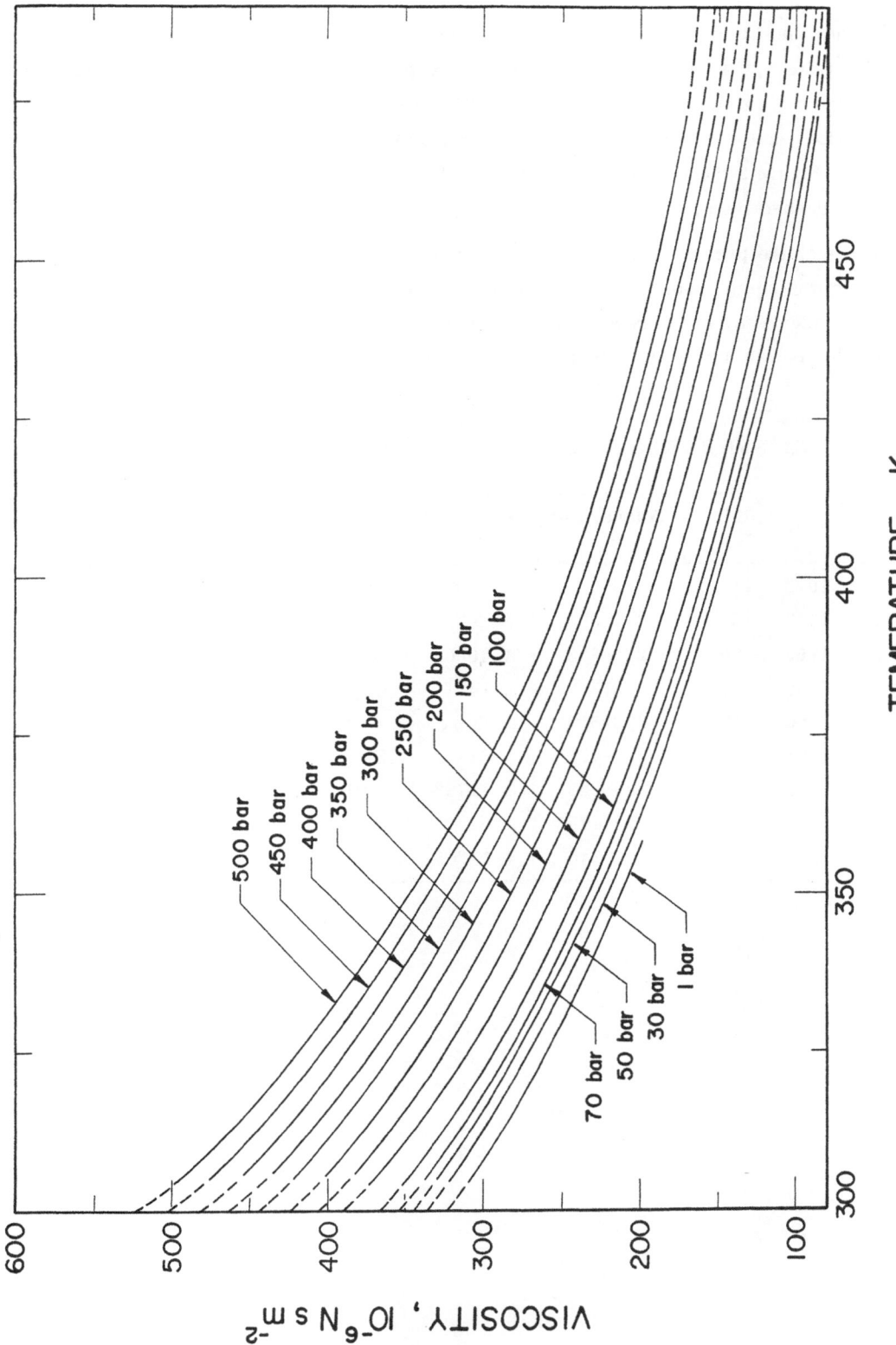

FIGURE 2. VISCOSITY OF n-HEPTENE [157].

n-HEXANE

The data of Agaev and Golubev [2], obtained by a capillary tube viscosimeter, were used to generate the recommended values presented in the table on the next page. The data are plotted as isotherms against pressure in Figure 1. Using the P-V-T data of [143], the dilute gas viscosities of [195] with a slight linear extrapolation and the residual viscosity concept, an extrapolation from 548 K to 1000 K was made up to 150 bar. In Figure 2 the values are shown as isobars against temperature. All extrapolated values are indicated as dashed lines.

Additional works on the pressure dependence of the viscosity of n-hexane are listed in the summary table below. Only a limited comparison with the recommended values was possible, revealing deviations of a few percent. The uncertainty in the recommended values is estimated to be ± 7% in the region of extrapolation and somewhat better in the experimental region.

ADDITIONAL REFERENCES ON THE VISCOSITY OF n-HEXANE

Authors	Year	Ref. No.	Temperature K	Pressure bar	Method	Departure % (no. points)
Kor et al.	1972	118	303	1-9807	Ultrasonic	–
Naziev et al.	1972	156	297-333	1-392	Capillary tube	1.3
Brazier and Freeman	1969	20	273-333	1-4000	Rolling ball	–
Kuss and Pollmann	1969	124	313	1-1471	Capillary tube	2.5
Kozlov et al.	1966	119	293-513	Isochores	Falling cylinder	–
Parisot and Johnson	1961	159	313-453	SL	Capillary tube	–
Bridgman	1949	22	303-348	1-11768	Falling weight	1.0
Khalilov	1939	113	293-473	SL	Capillary tube	–
Bridgman	1926	21	303-348	1-11768	Falling weight	1.0

VISCOSITY OF n-HEXANE

$[\mu, 10^{-6} \text{ N s m}^{-2}]$

T, K	\multicolumn{17}{c}{Pressure, bar}
	1	20	25	30.1*	35	40	45	50	60	70	80	100	150	200	300	400	500
380	8.1	155.0	156.2	157.4	158.6	159.8	161.0	162.2	164.7	167.2	169.7	174.3	185.4	197.3	–	–	–
400	8.5	132.5	133.8	135.0	136.3	137.5	138.8	140.0	142.4	144.8	147.2	151.1	162.0	173.5	195.0	–	–
420	9.0	113.0	114.2	115.5	116.7	117.9	119.2	120.4	122.8	125.1	127.5	131.7	142.5	153.4	174.1	195.0	–
440	9.5	94.6	96.0	97.3	98.7	100.0	101.7	103.4	105.6	107.7	109.9	115.0	125.6	136.2	155.6	174.7	193.0
460	10.0	77.7	79.2	80.8	82.5	84.3	86.3	88.3	89.8	92.1	95.2	100.0	110.5	121.1	139.5	157.1	173.8
480	10.4	61.1	63.2	66.0	67.9	69.9	72.2	74.5	76.6	79.5	82.4	87.1	97.8	107.9	125.5	142.1	157.6
500	10.8	13.6	16.0	49.0	52.4	56.2	58.3	61.3	64.4	67.5	70.5	76.0	86.6	96.7	113.7	129.5	144.0
520	11.2	13.4	14.7	17.9	24.9	38.6	43.0	47.0	52.2	56.4	59.6	65.5	76.6	86.6	103.6	118.9	132.8
525	11.3	13.4	14.5	17.2	21.5	33.4	39.2	43.6	49.4	53.7	57.2	63.0	74.4	84.4	101.2	116.4	130.2
530	11.4	13.4	14.5	16.8	20.0	30.2	36.3	40.6	46.8	51.2	54.9	60.7	72.0	82.0	99.0	114.0	127.6
540	11.6	13.4	14.4	16.2	18.5	25.8	31.5	35.2	42.2	46.8	50.5	56.4	67.7	77.8	94.5	109.4	122.8
550	11.9	13.5	14.4	15.9	17.8	22.7	27.5	30.6	38.2	43.0	46.7	52.5	63.8	73.6	90.4	104.9	118.0
560	12.0	13.6	14.4	15.7	17.3	20.7	24.3	26.9	34.6	39.4	43.1	49.0	60.2				
570	12.3	13.8	14.5	15.6	16.9	19.3	22.0	24.0	31.4	36.0	39.8	45.7	57.0				
580	12.5	14.0	14.6	15.6	16.6	18.5	20.7	22.1	28.5	33.1	36.8	42.8	54.1				
590	12.6	14.1	14.8	15.6	16.5	17.9	20.0	21.0	26.2	30.6	34.0	40.4	51.6				
600	12.9	14.3	14.9	15.6	16.5	17.5	19.4	20.4	24.4	28.5	31.6	38.3	49.4				
620	13.2	14.6	15.2	15.8	16.5	17.3	18.7	19.5	22.6	25.7	28.5	35.0	45.6				
640	13.6	14.9	15.4	16.0	16.6	17.4	18.4	19.2	21.7	24.4	26.8	32.5	42.7				
660	14.0	15.2	15.7	16.2	16.9	17.5	18.2	19.0	21.2	23.5	25.8	30.4	40.3				
680	14.4	15.5	16.0	16.5	17.0	17.7	18.3	19.0	21.0	23.0	25.0	28.9	38.3				
700	14.7	15.9	16.3	16.8	17.3	17.9	18.4	19.0	20.9	22.5	24.4	27.8	36.6				
750	15.6	16.6	17.0	17.5	18.0	18.4	18.9	19.3	20.9	22.0	23.5	26.0	33.4				
800	16.5	17.4	17.8	18.2	18.6	19.0	19.4	19.8	21.0	22.0	23.1	25.1	31.3				
850	17.4	18.1	18.5	19.0	19.4	19.7	20.1	20.4	21.3	22.1	23.0	24.8	30.0				
900	18.1	19.0	19.3	19.7	20.0	20.3	20.6	20.9	21.7	22.4	23.2	24.8	29.4				
1000	19.6	20.5	20.8	21.1	21.4	21.6	21.9	22.1	22.8	23.3	24.0	25.2	28.8				

*Critical pressure.

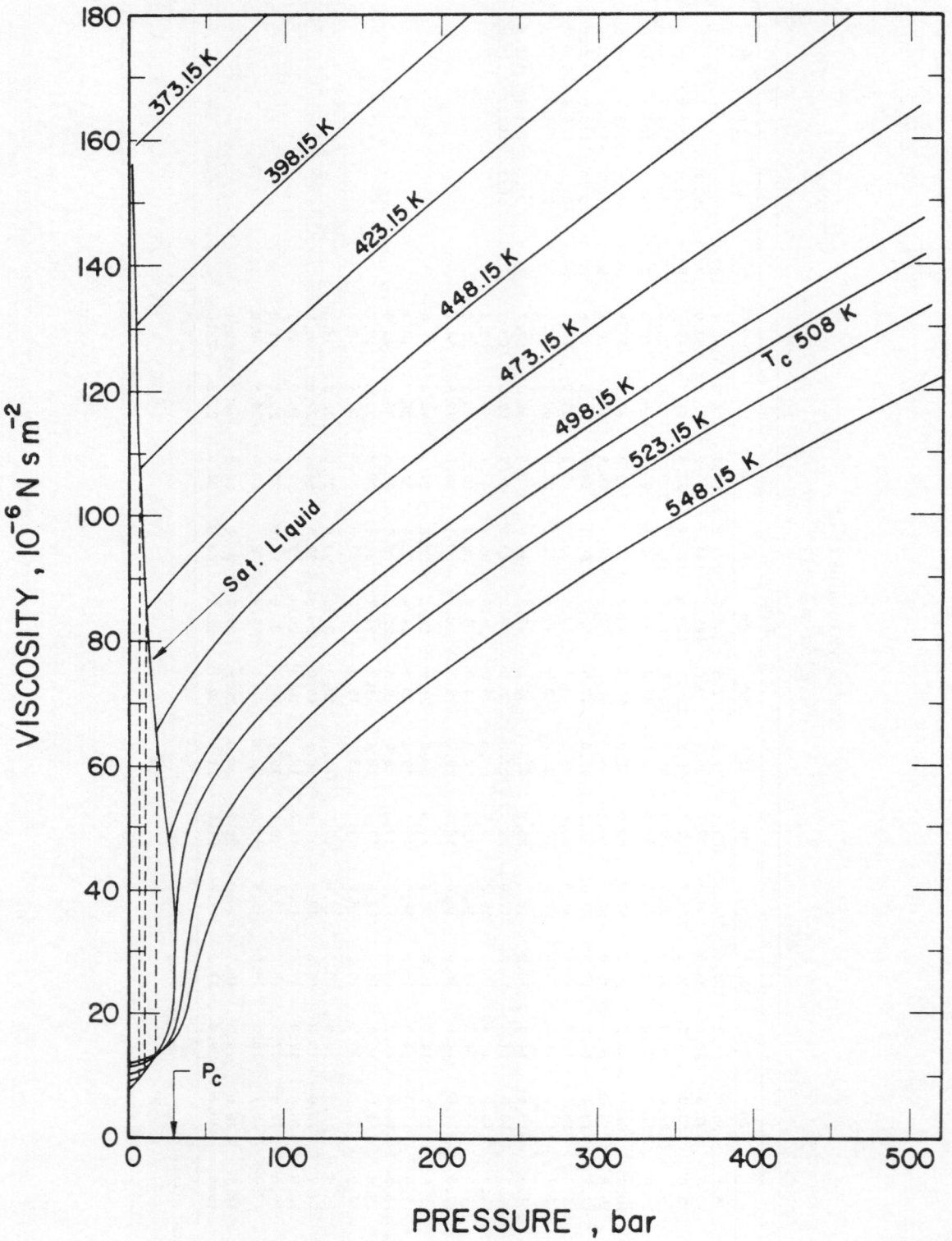

FIGURE 1. VISCOSITY OF n-HEXANE [2].

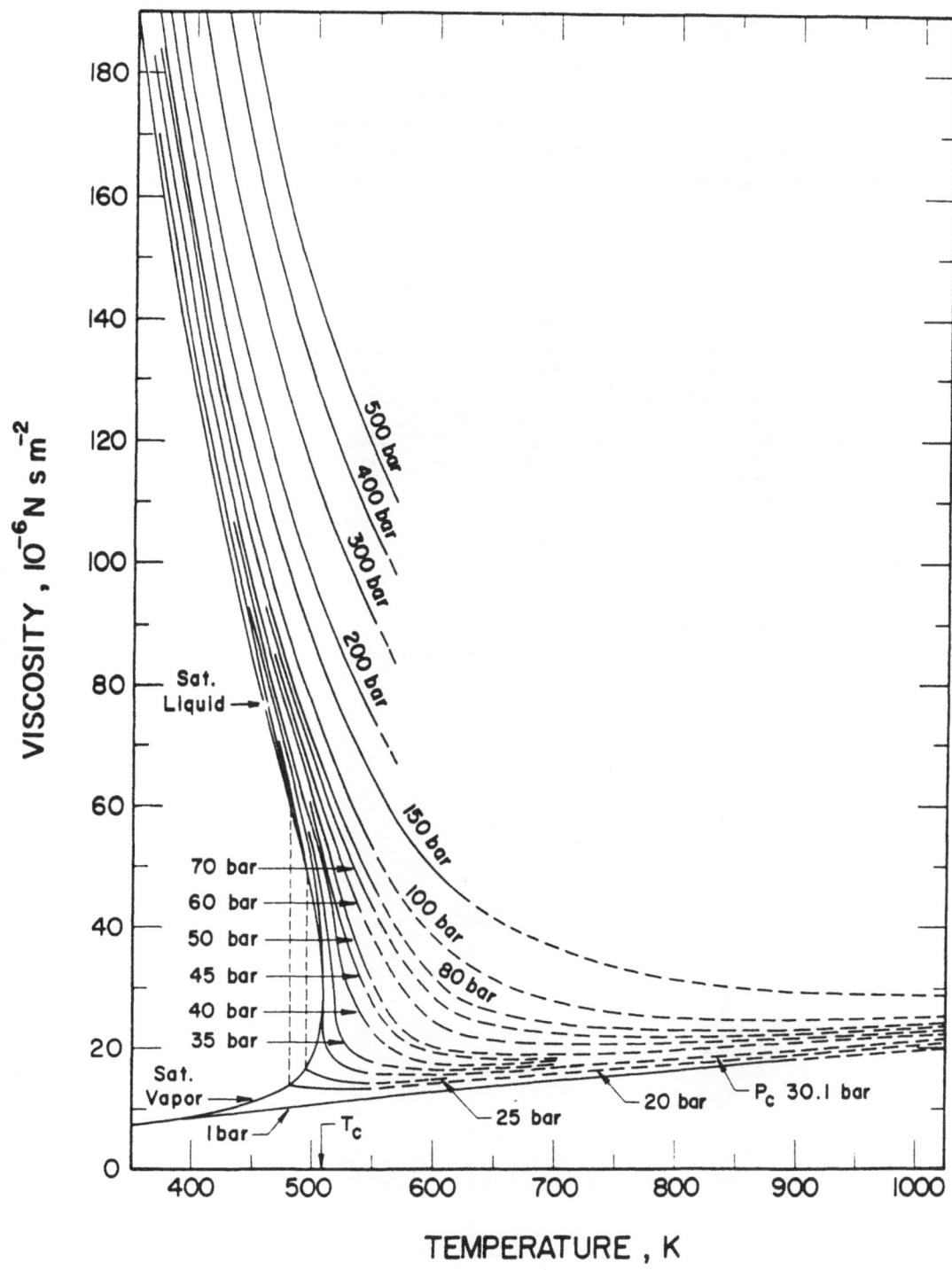

FIGURE 2. VISCOSITY OF n-HEXANE [2].

n-HEXENE

Only one set of data by Naziev et al. [156], obtained by a capillary tube viscosimeter, was found in the literature on the pressure dependence of the viscosity of n-hexene. The data are plotted in Figure 1 as isotherms against pressure, where smoothing modifications of maximum 3.2% had to be applied. Figure 2 shows isobars against temperature. The recommended values presented below were read from Figure 2. In the absence of further experimental data and the large smoothing modificating applied to the data in Figure 1, the estimate of ± 2% uncertainty reported by the authors is considered too optimistic. An uncertainty of ± 4% appears to be more realistic, though this, too, has to be considered a provisional value.

VISCOSITY OF n-HEXENE

$[\mu, \ 10^{-6} \ N \ s \ m^{-2}]$

T, K	Pressure, bar													
	1	10	20	30.7*	40	50	60	80	100	150	200	300	400	450
280	268	270	273	275	278	280	283	288	293	307	320	347	374	387
290	249	252	254	257	259	262	264	269	274	288	301	327	354	367
300	231	234	237	239	242	244	247	251	256	269	282	308	334	347
310	214	217	219	222	224	227	230	234	239	251	264	290	315	328
320	198	200	203	206	208	211	213	218	223	235	247	272	297	309
330	183	186	188	191	193	196	198	202	207	219	231	256	280	292
340	-	172	175	177	180	182	185	189	194	205	217	241	265	276
350	-	161	163	165	168	170	172	177	182	193	204	228	251	262
360	-	151	153	156	158	160	162	167	172	182	194	216	239	250
370	-	143	146	148	150	152	154	159	163	174	185	207	229	241
375	-	140	142	145	147	149	151	156	160	171	181	203	225	236

*Critical pressure.

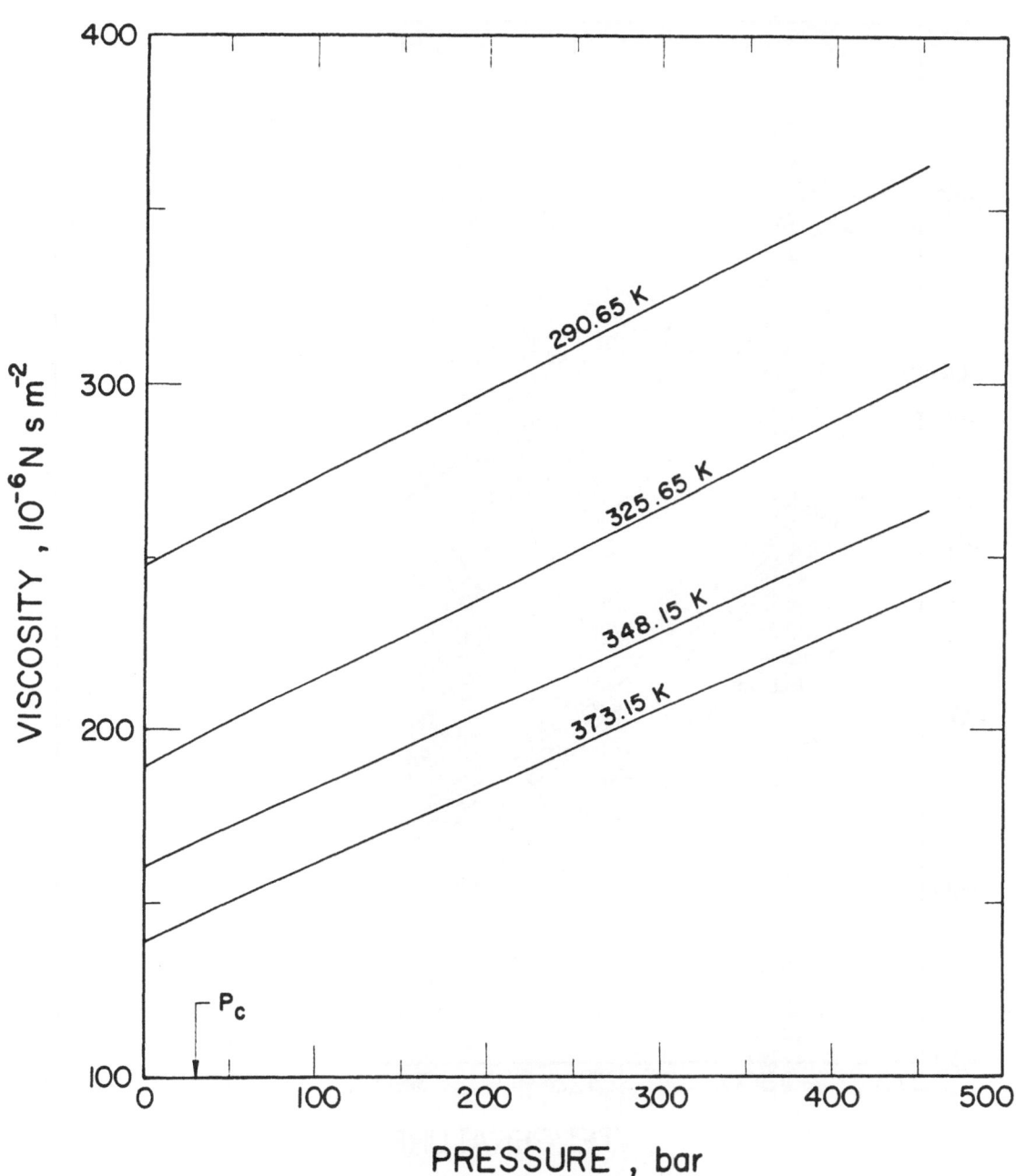

FIGURE 1. VISCOSITY OF n - HEXENE [156].

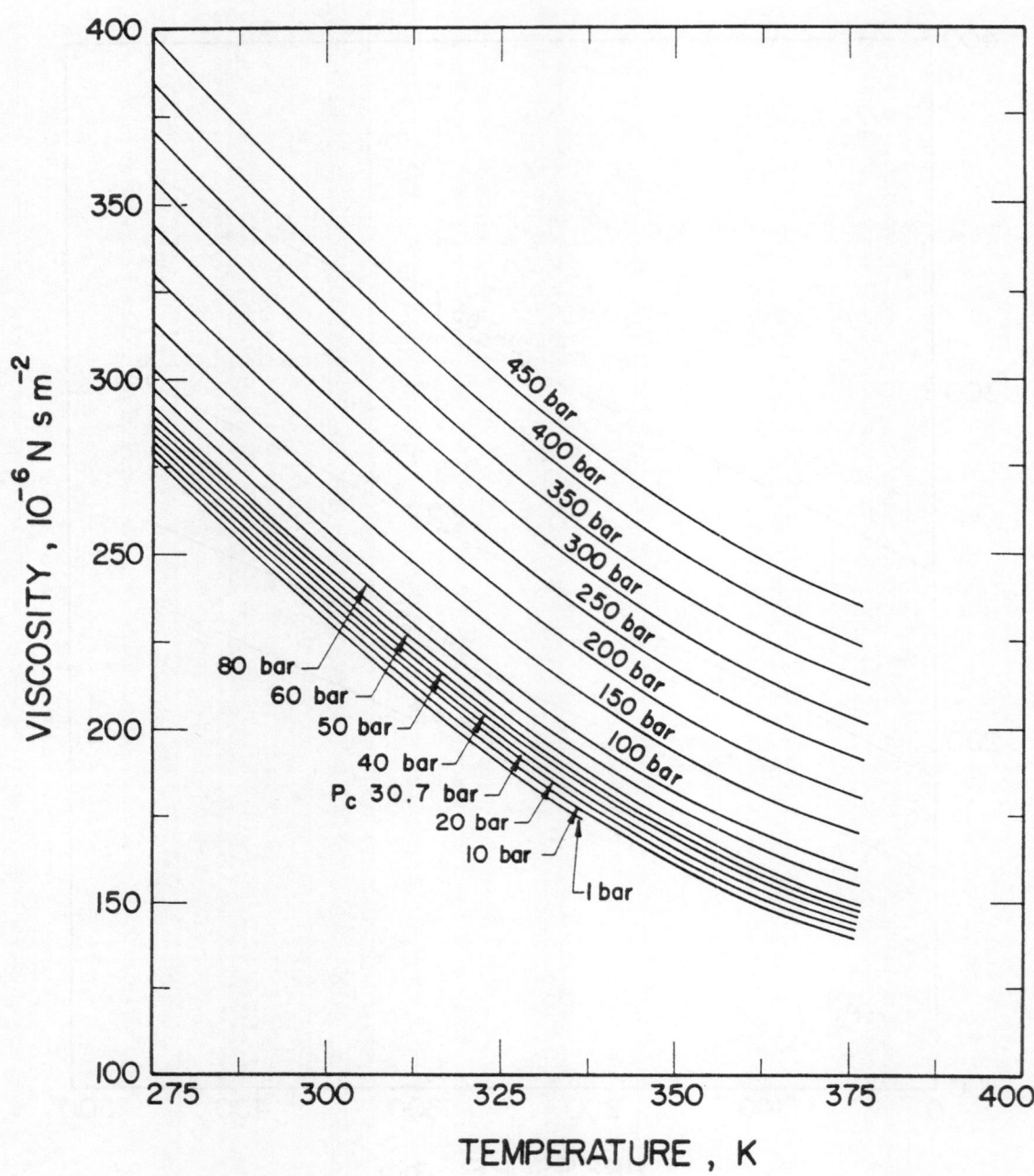

FIGURE 2. VISCOSITY OF n-HEXENE [156].

HYDROGEN

The values of [203] are used here as the recommended values for hydrogen, since they cover a very large region of states from 15 K to 1000 K, and from 1 to 1000 bar. They are plotted as isotherms against pressure in Figure 1 and as isobars against temperature in Figure 2. The recommended values are tabulated below.

A comparison was made with the data of Michels et al. [151], Golubev et al. [57, 60], Tsederberg et al. [199] and Kuss [122]. As seen from the departure plots in Figure 3, the agreement of less than 0.5% is excellent in the high temperature dense gas region. However, the agreement with the data with Rudenko et al. [173] is quite unsatisfactory, being greater than 10%.

A comparison with the recommended values for p-hydrogen presented in this work was also made. In the temperature region from 15 K to 100 K differences in the order of 5 to 10%, sometimes even larger, were found. This is in contrast to the statement of [203] that the viscosities of hydrogen and para-hydrogen differ only by 2-3% in the liquid region and may be considered equal in the dense gas region. It is presently not entirely clear to what extent these differences are due to inaccuracies of the recommended values presented here. For temperatures above room temperature, the uncertainty in the recommended values is estimated to be ± 1%. For lower temperatures, especially in the liquid region, it is felt that the uncertainty is of the order of 5-10%.

VISCOSITY OF HYDROGEN

$[\mu, 10^{-6} \text{ N s m}^{-2}]$

T, K	Pressure, bar											
	1	10	20	50	100	150	200	300	500	600	750	1000
15	21.00	22.39	24.05									
16	18.60	19.88	21.30	25.70								
17	16.80	17.84	19.11	22.86	29.40							
18	15.20	15.96	17.11	20.65	26.38							
19	14.00	14.84	15.88	18.95	24.07	29.95						
20	13.00	13.81	14.70	17.40	22.00	27.00	32.40					
21	1.16	12.65	13.50	16.09	20.38	24.75	29.55					
22	1.21	11.72	12.55	14.95	18.76	23.02	27.40					
23	1.25	11.00	11.75	13.90	17.60	21.25	25.25	33.50				
25	1.35	9.60	10.25	12.30	15.55	18.75	22.01	29.00				
27	1.45	8.11	9.00	10.91	13.79	16.54	19.52	25.39				
30	1.60	6.38	7.30	9.30	12.00	14.44	16.90	21.95				
33	1.75	2.20	5.98	8.05	10.56	12.75	14.81	19.01				
36	1.88	2.05	4.05	7.05	9.45	11.40	13.29	17.04				
40	2.07	2.22	2.75	5.85	8.28	10.00	11.70	15.06	23.30			
45	2.28	2.36	2.65	4.70	7.17	8.80	10.24	13.11	19.82			
50	2.49	2.54	2.80	4.20	6.25	7.90	9.25	11.77	17.41			
60	2.88	2.94	2.99	3.80	5.37	6.65	7.83	9.95	13.24	16.00	21.15	
80	3.58	3.62	3.68	4.00	4.80	5.70	6.55	8.10	11.31	12.90	15.61	20.92
100	4.21	4.23	4.24	4.42	5.00	5.62	6.34	7.63	9.95	11.10	13.07	16.80
150	5.60	5.62	5.64	5.74	6.00	6.33	6.69	7.46	8.98	9.70	10.82	12.88
200	6.81	6.82	6.85	6.91	7.06	7.23	7.50	8.02	9.13	9.60	10.47	11.86
300	8.96	8.96	8.98	9.02	9.10	9.21	9.33	9.62	10.30	10.60	11.21	12.10
500	12.64	12.64	12.65	12.67	12.71	12.75	12.83	12.96	13.29	13.50	13.79	14.34
750	16.60	16.60	16.60	16.62	16.64	16.67	16.70	16.78	16.97	17.10	17.26	17.59
1000	20.13	20.13	20.13	20.15	20.17	20.18	20.20	20.25	20.36	20.40	20.57	20.79

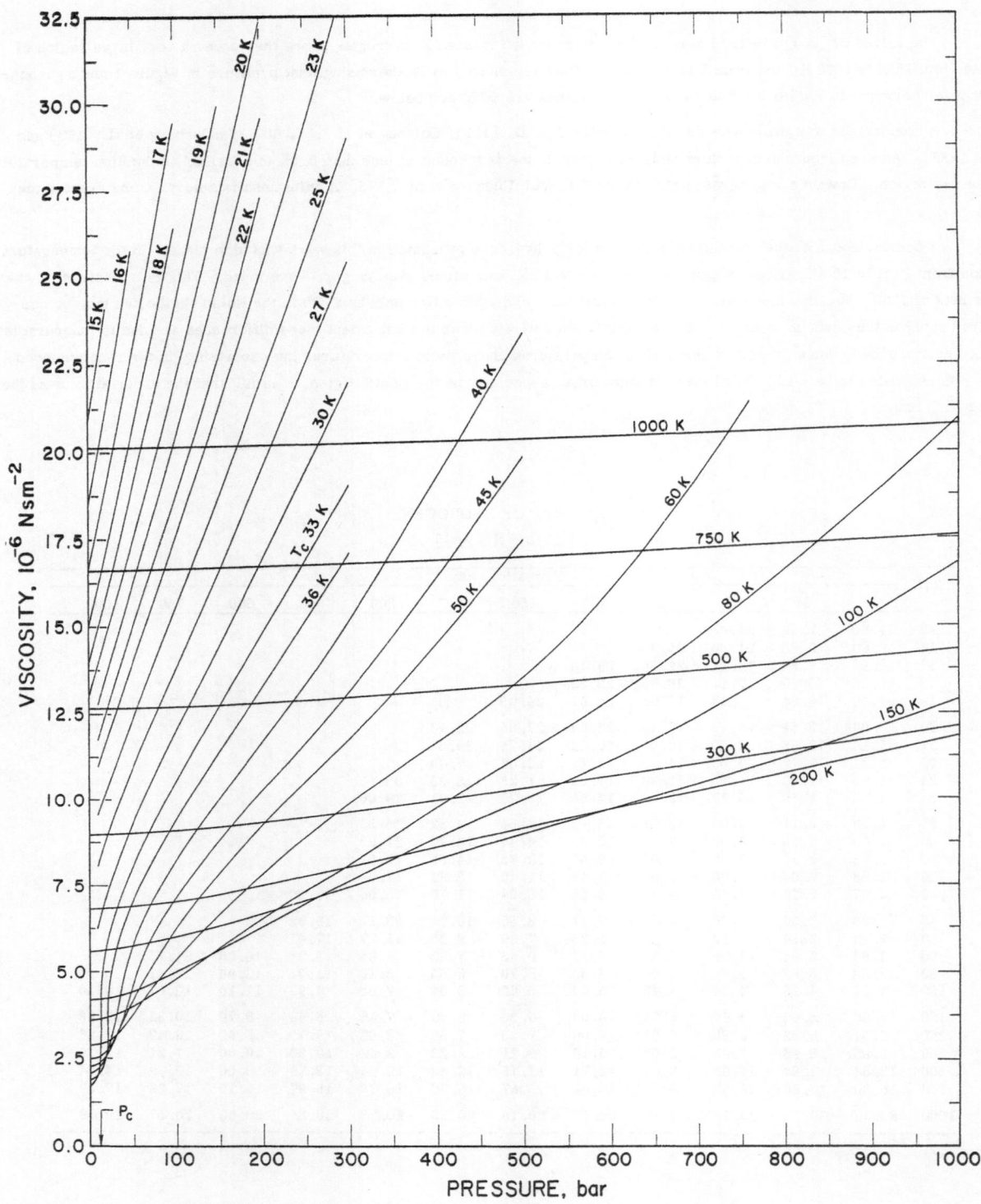

FIGURE 1. VISCOSITY OF HYDROGEN [203].

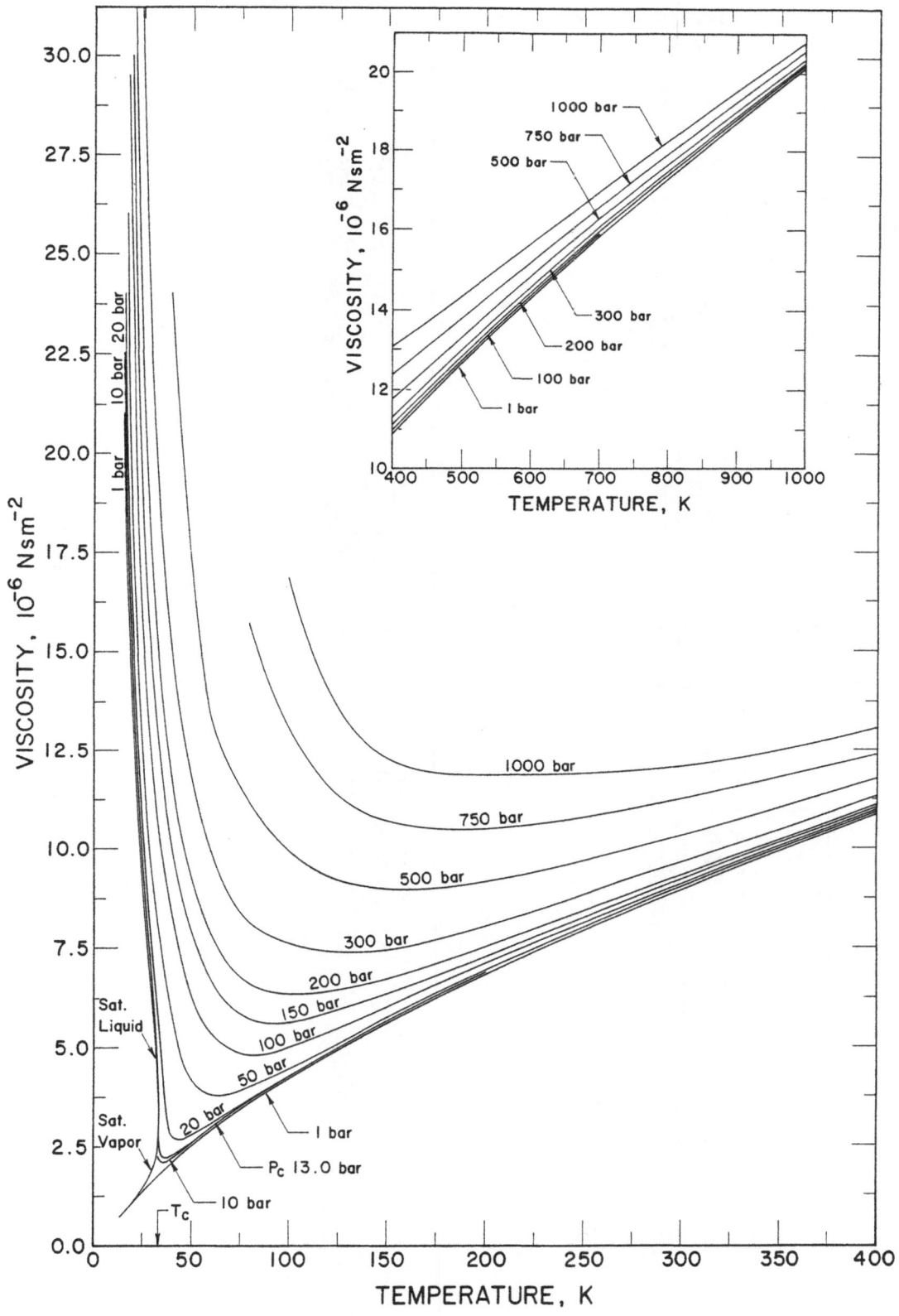

FIGURE 2. VISCOSITY OF HYDROGEN [203].

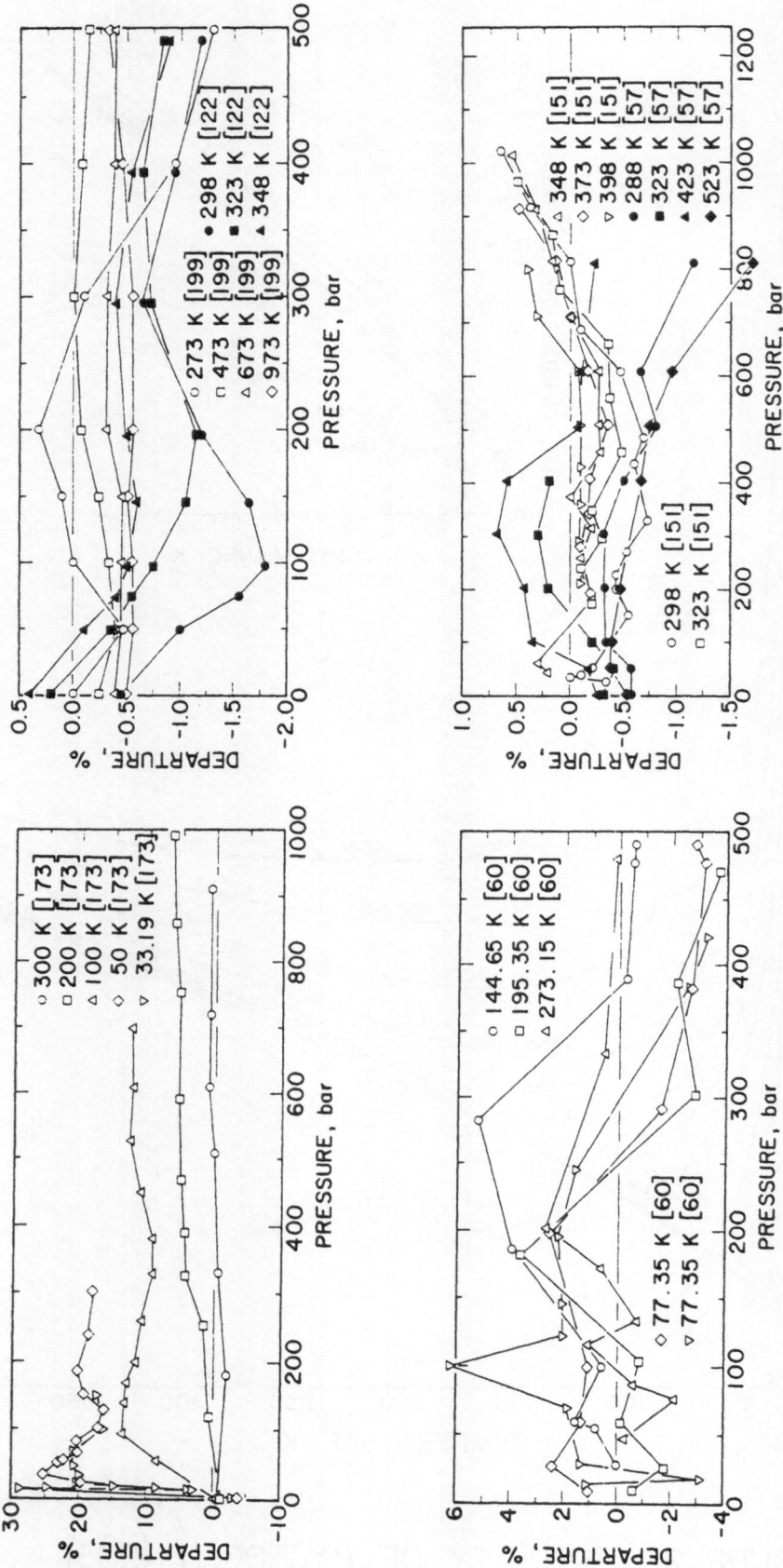

FIGURE 3. DEPARTURE PLOT ON THE VISCOSITY OF HYDROGEN.

p-HYDROGEN

Only one set of data by Diller [40] using the torsional crystal method, was found in the literature for the pressure dependence of the viscosity of para-hydrogen. The data are plotted as isotherms against pressure in Figure 1 where smoothing modifications up to 2% were applied. The dilute gas viscosity was taken from [126]. Figures 2A and 2B show isobars plotted against temperature. From 32 K to 36 K a minor graphical extrapolation to 350 bar was applied. The region of measurements for these isotherms can be seen from Figure 1. The recommended values tabulated below were read from Figure 2. The data have a high standard of precision, their uncertainty is estimated to be ± 1.5%. It has been stated [203] that the viscosity of para-hydrogen is not very different from that of hydrogen in the liquid phase (2 - 3%) and may be considered equal to that of the liquid above the critical temperature. Comparing these data with those recommended for hydrogen in this work reveals larger discrepancies up to 100 K. These may, however, be partly due to inaccuracies in the hydrogen data.

VISCOSITY OF p-HYDROGEN
$[\mu, 10^{-6} \text{ N s m}^{-2}]$

T,K	Pressure, bar																			
	1	10	12.9*	15	16	18	20	22	24	25	30	40	50	60	100	150	200	250	300	350
15	22.17	23.67	24.16	24.52	24.70	25.04	25.38	25.72	26.05	26.24	27.11	27.18	—	28.72	31.10	—	—	—	—	—
16	19.78	21.07	21.51	21.83	22.00	22.30	22.60	22.90	23.18	23.36	24.12	24.22	25.67	25.56	28.02	—	—	—	—	—
17	17.80	18.97	19.34	19.62	19.77	20.01	20.27	20.54	20.78	20.92	21.60	22.03	22.91	23.23	25.60	—	—	—	—	—
18	16.15	17.22	17.58	17.81	17.95	18.22	18.45	18.67	18.90	19.04	19.64	20.07	20.84	21.25	—	—	—	—	—	—
19	14.90	15.86	16.17	16.40	16.51	16.74	16.94	17.14	17.34	17.47	18.00	—	20.17	—	—	—	—	—	—	—
20	13.74	14.59	14.90	15.08	15.18	15.38	15.59	15.79	15.96	16.08	16.57	17.56	18.55	19.54	—	28.50	34.27	—	—	—
21	1.13	13.48	13.76	13.95	14.04	14.22	14.40	14.58	14.75	14.85	15.32	16.22	17.13	18.07	21.72	26.40	31.30	36.90	—	—
22	1.18	12.44	12.68	12.85	12.96	13.14	13.30	13.47	13.64	13.75	14.18	15.05	15.90	16.77	20.07	24.45	28.98	33.78	—	—
23	1.23	11.75	12.00	12.14	12.22	12.37	12.55	12.72	12.86	12.97	13.35	14.15	14.92	15.72	18.88	22.90	26.98	31.38	—	—
25	1.33	10.12	10.37	10.52	10.60	10.74	10.92	11.06	11.21	11.30	11.70	12.48	13.20	13.95	16.70	20.16	23.60	27.30	—	—
27	1.43	8.67	8.92	9.10	9.19	9.35	9.53	9.68	9.83	9.92	10.31	11.05	11.72	12.39	14.90	17.90	20.95	24.06	27.38	—
28	1.48	8.04	8.30	8.49	8.58	8.74	8.91	9.06	9.22	9.32	9.72	10.47	11.16	11.82	14.20	17.08	20.00	22.88	26.00	—
30	1.58	6.76	7.09	7.30	7.40	7.57	7.77	7.94	8.10	8.20	8.60	9.31	9.98	10.61	12.88	15.52	18.10	20.70	23.50	—
32	1.68	2.08	5.77	6.06	6.20	6.44	6.66	6.88	7.08	7.18	7.62	8.37	9.04	9.67	11.90	14.37	16.72	19.10	21.60	23.90
36	1.86	2.12	2.34	2.60	2.78	3.34	4.10	4.78	5.10	5.23	5.83	6.74	7.44	8.08	10.15	12.38	14.45	16.50	18.53	20.60
37	1.91	2.15	2.31	2.51	2.63	2.96	3.46	4.15	4.60	4.78	5.45	6.37	7.08	7.73	9.80	12.00	14.06	16.02	18.02	20.00
38	1.95	2.18	2.31	2.47	2.56	2.77	3.12	3.50	4.03	4.31	5.07	6.02	6.75	7.40	9.47	11.62	13.63	15.52	17.48	19.35
39	2.00	2.21	2.33	2.46	2.53	2.69	2.92	3.22	3.63	3.91	4.72	5.71	6.45	7.10	9.15	11.24	13.24	15.08	16.96	18.80
40	2.04	2.25	2.37	2.47	2.53	2.66	2.84	3.07	3.38	3.57	4.38	5.40	6.16	6.80	8.85	10.94	12.85	14.67	16.47	18.28
42	2.13	2.32	2.43	2.51	2.57	2.67	2.82	2.97	3.18	3.27	3.84	4.86	5.65	6.29	8.33	10.33	12.18	13.92	15.58	17.25
44	2.22	2.40	2.50	2.58	2.63	2.71	2.83	2.94	3.10	3.18	3.56	4.40	5.22	5.83	7.87	9.80	11.58	13.22	14.80	16.40
46	2.30	2.47	2.56	2.64	2.69	2.76	2.85	2.95	3.07	3.14	3.41	4.08	4.87	5.46	7.42	9.30	11.02	12.60	14.08	15.65
48	2.39	2.55	2.63	2.70	2.75	2.81	2.89	2.96	3.05	3.12	3.32	3.87	4.58	5.16	7.03	8.87	10.52	12.03	13.47	14.95
50	2.47	2.62	2.71	2.77	2.80	2.86	2.93	2.99	3.06	3.10	3.29	3.75	4.36	4.91	6.70	8.48	10.05	11.50	12.88	14.30
55	2.66	2.80	2.86	2.91	2.93	2.97	3.02	3.08	3.12	3.15	3.30	3.65	4.08	4.49	6.10	7.70	9.13	10.48	11.76	13.02
60	2.85	2.97	3.01	3.05	3.07	3.11	3.15	3.19	3.23	3.25	3.37	3.65	3.97	4.29	5.66	7.13	8.45	9.70	10.87	12.05
65	3.03	3.14	3.18	3.22	3.23	3.27	3.29	3.34	3.37	3.39	3.49	3.70	3.94	4.20	5.36	6.70	7.92	9.09	10.18	11.26
70	3.21	3.30	3.35	3.38	3.39	3.42	3.45	3.48	3.51	3.53	3.61	3.79	3.98	4.20	5.18	6.38	7.50	8.55	9.59	10.63
80	3.55	3.61	3.65	3.67	3.68	3.71	3.73	3.76	3.78	3.80	3.86	3.99	4.14	4.30	5.04	6.01	6.95	7.85	8.75	9.65
90	3.88	3.95	3.98	4.00	4.01	4.03	4.05	4.07	4.08	4.09	4.14	4.23	4.35	4.48	5.05	5.85	6.65	7.40	8.21	9.00
100	4.20	4.28	4.30	4.32	4.32	4.33	4.35	4.36	4.37	4.38	4.41	4.50	4.59	4.70	5.17	5.84	6.52	7.18	7.86	8.55

*Critical pressure.

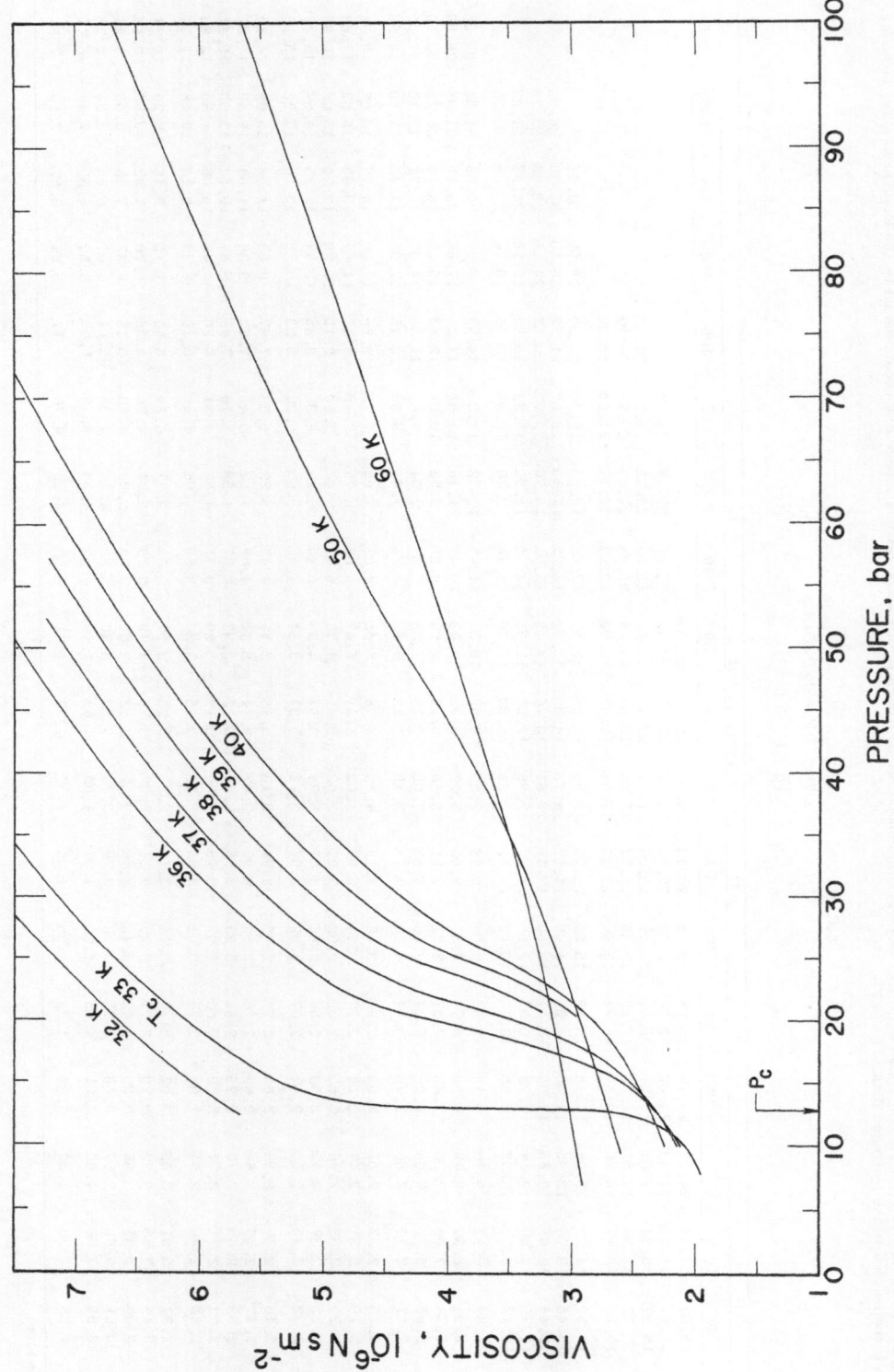

FIGURE I A. VISCOSITY OF p - HYDROGEN [40].

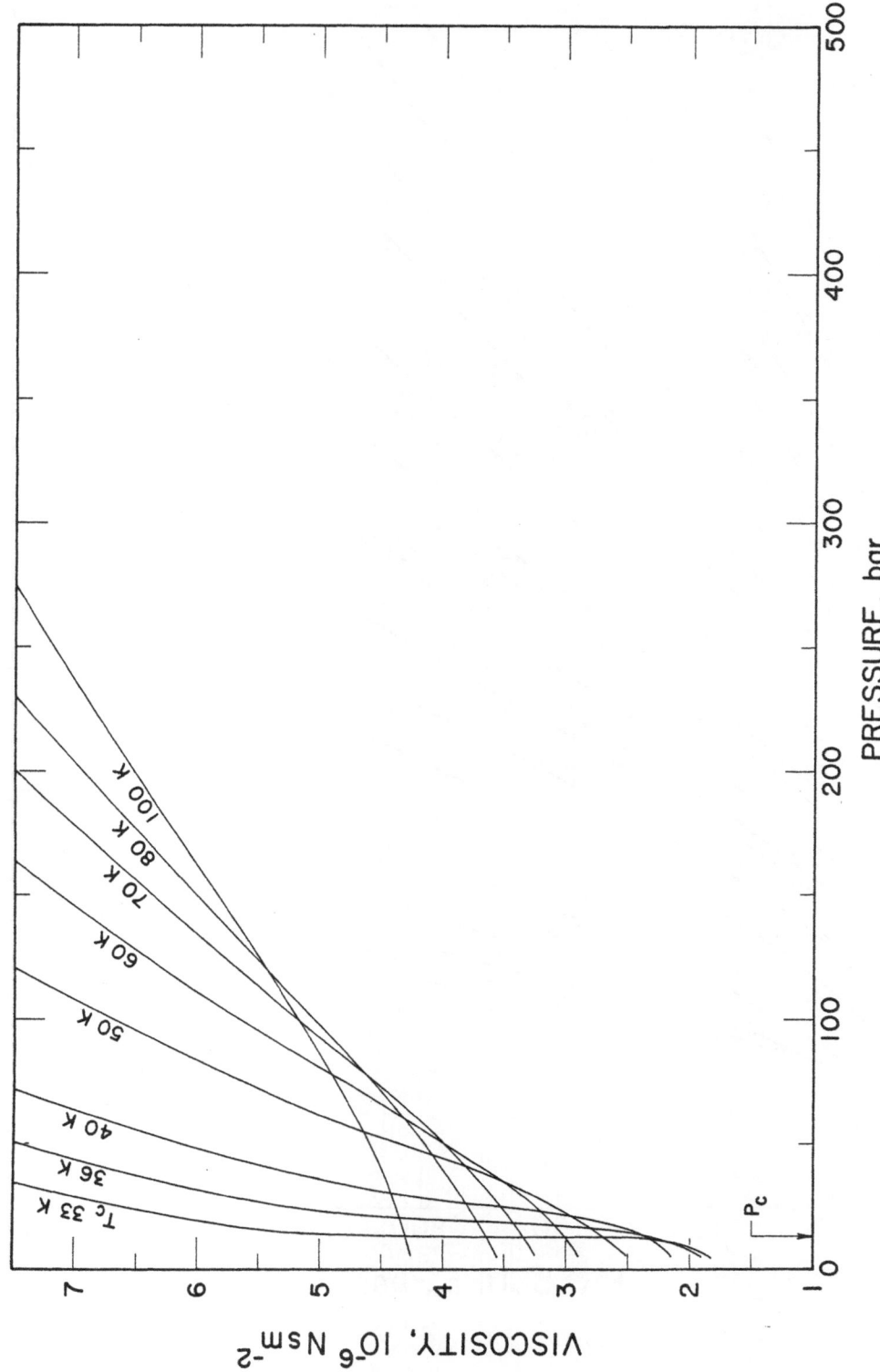

FIGURE IB. VISCOSITY OF p - HYDROGEN [40].

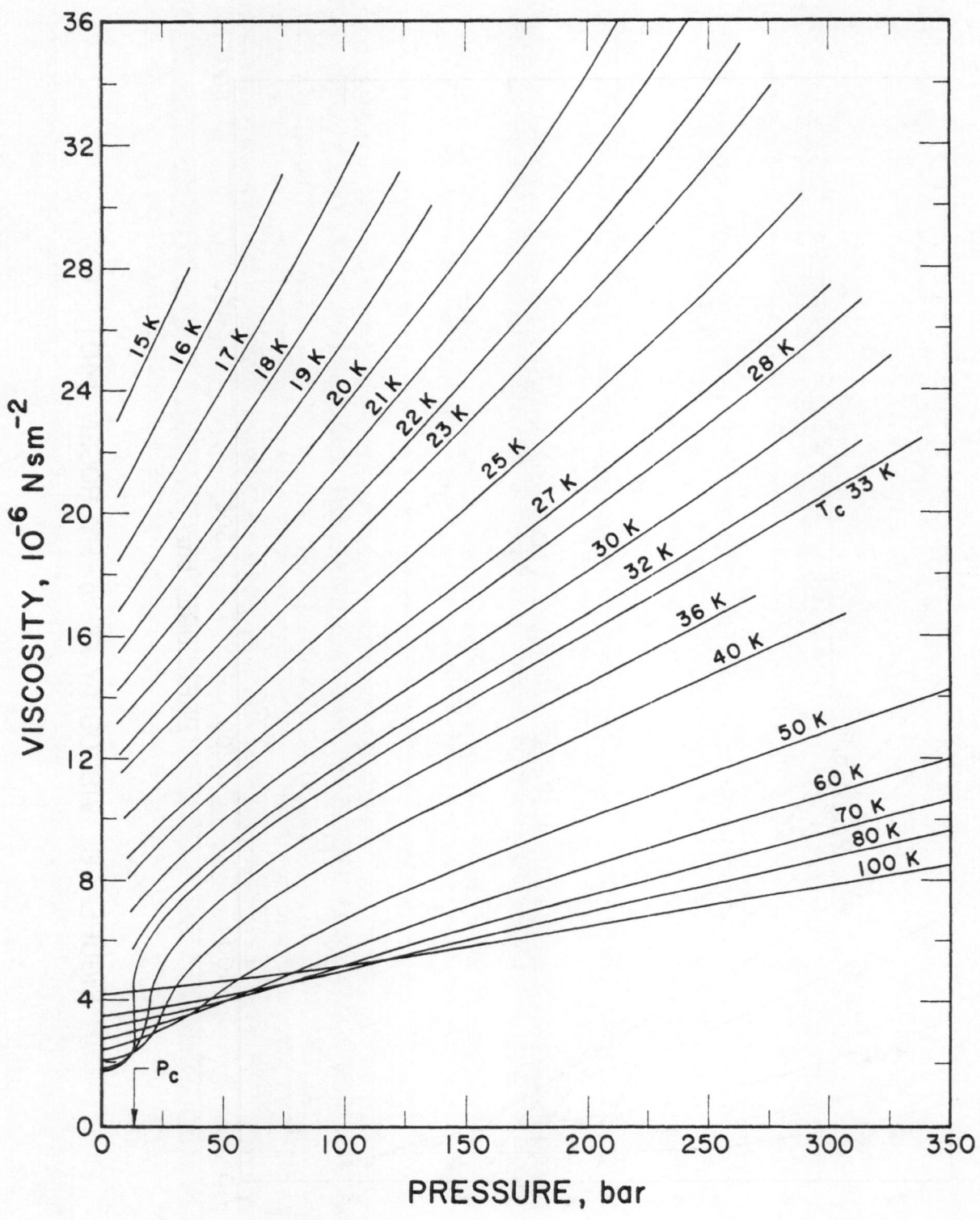

FIGURE 1C. VISCOSITY OF p-HYDROGEN [40].

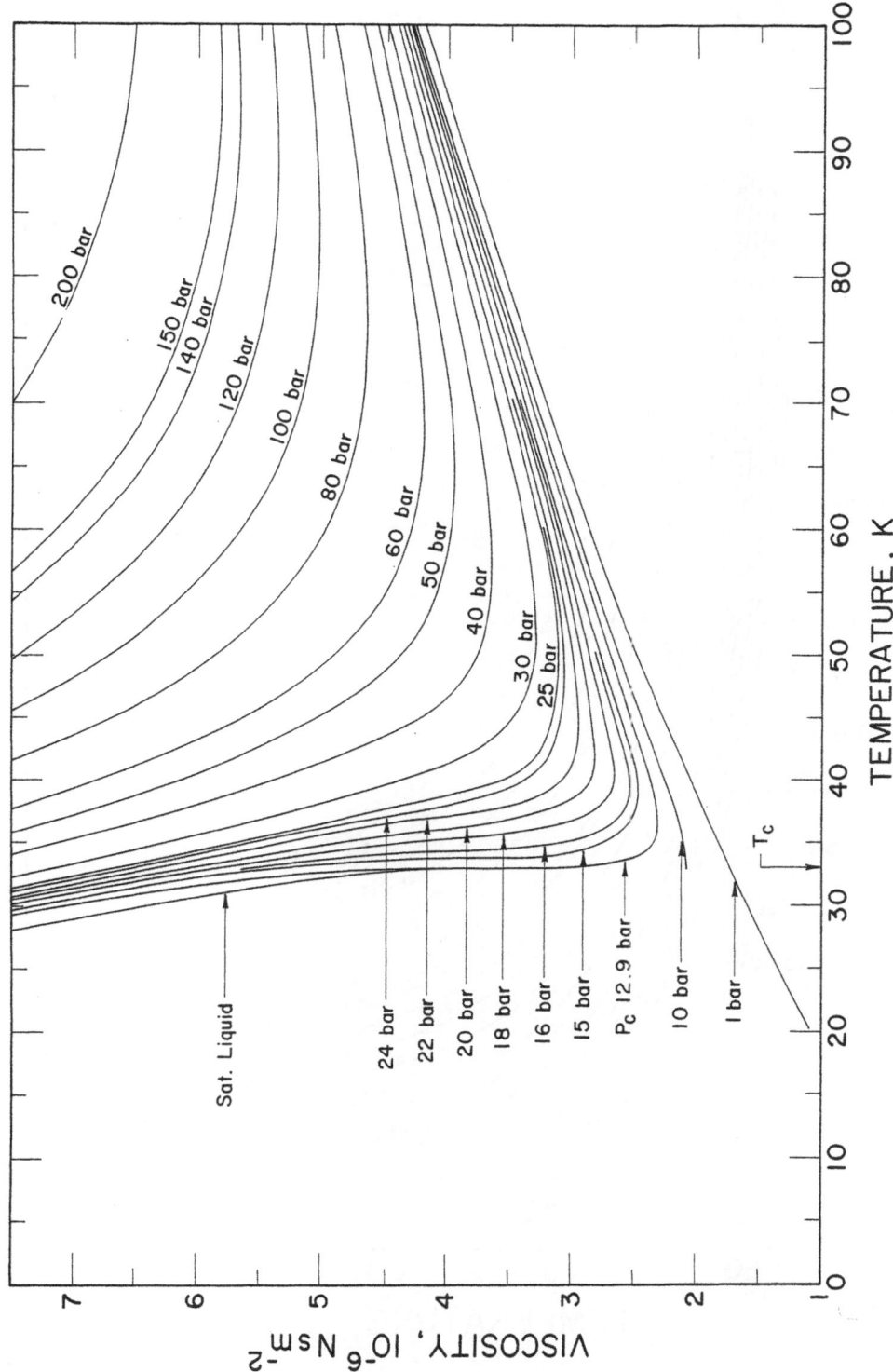

FIGURE 2A. VISCOSITY OF p - HYDROGEN [40].

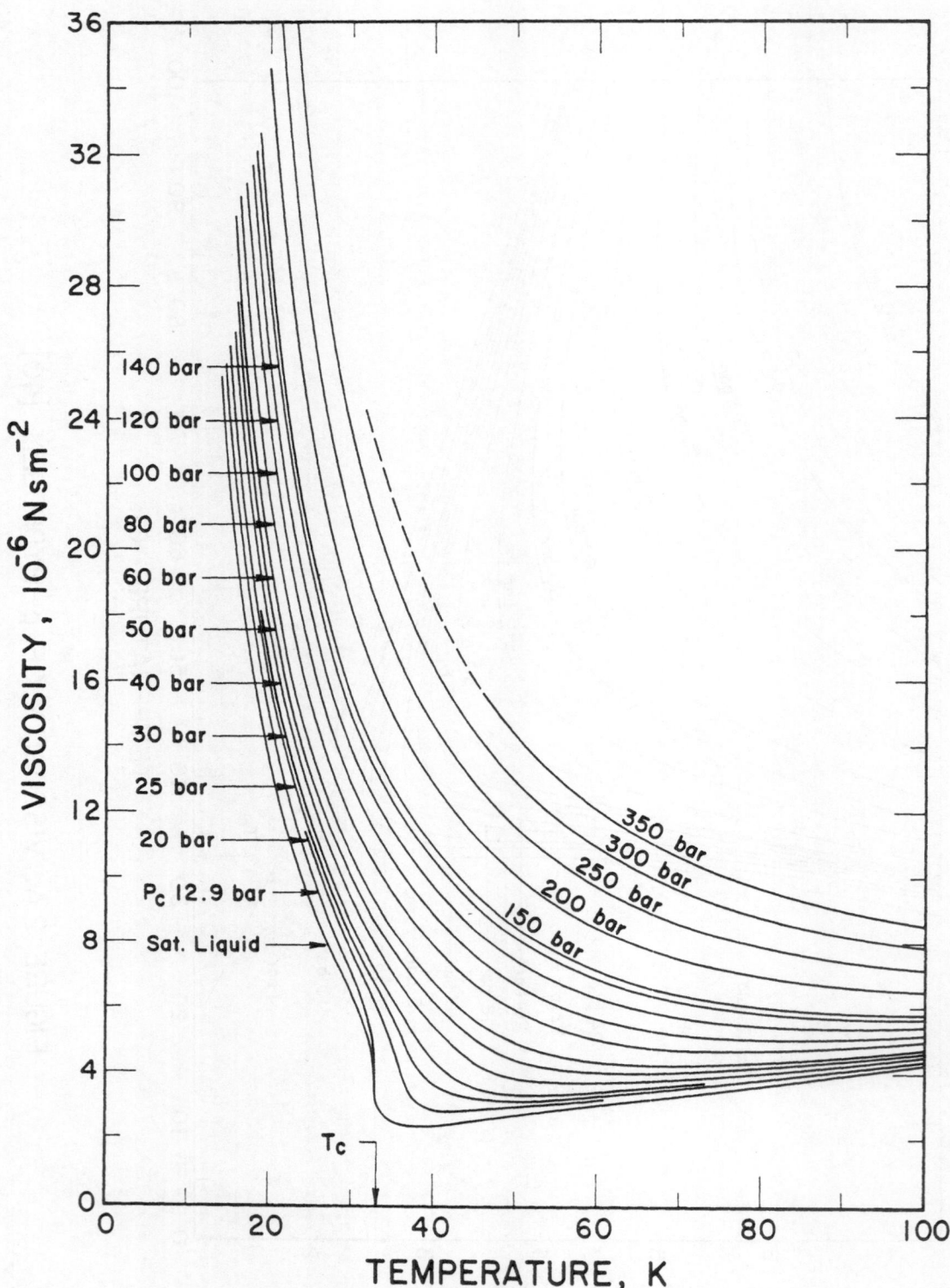

FIGURE 2B. VISCOSITY OF ρ-HYDROGEN [40].

KRYPTON

The data of Trappeniers et al. [198] were used to generate the recommended values tabulated on the next page. These data were obtained by a capillary tube method and appear to have a high standard of precision. Using the P-V-T data of [13, 199] the 1 bar data of [135], and the residual viscosity concept, it was possible to extrapolate over a wide region of states, from a low of 270 K up to a high of 600 K and up to 500 bar. The values are plotted as isotherms against pressure in Figure 1 and as isobars against temperature in Figure 2. In both figures, the extrapolated values may be recognized by dashed lines. The recommended values were read from Figure 2.

Additional works on the pressure dependence of the viscosity of krypton are listed in the summary table below. As shown in Figure 3, excellent agreement can be reported for the data of Reynes et al. [169], which were also obtained by the capillary tube method. Good agreement can also be observed for the data at lower pressures of Kestin et al. [105]. The uncertainty in the recommended values is estimated to be ± 1.5%. This is substantiated by comparison with a set of recommended values by [76], which was derived from the principle of corresponding states using experimental data of argon. These latter values include the liquid region as well and are recommended for use in that region.

ADDITIONAL REFERENCES ON THE VISCOSITY OF KRYPTON

Authors	Year	Ref. No.	Temperature K	Pressure bar	Method	Departure % (no. points)	
Slyusar and Rudenko	1972	181	117-209	SL	Falling cylinder	-	
Kestin and Yata	1968	112	293-303	1-21	Oscillating disk	0.4	(8)
Boon et al.	1967	17	116-123	SL	Capillary tube	-	
Reynes	1964	167	373-473	71-830	Capillary tube	0.00 ± 0.76	(20)
Reynes and Thodos	1964	169	373-473	71-830	Capillary tube	0.00 ± 0.76	(20)
Boon and Thomaes	1963	18	116-123	SL	Capillary tube	-	
Kestin and Leidenfrost	1959	105	293	1-22	Oscillating disk	0.8	(6)

VISCOSITY OF KRYPTON

$[\mu,\ 10^{-6}\ N\ s\ m^{-2}]$

T, K	Pressure, bar																	
	1	20	40	50	54.3*	60	80	100	120	140	150	160	180	200	250	300	400	500
150	13.1	–	–	–	–	–	–	–	–	–	–	–	–	–	–	–	–	–
200	17.4	–	–	–	–	–	–	–	–	–	–	–	–	–	–	–	–	–
250	21.6	–	–	–	–	–	–	–	–	–	–	–	–	–	–	–	–	–
270	23.2	24.2	25.4	26.3	26.7	27.2	29.8	33.3	38.0	43.4	46.4	49.4	55.4	60.5	72.4	82.4	98.7	111.8
280	24.0	24.9	26.0	26.8	27.2	27.7	29.9	32.8	36.6	41.1	43.4	45.9	50.8	55.7	67.4	77.2	93.0	106.2
290	24.8	25.6	26.7	27.4	27.7	28.2	30.1	32.6	35.7	39.6	41.4	43.5	47.8	52.2	63.0	72.4	87.7	101.0
300	25.6	26.4	27.3	28.0	28.4	28.7	30.4	32.6	35.2	38.4	40.1	41.9	45.6	49.4	59.2	68.1	83.0	96.2
310	26.4	27.1	28.1	28.6	29.0	29.2	30.8	32.8	35.1	37.8	39.3	40.8	44.0	47.4	56.2	64.5	78.8	91.8
320	27.1	27.8	28.8	29.2	29.6	29.8	31.3	33.0	35.2	37.5	38.8	40.5	43.0	46.0	53.8	61.5	75.4	88.0
340	28.6	29.2	30.1	30.6	30.9	31.1	32.4	33.8	35.6	37.4	38.5	39.6	41.8	44.2	50.7	57.2	69.8	81.2
360	30.1	30.6	31.4	31.9	32.2	32.4	33.6	34.8	36.2	37.8	38.6	39.6	41.4	43.4	49.0	54.5	65.6	76.0
380	31.6	32.1	32.8	33.2	33.4	33.8	34.8	35.8	37.0	38.4	39.1	40.0	41.4	43.2	47.9	52.8	62.6	72.1
400	33.0	33.5	34.2	34.6	34.8	35.0	35.9	36.9	38.0	39.2	39.8	40.5	41.8	43.4	47.4	51.7	60.6	69.2
420	34.4	34.9	35.6	35.9	36.0	36.2	37.0	38.0	39.0	40.0	40.6	41.1	42.4	43.7	47.3	51.1	59.1	67.0
440	35.7	36.2	36.8	37.2	37.3	37.5	38.2	39.0	40.0	40.8	41.4	41.8	43.1	44.3	47.5	51.0	58.2	–
450	36.4	36.9	37.4	37.8	37.9	38.1	38.8	39.5	40.5	41.3	41.8	42.3	43.4	44.6	47.6	50.9	57.8	–
500	39.6	40.1	40.5	40.8	40.9	41.0	41.7	42.3	43.1	43.8	44.2	44.6	45.5	46.4	48.9	51.4	56.9	–
550	42.7	43.1	43.5	43.7	43.8	43.9	44.6	45.1	45.7	46.3	46.6	47.0	47.7	48.6	50.5	52.6	57.2	–
600	45.7	46.0	46.4	46.6	46.7	46.8	47.4	47.9	48.2	48.8	49.2	49.5	50.1	50.8	52.6	54.3	58.4	–

*Critical pressure.

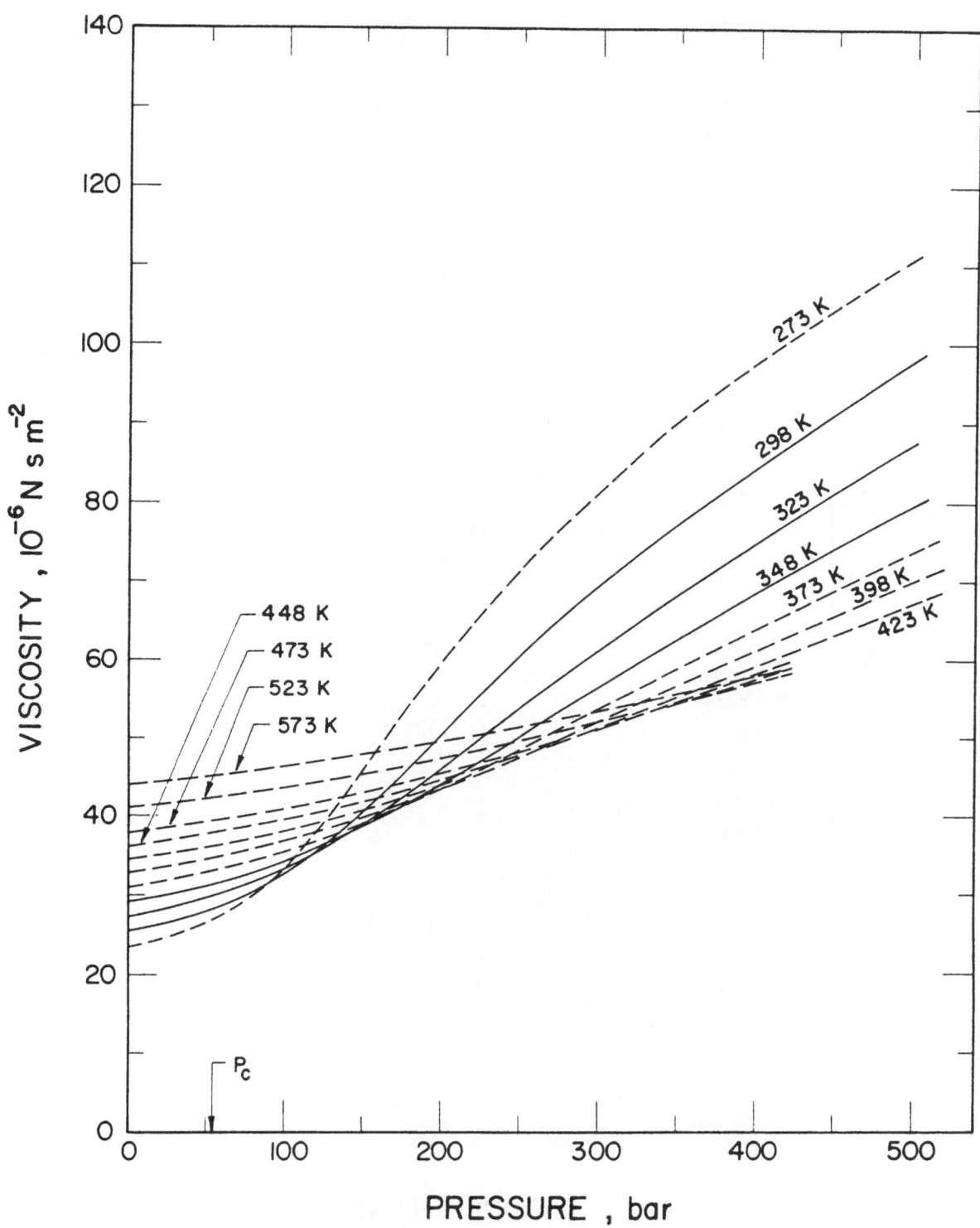

FIGURE I. VISCOSITY OF KRYPTON [198].

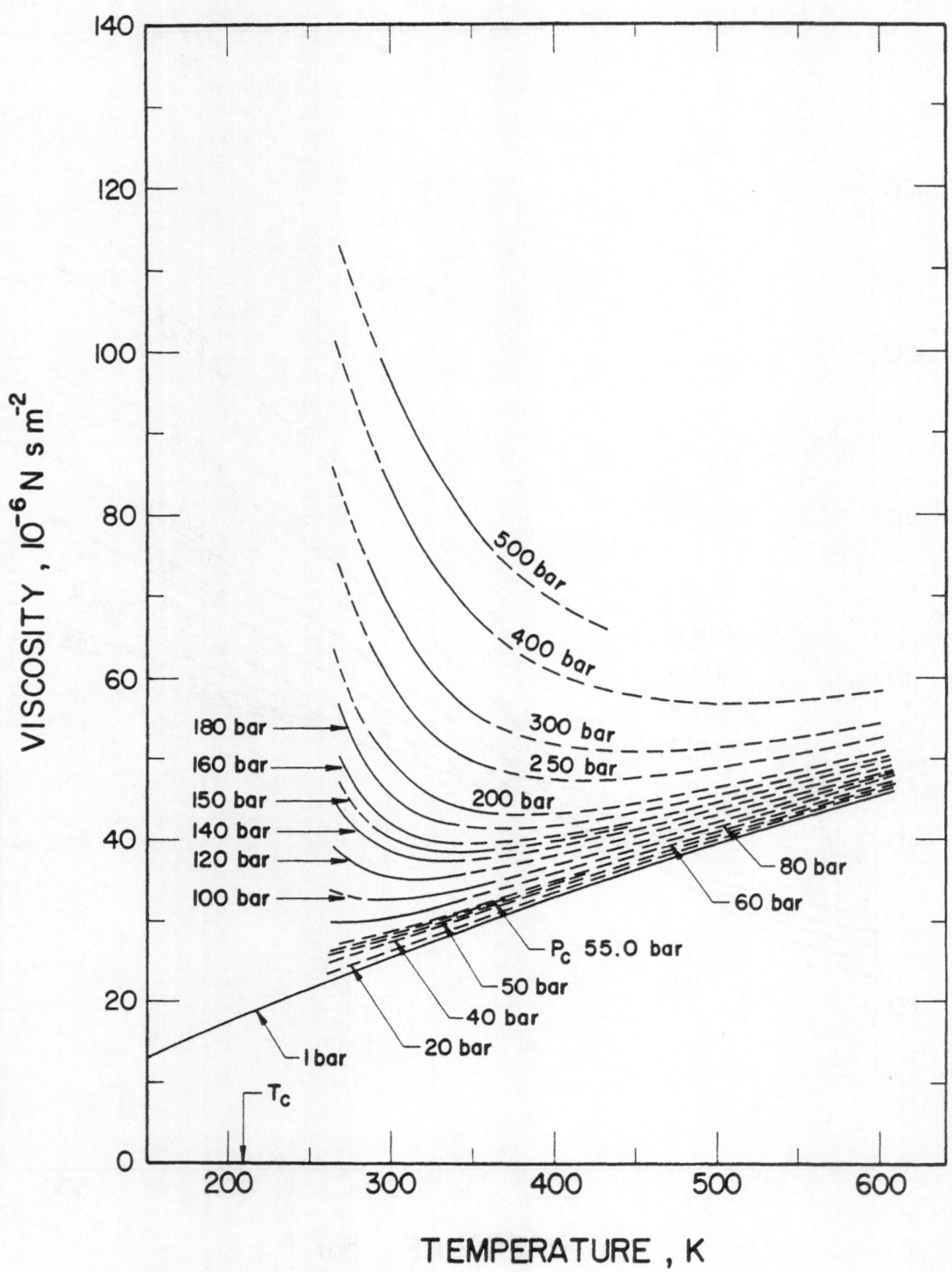

FIGURE 2. VISCOSITY OF KRYPTON [198].

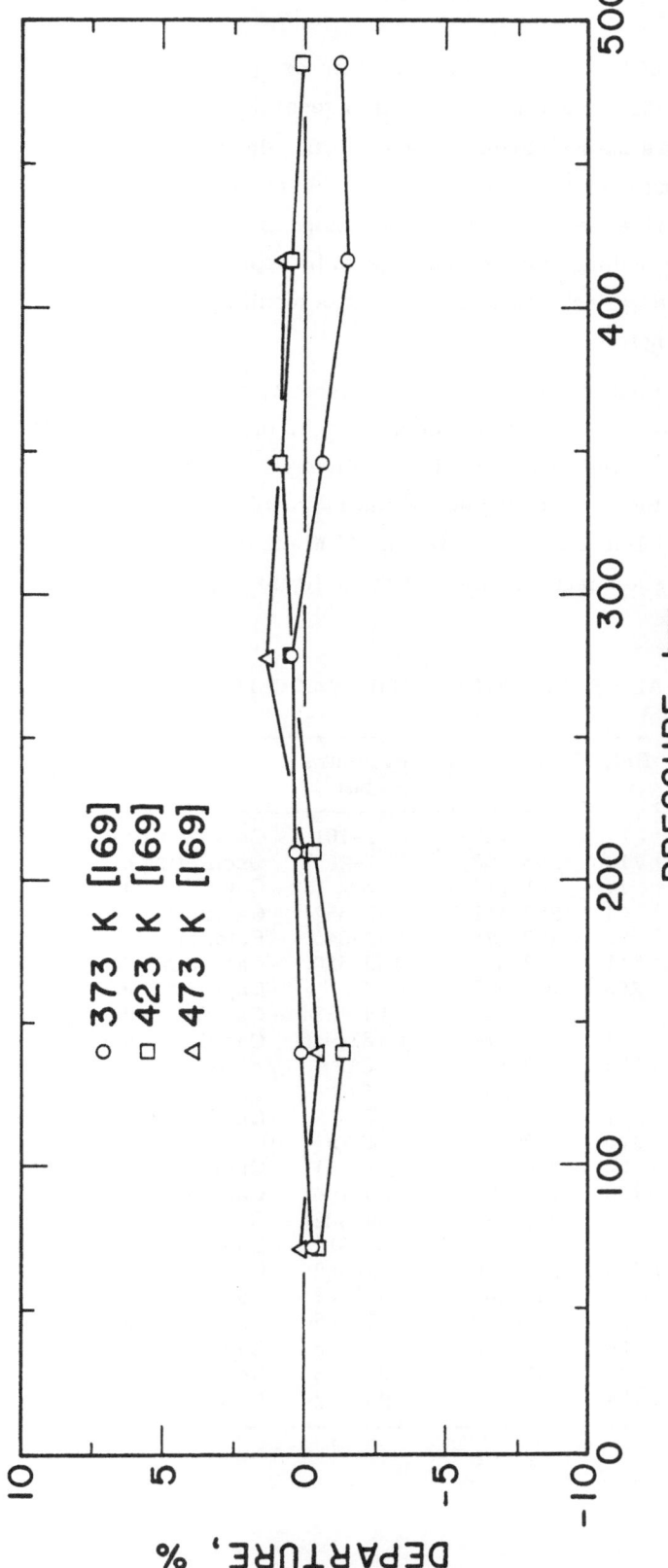

FIGURE 3. DEPARTURE PLOT ON THE VISCOSITY OF KRYPTON.

METHANE

The data of Huang et al. [85] obtained by a falling cylinder apparatus, and those of Gonzales et al. [62] obtained by a capillary tube viscosimeter, were used to generate the recommended values presented on the next page. Figure 1A shows the isotherms of Huang et al. [85] plotted against pressure, where smoothing modifications of 1% maximum have been applied. Figure 1B shows isotherms of Gonzalez et al. [62] plotted against pressure, where the smoothing modifications amounted up to 2.5%. In the temperature overlap of both sources, again smoothing modifications had to be applied, as can be seen from Figures 2A and 2B, where isobars are plotted against temperature. These modifications are about 3%. The recommended values were read from Figures 2A and 2B.

Additional works on the pressure dependence of the viscosity of methane are listed in the summary table below. With few exceptions, especially in the liquid region, the agreement with the recommended values is satisfactory, as seen from the departure plots in Figure 3. The uncertainty in the recommended values is estimated to be ± 3% in the gaseous region and the majority of the liquid region. The large deviations with respect to the data of Hellemans et al. between 145 K and the critical temperature make it difficult to make a reliable accuracy statement for this part of the liquid region.

ADDITIONAL REFERENCES ON THE VISCOSITY OF METHANE

Authors	Year	Ref. No.	Temperature K	Pressure bar	Method	Departure % (no. points)
Hellemans et al.	1970	81	97-187	1-101	Oscillating disk	-1.97 ± 7.10 (23)
Kestin and Yata	1968	112	293-303	1-25	Oscillating disk	1.2
Boon et al.	1967	17	91-114	SL	Capillary tube	–
Giddings et al.	1966	54	283-411	1-552	Capillary tube	- 95 ± 1.14 (39)
Carmichael et al.	1965	25	278-478	1-345	Rotating cylinder	-1.85 ± 1.17 (33)
Martin et al.	1965	141	303	150-550	Capillary tube	–
Vernet and Kniazeff	1964	205	113-273	75	Capillary tube	–
Barua et al.	1964	12	223-423	10-423	Capillary tube	-0.75 ± 91 (33)
Boon and Thomaes	1963	18	91-114	SL	Capillary tube	–
Swift et al.	1960	190	133-191	SL	Falling body	–
Swift et al.	1959	189	123-190	SL	Falling body	–
Baron et al.	1959	11	325-408	7-552	Capillary tube	-0.02 ± 0.01 (40)
Kestin and Leidenfrost	1959	105	293	1-79	Oscillating disk	1.9
Iwasaki and Takahashi	1959	92	298-348	1-511	Oscillating disk	4.8
Pavlovich and Timrot	1958	160	112-323	20-196	Capillary tube	31
Ross and Brown	1957	171	223-298	69-689	Capillary tube	3.38 ± 2.32 (16)
Meshcheryakov and Golubev	1954	57	258-523	1-811	Capillary tube	-2.88 ± 2.54 (36)
Stewart	1952	186	294-340	1-552	Capillary tube	2.3
Carr	1952	31	295-366	1-552	Capillary tube	4
Kuss	1952	122	293-353	1-608	Capillary tube	12.8
Comings et al.	1944	36	303-368	1-173	Capillary tube	1.5
Bicher and Katz	1943	15	298-498	28-345	Rolling ball	12
Sage and Lacey	1938	174	311-378	28-172	Rolling ball	–

VISCOSITY OF METHANE

$$[\mu,\ 10^{-6}\ N\ s\ m^{-2}]$$

T, K	Pressure, bar																			
	1	20	30	40	46.1*	55	60	70	80	90	100	120	140	150	200	300	400	500	600	700
100	157.2	160.8	162.7	164.6	165.8	167.6	168.7	170.9	173.1	175.3	177.5	181.1	184.7	186.5						
105	136.5	139.6	141.3	142.9	144.0	145.4	146.2	147.8	149.5	151.1	152.7	155.9	159.2	160.8	169.5	186.5				
110	121.4	124.2	125.6	127.1	128.0	129.2	130.0	131.5	132.9	134.4	135.9	138.7	141.6	143.0	150.8	165.6				
115	5.3	112.5	113.8	115.0	115.8	117.0	117.7	119.1	120.4	121.8	123.2	125.9	128.6	130.0	137.0	150.5				
120	5.4	101.5	102.9	104.3	105.2	106.4	107.1	108.4	109.8	111.1	112.5	115.1	117.6	118.9	125.5	138.8				
130	5.8	83.5	84.8	85.8	86.6	87.6	88.2	89.4	90.6	91.8	93.0	95.3	97.6	98.7	104.4	115.8				
140	6.0	69.2	70.3	71.3	72.1	73.0	73.5	74.5	75.5	76.8	78.0	79.8	81.6	82.5	87.2	96.8				
150	6.3	58.1	59.1	60.3	60.6	61.5	62.0	63.0	64.0	65.2	66.3	68.1	69.8	70.7	75.3	84.1				
160	6.6	49.0	49.9	50.7	51.4	52.3	52.8	53.8	54.8	55.9	56.9	58.7	60.6	61.5	66.1	74.8				
170	6.9	7.4	41.5	42.7	43.5	44.5	45.0	46.2	47.3	48.4	49.5	51.5	53.3	54.4	59.0	67.0				
180	7.2	7.6	8.0	34.6	35.5	37.2	38.0	39.3	40.5	41.7	42.9	45.0	47.0	48.0	52.5	60.2				
200	7.8	8.2	8.5	9.0	9.8	12.2	15.8	22.0	25.5	27.7	29.6	32.6	35.3	36.4	41.5	49.1				
205	8.0	8.4	8.7	9.0	9.6	11.0	12.4	16.5	21.0	24.0	26.4	29.9	32.6	33.8	39.1	46.8				
210	8.1	8.5	8.7	9.0	9.6	10.4	11.2	13.6	17.4	21.0	23.5	27.2	30.2	31.5	37.0	44.8				
215	8.3	8.6	8.9	9.2	9.8	10.3	10.9	12.5	15.2	18.5	21.0	24.8	28.0	29.3	35.1	43.0				
220	8.5	8.8	9.0	9.4	9.9	10.4	10.8	12.0	14.0	16.3	18.8	22.6	26.0	27.4	33.3	41.2				
225	8.6	9.0	9.1	9.5	10.0	10.4	10.8	11.7	13.4	15.3	17.1	20.8	24.1	25.5	31.6	39.6				
230	8.8	9.2	9.4	9.6	10.1	10.5	10.8	11.5	13.0	14.5	16.0	19.3	22.5	23.9	30.0	38.1				
240	9.1	9.5	9.6	10.0	10.3	10.7	10.9	11.5	12.5	13.6	14.6	17.2	19.9	21.2	27.1	35.3				
250	9.4	9.8	10.0	10.3	10.5	11.0	11.1	11.7	12.3	13.2	13.9	16.0	18.2	19.3	24.8	32.7				
260	9.8	10.1	10.2	10.6	10.8	11.2	11.4	11.9	12.3	13.0	13.6	15.3	17.1	18.1	22.9	30.5	–	–	–	–
280	10.4	10.8	10.9	11.3	11.4	11.6	11.9	12.3	12.6	13.1	13.6	14.8	16.7	16.7	20.4	26.8	–	–	–	–
300	11.0	11.5	11.7	12.0	12.1	12.4	12.5	12.9	13.2	13.6	13.9	14.9	16.0	16.5	19.2	24.5	30.3	34.9	38.8	42.6
320	11.7	12.1	12.3	12.6	12.7	13.0	13.1	13.4	13.7	14.0	14.3	15.1	15.9	16.3	18.7	23.3	28.5	32.6	36.2	39.5
340	12.3	12.7	12.9	13.2	13.3	13.5	13.6	13.9	14.1	14.4	14.6	15.3	16.0	16.4	18.4	22.5	27.1	30.9	34.3	37.3
360	13.0	13.4	13.6	13.8	13.9	14.1	14.2	14.4	14.6	14.8	15.0	15.6	16.2	16.5	18.3	22.0	26.1	29.6	32.9	35.7
380	13.5	14.0	14.2	14.4	14.5	14.7	14.8	15.0	15.1	15.3	15.5	16.0	16.5	16.8	18.3	21.6	25.3	28.6	31.7	34.5
400	14.2	14.6	14.8	15.0	15.1	15.3	15.4	15.5	15.7	15.8	16.0	16.4	16.9	17.1	18.5	21.5	24.7	27.7	30.7	33.5
420	14.7	15.3	15.4	15.6	15.7	15.8	15.9	16.0	16.2	16.3	16.5	16.9	17.3	17.5	18.7	21.4	24.4	27.3	30.0	32.6
440	15.4	16.0	16.1	16.2	16.3	16.4	16.4	16.6	16.7	16.9	17.0	17.4	17.8	18.0	19.0	21.5	24.1	26.8	29.4	31.8
460	16.0	16.6	16.7	16.8	16.9	17.0	17.0	17.2	17.3	17.5	17.6	18.0	18.3	18.5	19.5	21.7	24.1	26.6	28.9	31.3
480	16.6	17.2	17.3	17.4	17.5	17.6	17.6	17.8	17.9	18.1	18.2	18.5	18.8	18.8	19.9	21.9	24.1	26.4	28.6	30.7
500	17.2	17.8	17.9	18.0	18.1	18.2	18.2	18.3	18.5	18.6	18.7	19.0	19.3	19.5	20.4	22.2	24.3	26.3	28.3	30.3
520	17.7	18.4	18.5	18.6	18.7	18.8	18.8	18.9	19.1	19.2	19.3	19.6	19.9	20.1	20.9	22.5	24.5	26.3	28.2	30.0

*Critical pressure.

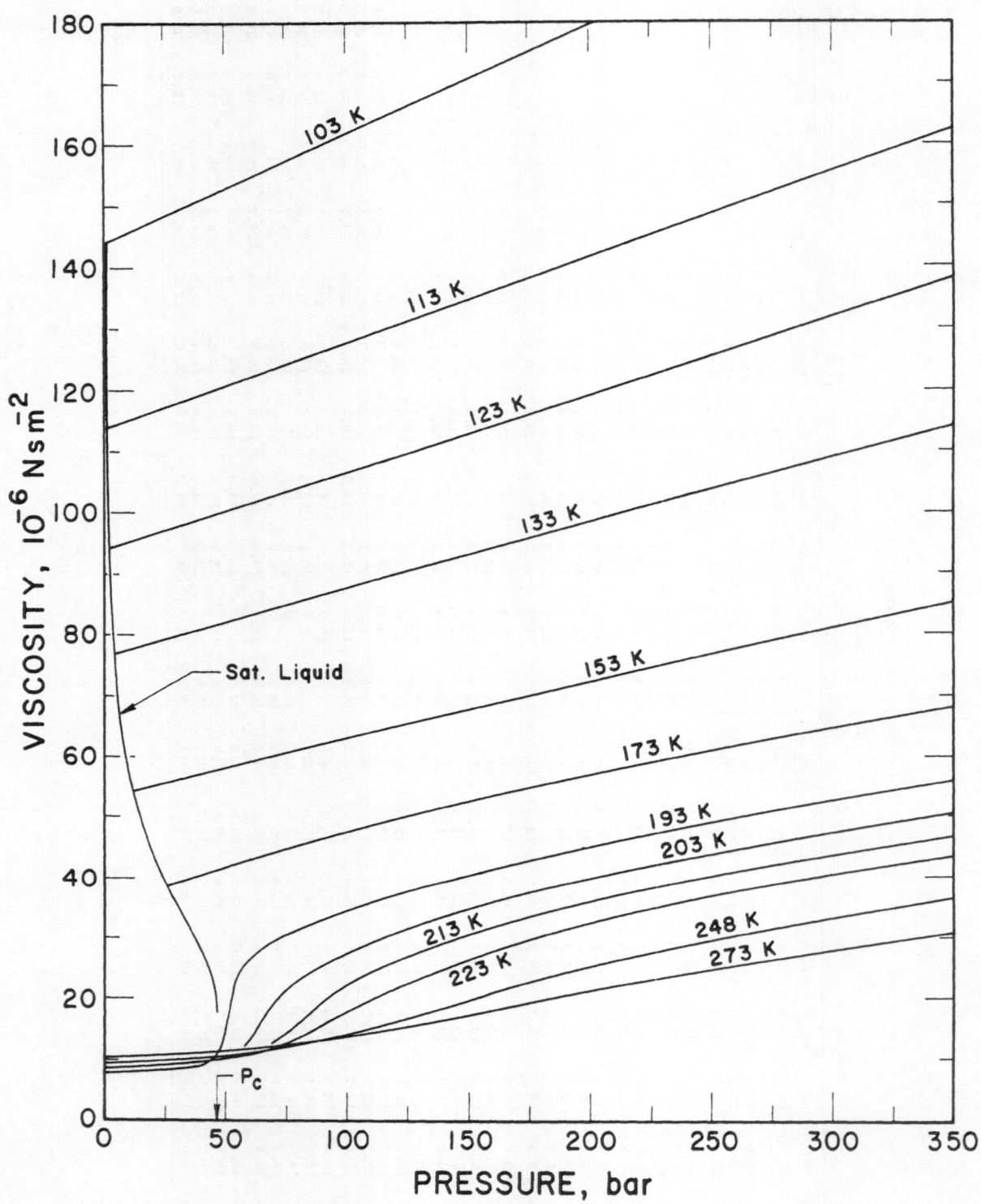

FIGURE I A. VISCOSITY OF METHANE [85].

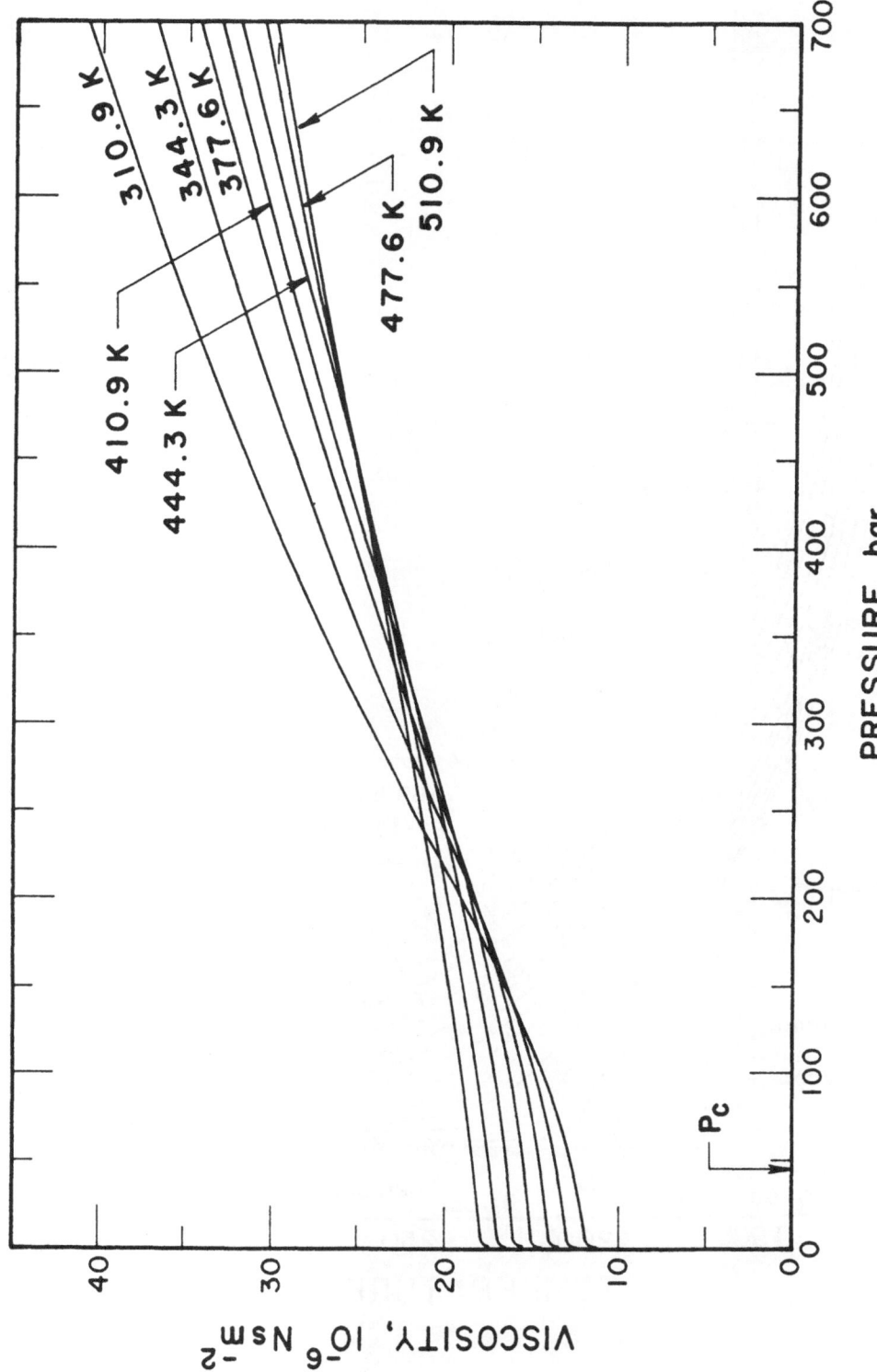

FIGURE 1B. VISCOSITY OF METHANE [62].

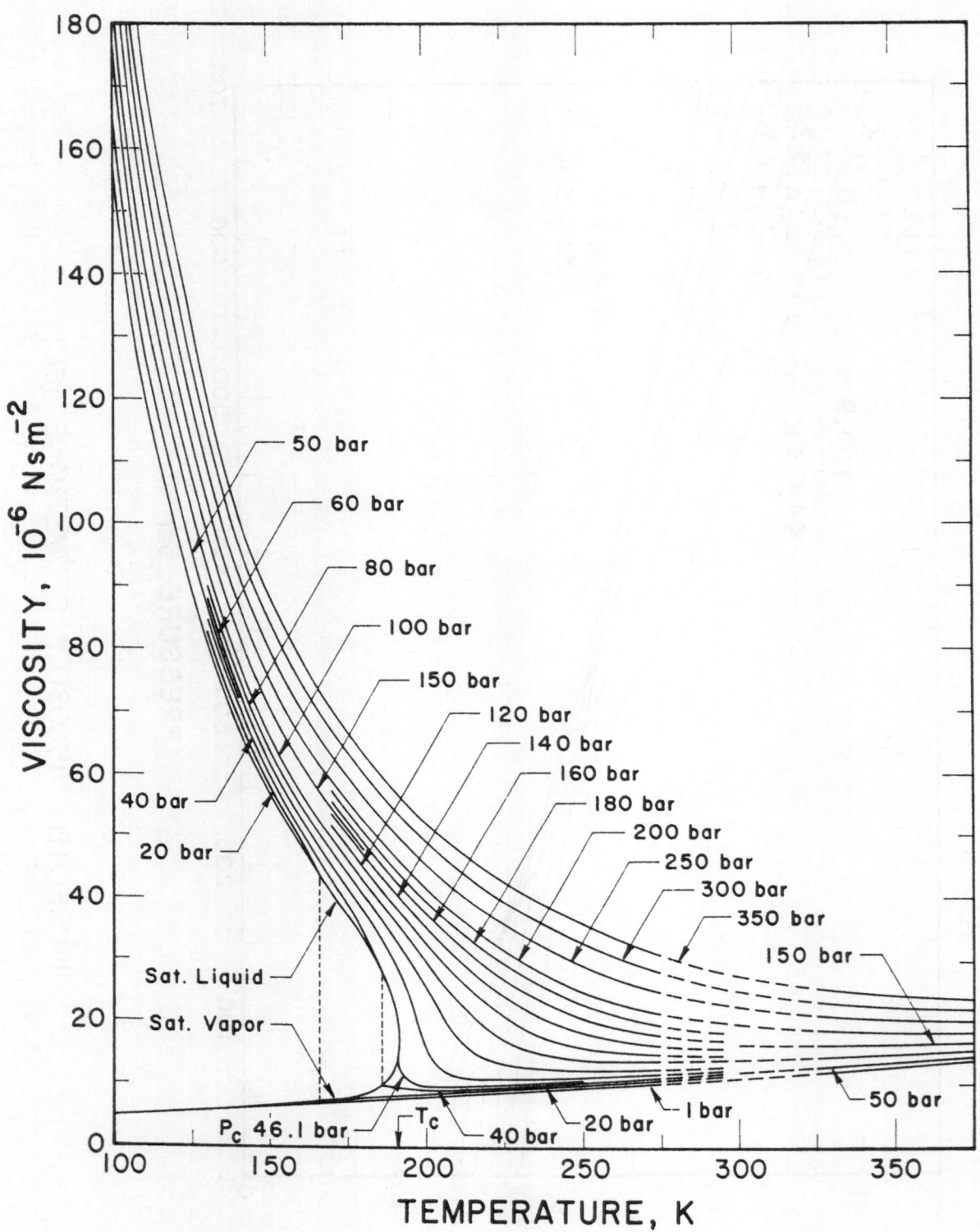

FIGURE 2A. VISCOSITY OF METHANE [62, 85].

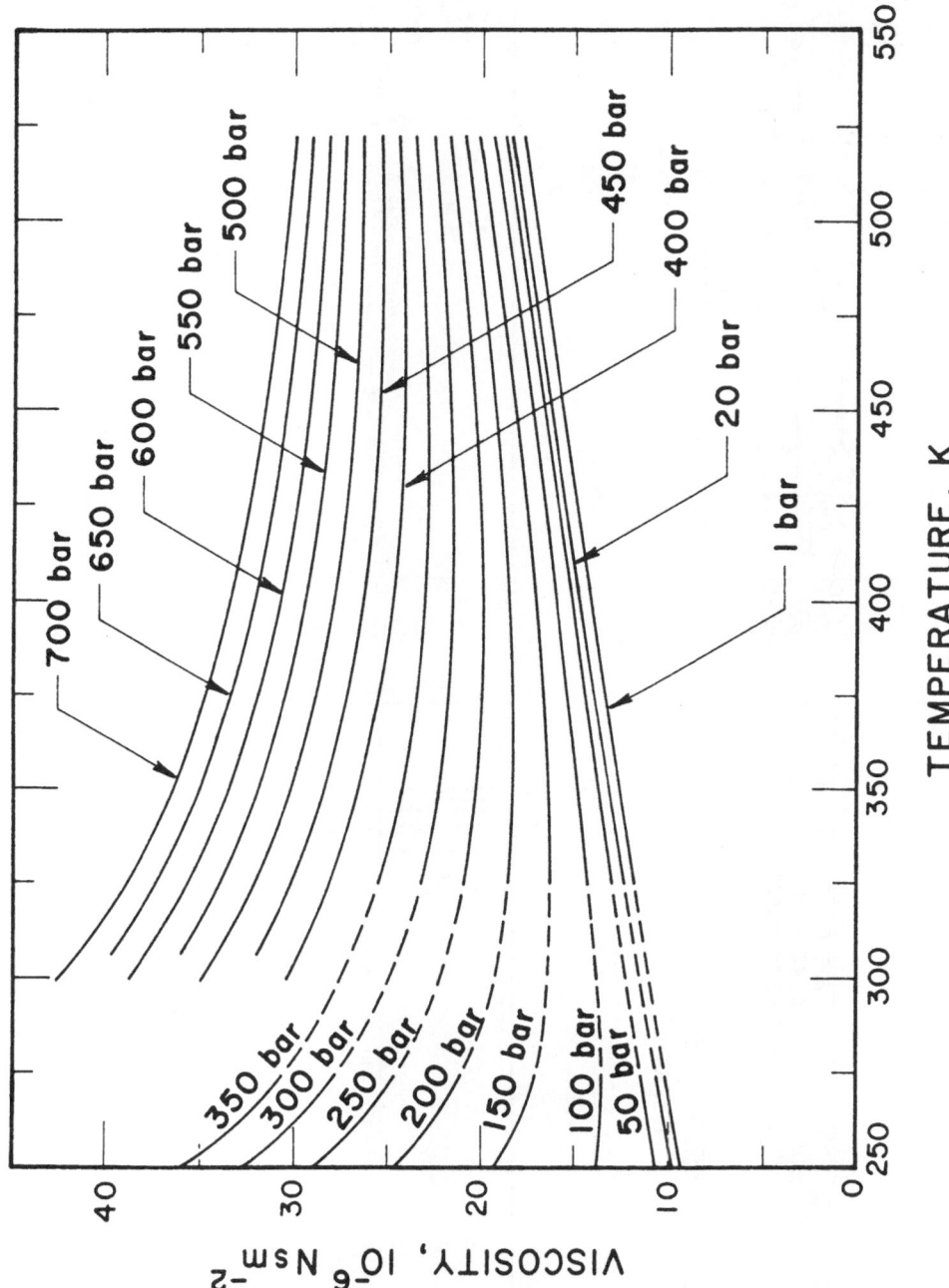

FIGURE 2B. VISCOSITY OF METHANE [62].

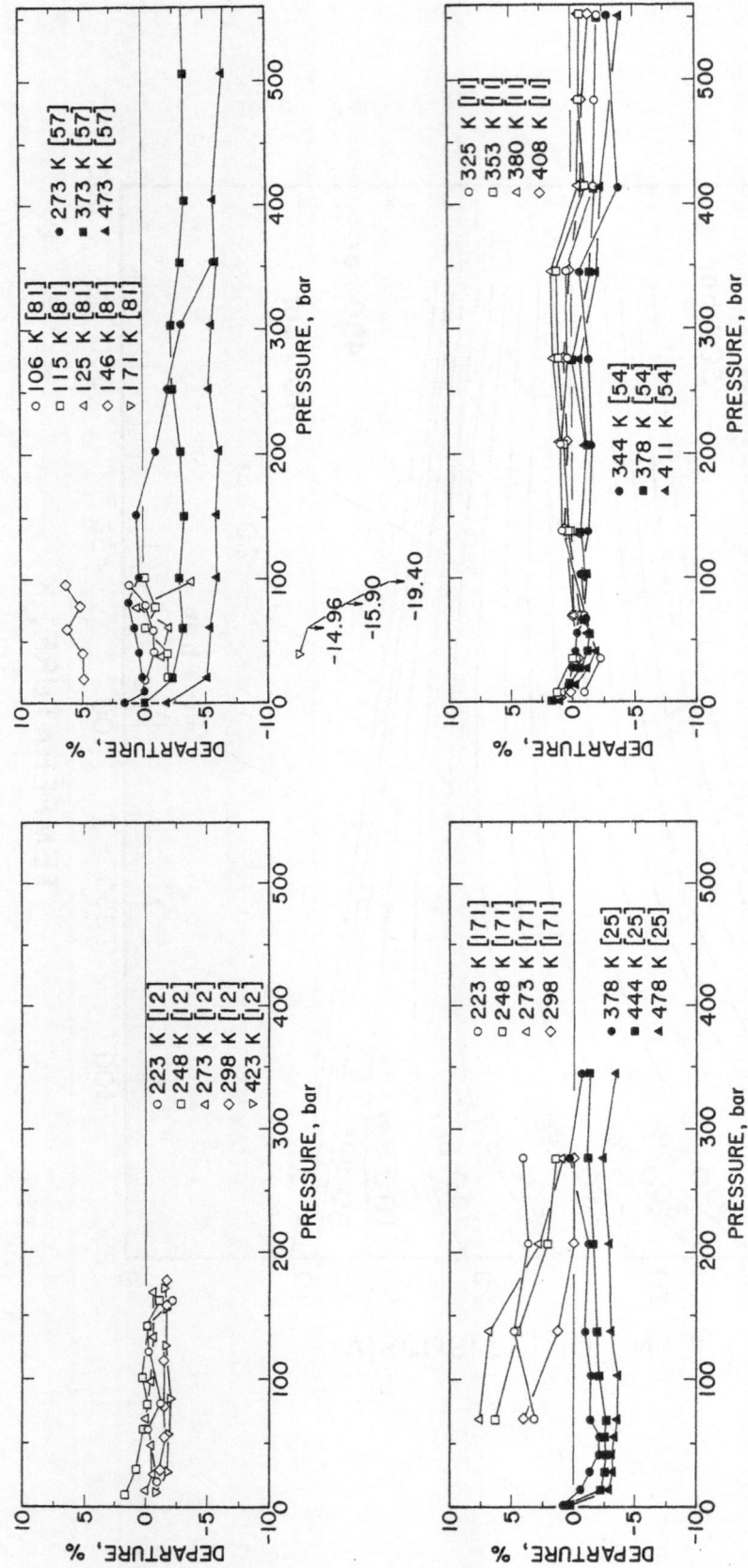

FIGURE 3. DEPARTURE PLOT ON THE VISCOSITY OF METHANE.

METHANOL

The data of [57] and [87], both obtained by a capillary tube viscosimeter, were used to generate the recommended values given below. Figure 1 shows the data of the two authors as isotherms against pressure. The full lines up to 800 bar are the data of [57], while those at lower temperatures extending only to about 245 bar are reported by [87]. In the low temperature region considered, the isotherms have only a very slight curvature. Therefore, the measured isotherms of [87] have been extrapolated to high pressures, as shown by the dashed lines in Figure 1. Slight extrapolations to the vapor pressure from [127] gave the saturated liquid line, while the saturated vapour line was estimated from the viscosity of the dilute gas at the boiling temperature and the critical viscosity as estimated from the saturated liquid line. Dilute gas values were taken from [195]. Saturated liquid and saturated vapour lines thus estimated are not highly reliable and are therefore included as dashed lines in the figures. In Figure 2, the data of Figure 1 are plotted as isobars against temperature. Extrapolated values are again identified by dashed lines. The recommended values were read from this diagram. The uncertainty is estimated to be ± 6%.

VISCOSITY OF METHANOL
$[\mu, 10^{-6} \text{ N s m}^{-2}]$

T,K	Pressure, bar													
	1	50	100	150	200	250	300	350	400	450	500	600	700	800
290	605	622	632	650	662	672	685	700	715	727	740	755	775	795
300	535	547	560	572	585	600	612	625	637	652	665	682	700	720
310	470	482	492	502	515	527	540	552	562	575	587	605	625	650
320	412	425	432	442	455	467	475	487	500	507	575	530	550	572
330	365	375	382	392	402	412	422	432	442	450	460	475	492	515
340	11.2	332	340	350	360	367	377	385	394	402	412	425	440	457
350	11.5	292	300	310	320	327	337	345	352	362	370	385	400	415
375	12.3	222	227	237	245	252	260	270	275	280	287	297	310	325
400	13.2	170	175	185	192	200	205	210	211	222	227	240	250	262
425	14.0	132	140	147	157	165	170	175	180	185	190	200	210	222
450	14.8	107	112	120	130	135	142	147	152	157	162	170	180	190
475	15.6	87.5	92.5	100	107	112	120	125	130	132	137	147	155	165
500	16.5	17.0	70.0	80.0	87.5	92.5	100	105	110	112	117	125	135	142
525	17.3	18.0	33.0	58.0	70.0	77.5	82.5	87.5	92.5	95.0	100	110	120	127
550	18.1	19.0	21.0	40.0	55.0	60.0	67.5	72.5	77.5	82.5	87.5	97.5	107	115

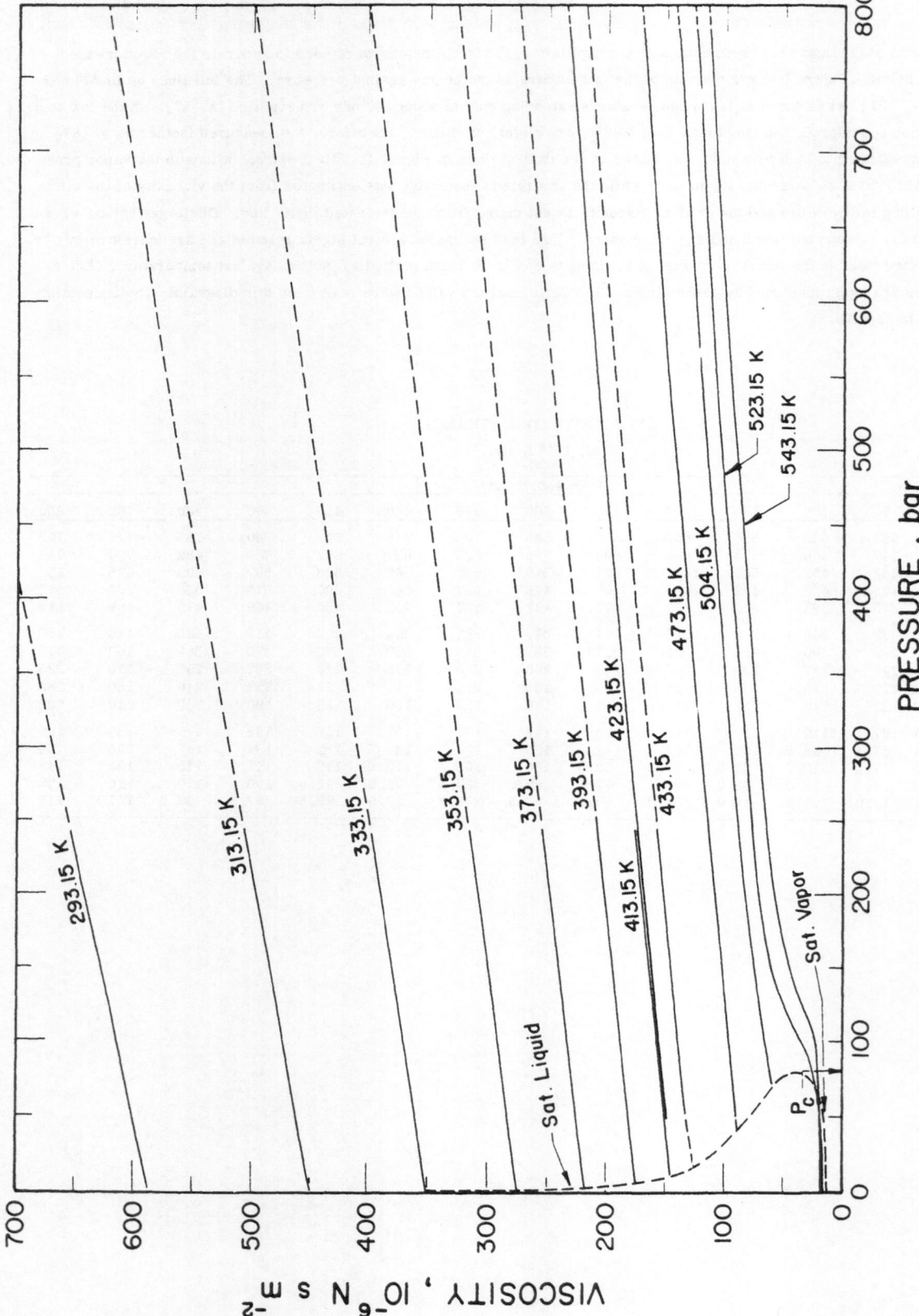

FIGURE 1. VISCOSITY OF METHANOL [57, 87].

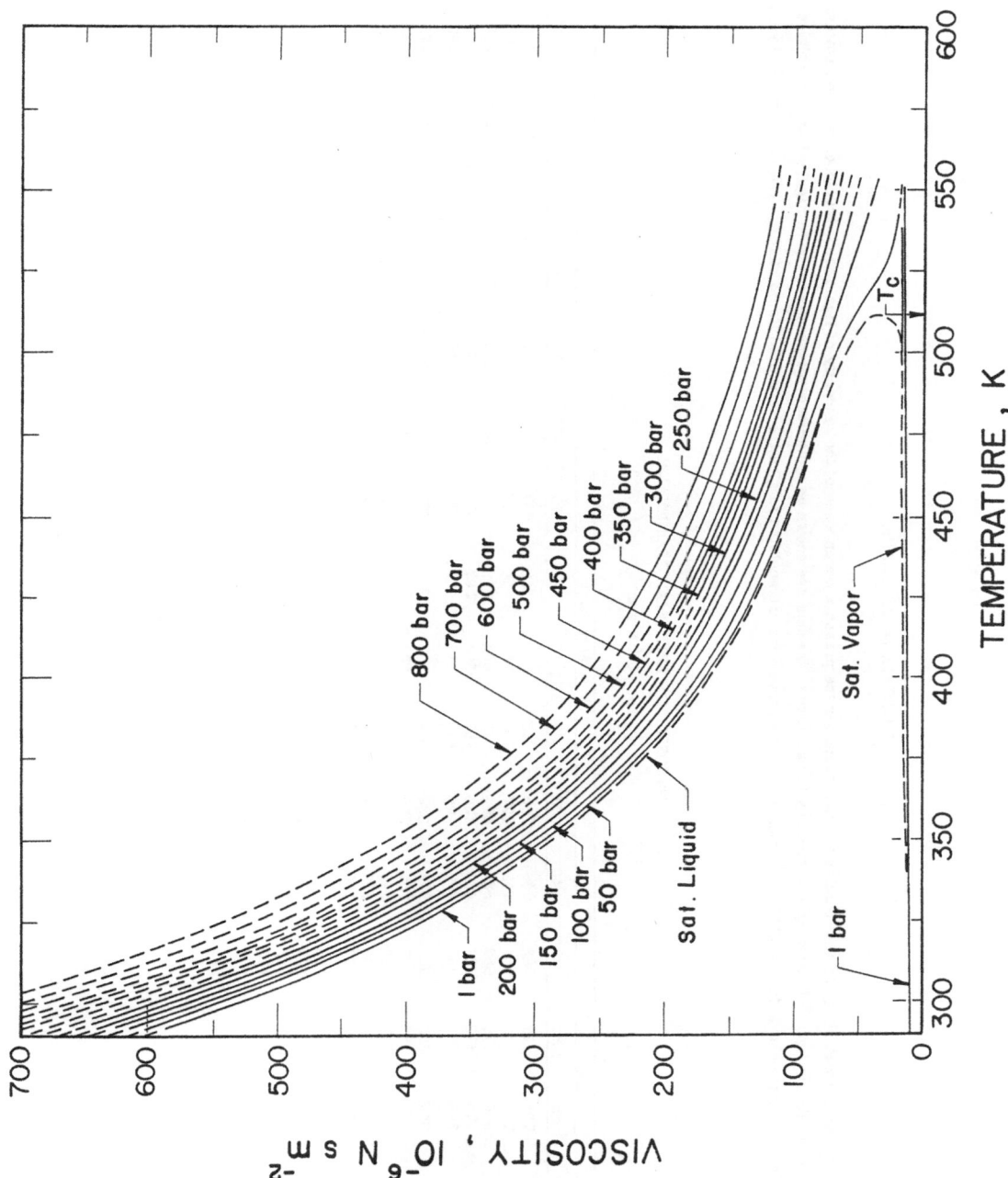

FIGURE 2. VISCOSITY OF METHANOL [57, 87].

METHYLCYCLOHEXANE

Only one set of data by Guseinov et al. [73] was found in the literature for the pressure dependence of the viscosity of liquid methylcyclohexane. No statements about the experimental procedure used nor the estimate of accuracy are given by the authors. The data are shown as isotherms against pressure in Figure 1 and as isobars against temperature in Figure 2. Using the vapour pressure data of [127], the measured isotherms were extrapolated to give the saturated liquid states. The recommended values are presented below.

VISCOSITY OF METHYLCYCLOHEXANE

$[\mu,\ 10^{-6}\ N\ s\ m^{-2}]$

T, K	Pressure, bar																			
	10	15	20	25	30	35	40	50	60	70	80	90	100	150	200	250	300	350	400	500
290	757	762	767	772	777	783	786	795	805	813	822	837	840	882	925	967	1010	1050	1095	11625
300	687	695	702	706	710	714	719	727	735	744	754	763	772	815	852	895	925	965	1010	1090
320	557	565	572	576	579	583	586	593	600	609	619	628	637	680	712	750	780	815	852	930
340	455	467	467	470	473	476	480	486	492	500	507	515	522	550	580	612	640	670	700	772
360	377	382	387	390	393	396	399	404	410	416	422	428	435	457	480	512	537	562	587	640
380	317	321	325	327	329	332	334	338	342	347	352	357	362	382	402	430	450	475	497	545
400	267	271	275	277	280	282	285	290	295	298	301	304	307	325	345	367	387	407	427	470
420	225	228	232	235	237	240	242	247	252	257	261	265	270	285	300	320	340	357	375	412
440	190	193	197	200	203	206	209	214	220	224	227	231	235	252	265	285	300	317	332	367
460	162	166	170	172	175	177	180	185	190	194	199	203	207	222	235	252	270	285	300	330
480	–	143	147	150	152	155	157	162	167	171	175	178	182	197	212	227	242	257	272	300
500	–	125	130	131	133	135	137	141	145	149	154	158	162	175	190	210	222	237	250	275
520	–	–	112	115	117	119	121	125	130	133	137	141	145	160	177	192	207	220	232	255
530	–	–	105	107	110	112	115	120	125	128	131	134	137	155	170	187	200	215	227	247

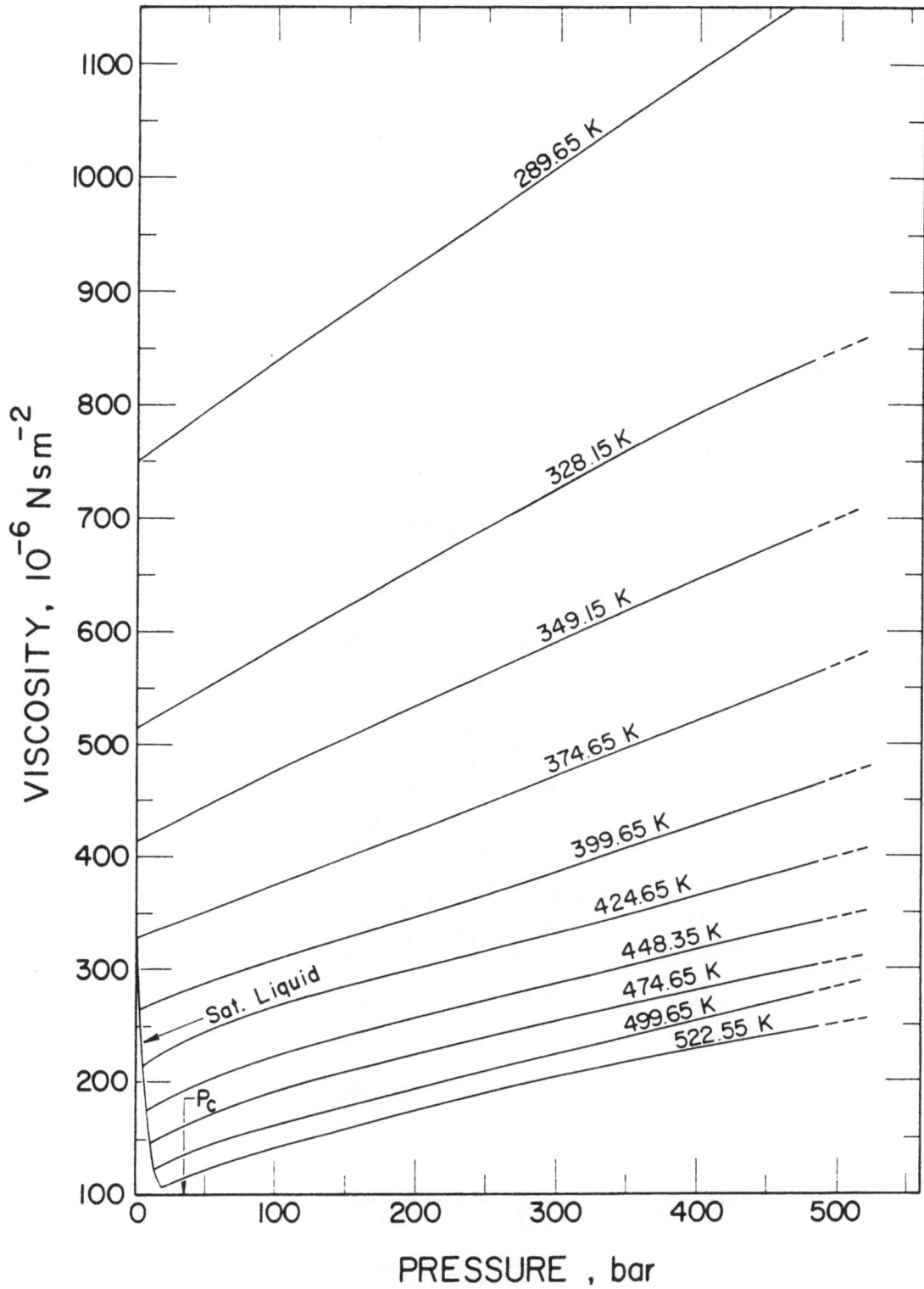

FIGURE 1. VISCOSITY OF METHYLCYCLOHEXANE [73].

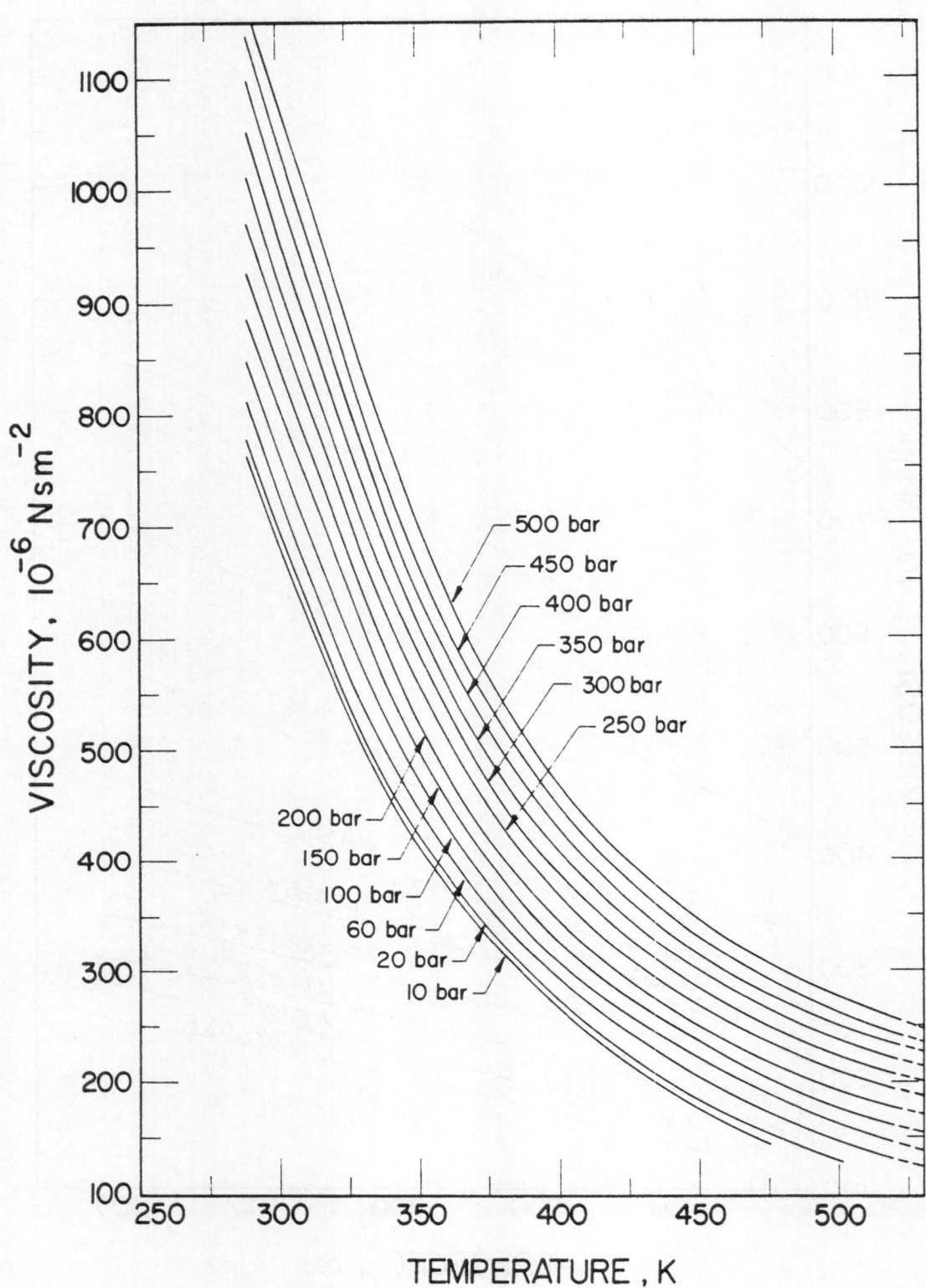

FIGURE 2. VISCOSITY OF METHYLCYCLOHEXANE [73].

NEON

Of the several sets of data found in the literature on the pressure dependence of the viscosity of gaseous neon, only one reference appears to exist in the liquid state. The data of [49, 82 and 197], were used along with the slightly adjusted dilute gas viscosities of [195], to generate a residual viscosity curve. The residual concept is obeyed within a few percent. Using the P-V-T data of [53, 144 and 187], viscosity values as a function of temperature and pressure were generated over a large region of states. A selection of the data are plotted as isotherms against pressure in Figure 1. Here the saturated liquid line was found by extrapolation of the liquid isotherms to the vapor pressure. Representations of viscosity against temperature for constant pressures are shown in Figures 2A and 2B. The saturated vapor line was taken from [195]. In the liquid region, some extrapolation to higher pressures was made as shown by the dashed lines. The final set of recommended values are tabulated on the next two pages.

A comparison was made of the recommended values with the data of [49, 168 and 197]. All these values are in the gaseous region and have only a minor pressure dependence. It is thus not too surprising that the agreement is excellent. In the low temperature region, where the pressure dependence is significant, only the data of [182] are available and thus no comparison with other authors is possible. In the temperature region above 200 K, uncertainty of the recommended values is believed to be within ± 2%. Below 200 K, and especially in the liquid region, it is at best difficult to make a reliable estimate of uncertainties and a provisional figure of ± 10% is assumed at this time.

VISCOSITY OF NEON

$[\mu, 10^{-6} \text{ N s m}^{-2}]$

T, K	Pressure, bar												
	1	10	15	20	25	27.2*	30	35	40	45	50	60	70
30	5.0	109	111	112	114	115	115	117	118.2	120	120	123	126
35	6.0	73.2	74.4	76.0	77.8	78.7	79.6	80.8	82.0	83.5	85.0	88.0	91.5
40	6.8	7.0	44.0	46.0	48.0	49.0	50.0	51.5	53.0	55.5	58.0	60.0	63.6
45	7.5	7.8	8.1	8.8	11.0	13.2	22.6	29.3	33.0	36.3	37.0	43.5	46.7
50	8.3	8.5	8.8	9.2	9.7	10.0	10.4	12.0	13.8	16.6	18.8	25.1	27.8
55	9.0	9.0	9.1	9.3	9.9	10.1	10.1	10.9	11.7	12.8	14.6	16.6	20.8
60	9.8	9.8	10.0	10.2	10.4	10.6	10.7	11.1	11.5	12.2	13.4	14.6	16.6
65	10.6	10.6	10.8	10.9	11.1	11.2	11.3	11.6	11.8	12.3	12.8	14.0	15.4
70	11.2	11.2	11.2	11.4	11.6	11.7	11.8	12.1	12.3	12.8	13.0	14.0	15.0
75	11.8	11.8	11.9	12.0	12.2	12.3	12.4	12.6	12.8	13.1	13.4	14.2	14.8
80	12.4	12.5	12.6	12.7	12.8	12.8	13.0	13.2	13.3	13.7	13.8	14.5	15.0
85	13.0	13.0	13.0	13.7	13.2	13.4	13.4	13.6	13.8	14.0	14.2	14.6	15.2
90	13.6	13.6	13.6	13.6	13.7	13.9	13.9	14.0	14.3	14.4	14.6	14.9	15.4
95	14.1	14.1	14.1	14.2	14.3	14.4	14.4	14.6	14.8	14.9	15.1	15.4	15.7
100	14.6	14.6	14.6	14.7	14.8	14.9	14.9	15.1	15.2	15.4	15.5	15.9	16.0
110	15.6	15.7	15.7	15.7	15.8	15.9	15.9	16.0	16.1	16.2	16.4	16.6	16.8
120	16.7	16.7	16.7	16.7	16.8	16.9	16.9	16.9	17.0	17.1	17.3	17.5	17.8
130	17.8	17.8	17.9	17.9	18.0	18.0	18.1	18.1	18.2	18.2	18.3	18.5	18.6
140	18.7	18.7	18.7	18.7	18.8	18.8	18.9	18.9	19.0	19.1	19.2	19.3	19.5
150	19.6	19.7	19.7	19.8	19.8	19.8	19.9	19.9	20.0	20.0	20.1	20.2	20.4
160	20.6	20.6	20.7	20.7	20.8	20.8	20.8	20.9	20.9	21.0	21.0	21.1	21.2
170	21.5	21.6	21.7	21.7	21.7	21.7	21.8	21.8	21.9	21.9	22.0	22.0	22.1
180	22.4	22.5	22.5	22.6	22.6	22.6	22.6	22.7	22.7	22.7	22.8	22.8	22.9
190	23.2	23.3	23.3	23.3	23.4	23.4	23.4	23.4	23.5	23.5	23.6	23.7	23.8
200	24.1	24.1	24.1	24.1	24.1	24.1	24.2	24.2	24.2	24.3	24.4	24.5	24.6
220	25.7	25.7	25.7	25.7	25.8	25.8	25.8	25.9	25.9	26.0	26.0	26.0	26.1
240	27.3	27.3	27.3	27.4	27.4	27.4	27.4	27.4	27.5	27.5	27.6	27.7	27.8
260	28.7	28.8	28.8	28.9	28.9	28.9	29.0	29.1	29.1	29.2	29.2	29.2	29.3
280	30.3	30.3	30.3	30.4	30.4	30.4	30.4	30.4	30.5	30.5	30.6	30.6	30.7
300	31.7	31.7	31.8	31.8	31.8	37.9	31.9	31.9	32.0	32.0	32.0	32.0	32.0
325	33.4	33.4	33.5	33.5	33.6	33.6	33.7	33.7	33.7	33.7	33.8	33.8	33.9
350	35.0	35.0	35.1	35.1	35.2	35.2	35.2	35.2	35.3	35.3	35.3	35.3	35.4
375	36.7	36.8	36.9	37.0	37.1	37.1	37.1	37.1	37.2	37.2	37.2	37.2	37.3
400	38.4	38.4	38.4	38.5	38.5	38.5	38.5	38.5	38.6	38.6	38.6	38.6	38.7
450	41.6	41.6	41.6	41.6	41.6	41.6	41.7	41.7	41.7	41.7	41.7	41.7	41.8
500	44.6	44.6	44.6	44.6	44.6	44.7	44.7	44.7	44.7	44.7	44.7	44.7	44.7

*Critical pressure.

VISCOSITY OF NEON (continued)

T,K	Pressure, bar												
	80	90	100	110	120	130	140	150	160	170	180	190	200
30	129	132	135	137	139	141	143	145	147	149	151	153	155
35	95.0	97.5	100	102	104	106	108	110	112	114	116	118	120
40	66.1	68.6	71.6	74.0	76.0	78.0	80.0	82.0	84.0	86.0	88.0	90.0	92.5
45	49.7	52.2	54.0	56.8	59.0	61.1	63.2	65.1	67.0	69.0	71.0	73.0	75.0
50	33.0	36.3	38.4	41.7	44.0	46.4	48.7	51.0	53.0	55.0	57.2	59.4	60.5
55	20.6	25.8	26.4	30.9	32.8	35.6	37.8	39.0	42.0	44.0	46.0	47.6	50.0
60	18.6	20.8	21.4	24.8	27.0	28.4	30.1	32.0	33.6	35.0	37.0	38.6	40.0
65	17.0	18.4	19.6	21.4	23.0	24.4	25.9	27.0	28.8	30.4	31.6	33.0	34.6
70	16.0	17.0	18.3	19.4	20.8	21.9	23.2	24.4	25.6	27.0	27.9	29.0	30.4
75	15.6	16.5	17.6	18.4	19.4	20.5	21.5	22.6	23.6	24.6	25.6	26.6	27.4
80	15.6	16.4	17.4	17.9	18.8	19.6	20.5	21.4	22.3	23.2	24.0	24.9	25.4
85	15.8	16.3	17.2	17.6	18.4	19.0	19.9	20.6	21.3	22.0	22.8	23.6	24.4
90	16.0	16.2	17.1	17.6	18.2	18.7	19.2	20.0	20.6	21.4	22.0	22.7	23.4
95	16.3	16.7	17.2	17.7	18.2	18.6	19.1	19.8	20.4	21.0	21.6	22.1	22.7
100	16.6	17.0	17.4	17.8	18.3	18.8	19.3	19.6	20.3	20.8	21.5	21.8	22.2
110	17.2	17.6	18.0	18.3	18.7	19.1	19.4	19.8	20.3	20.7	21.2	21.5	21.8
120	18.0	18.3	18.6	18.9	19.2	19.6	19.9	20.2	20.7	21.0	21.4	21.6	21.8
130	18.9	19.1	19.4	19.7	19.8	20.2	20.5	20.7	21.2	21.5	21.8	21.9	22
140	19.7	20.0	20.1	20.4	20.5	20.9	21.1	21.4	21.7	22.0	22.3	22.4	22.5
150	20.6	20.8	27.0	21.2	21.4	21.6	21.8	22.0	22.3	22.6	22.8	22.9	23.0
160	21.4	21.6	21.7	21.9	22.0	22.3	22.6	22.8	23.0	23.2	23.4	23.5	23.6
170	22.3	22.4	22.6	22.7	22.8	23.1	23.3	23.5	23.7	23.8	24.0	24.2	24.3
130	23.1	23.2	23.3	23.4	23.6	23.8	24.0	24.2	24.4	24.6	24.7	24.9	25.0
190	23.9	24.0	24.0	24.1	24.2	24.4	24.6	24.8	25.0	25.2	25.3	25.5	25.6
200	24.6	24.7	24.8	25.0	25.2	25.3	25.4	25.8	25.9	26.0	26.1	26.2	26.4
220	26.2	26.3	26.4	26.5	26.7	26.8	26.9	27.0	27.2	27.3	27.4	27.6	27.7
240	27.9	28.0	28.0	28.1	28.1	23.2	28.3	28.9	28.5	28.7	28.8	28.9	29.0
260	29.4	29.4	29.4	29.6	29.6	29.7	29.8	29.8	30.0	30.1	30.2	30.3	30.4
280	30.7	30.8	30.8	30.9	31.0	31.0	31.1	31.2	31.3	31.5	31.6	31.7	31.8
300	32.1	32.2	32.2	32.3	32.4	32.4	32.5	32.6	32.7	32.8	32.9	33.0	33.0
325	33.9	34.0	34.0	34.7	34.1	34.2	34.3	34.4	34.4	34.5	34.6	34.7	34.7
350	35.4	35.5	35.6	35.6	35.7	35.8	35.8	35.9	36.0	36.1	36.1	36.2	36.2
375	37.3	37.3	37.4	37.4	37.4	37.4	37.4	37.5	37.5	37.6	37.6	37.7	37.7
400	38.7	38.7	38.8	38.8	38.8	38.9	39.0	39.0	39.1	39.1	39.2	39.2	39.2
450	41.8	41.8	41.8	41.8	41.8	41.9	41.9	41.9	41.9	41.9	42.0	42.0	42.0
500	44.7	44.7	44.7	44.7	44.7	44.7	44.7	44.7	44.7	44.8	44.8	44.8	44.8

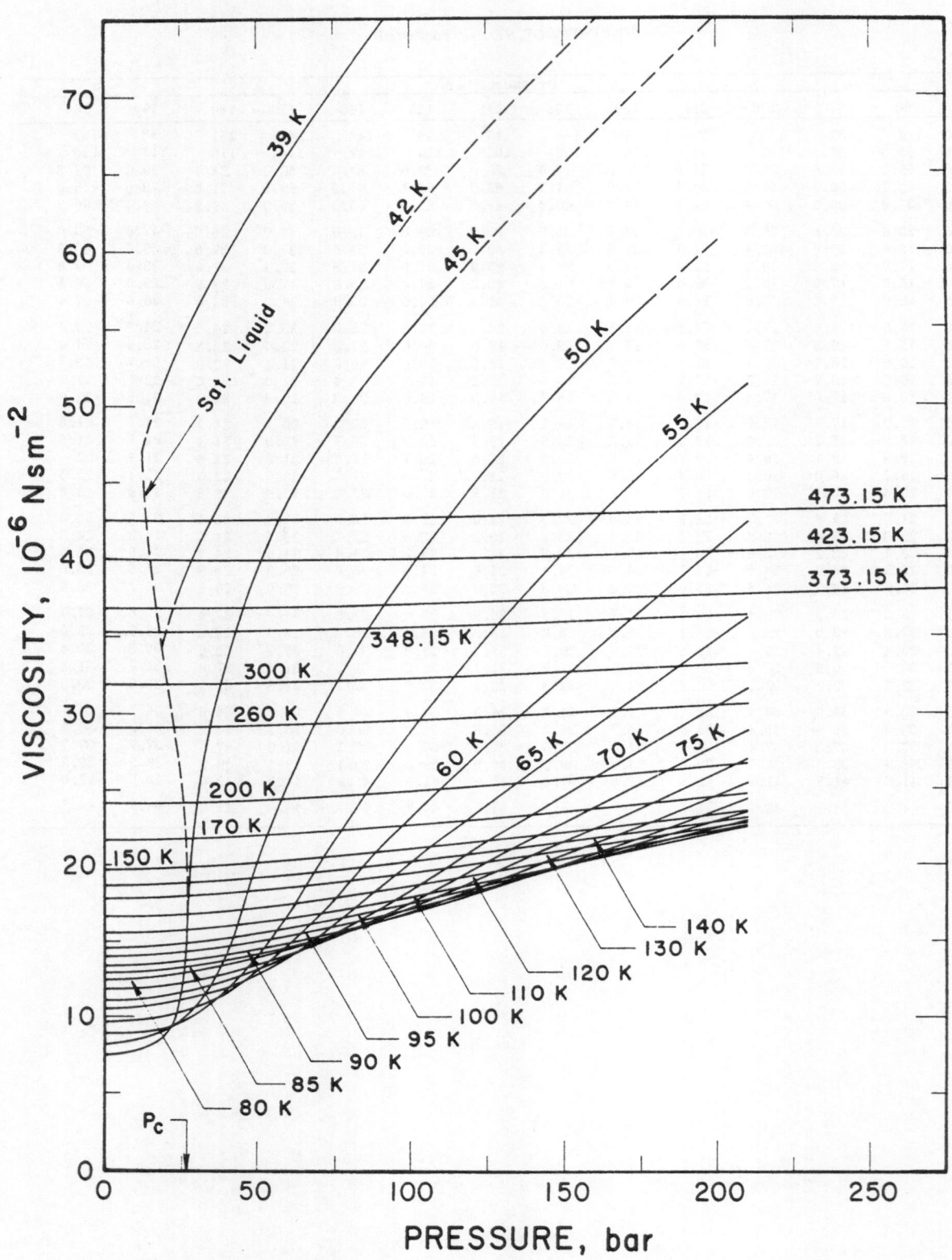

FIGURE 1. VISCOSITY OF NEON [49, 82, 197].

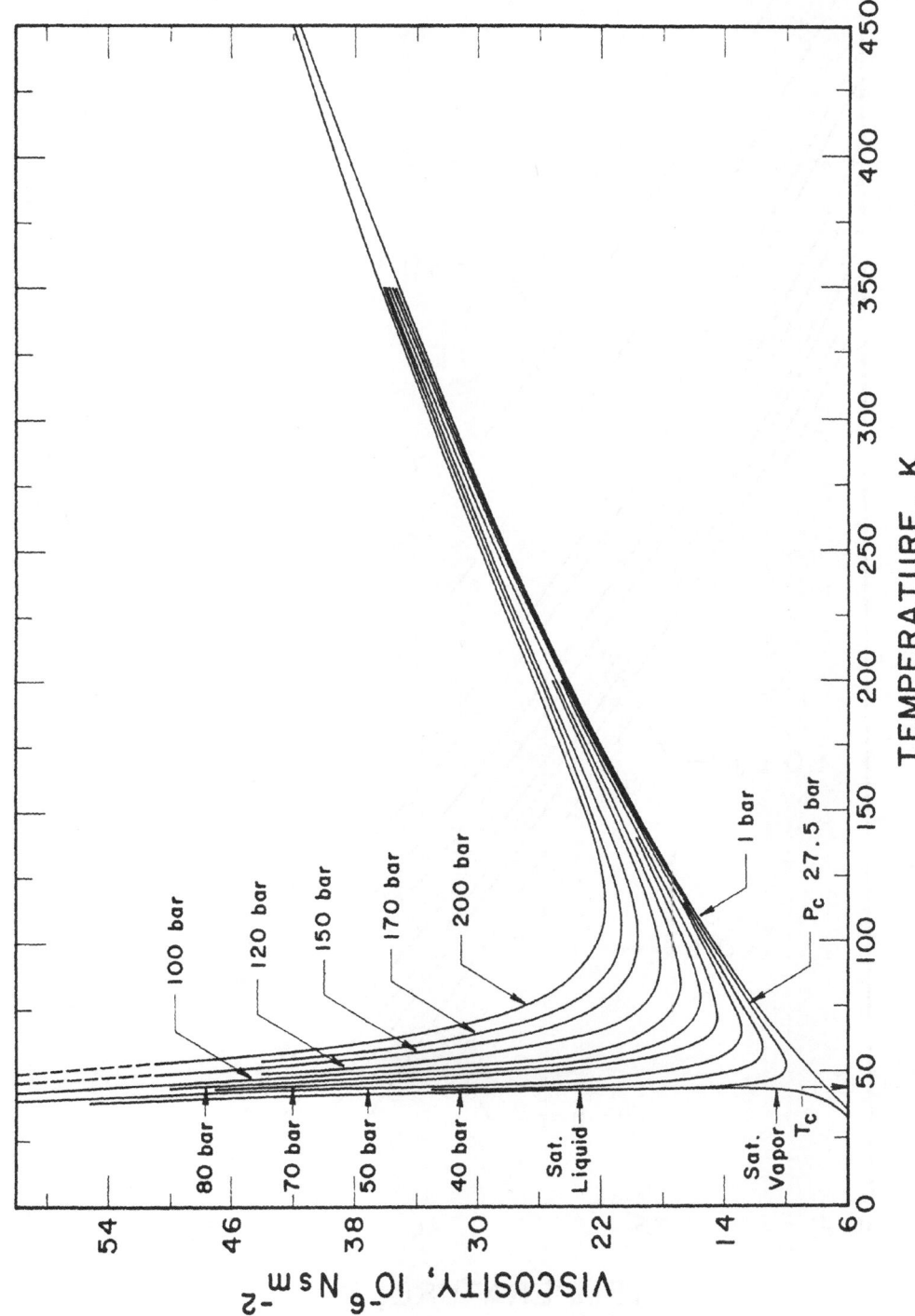

FIGURE 2A. VISCOSITY OF NEON [49, 82, 197].

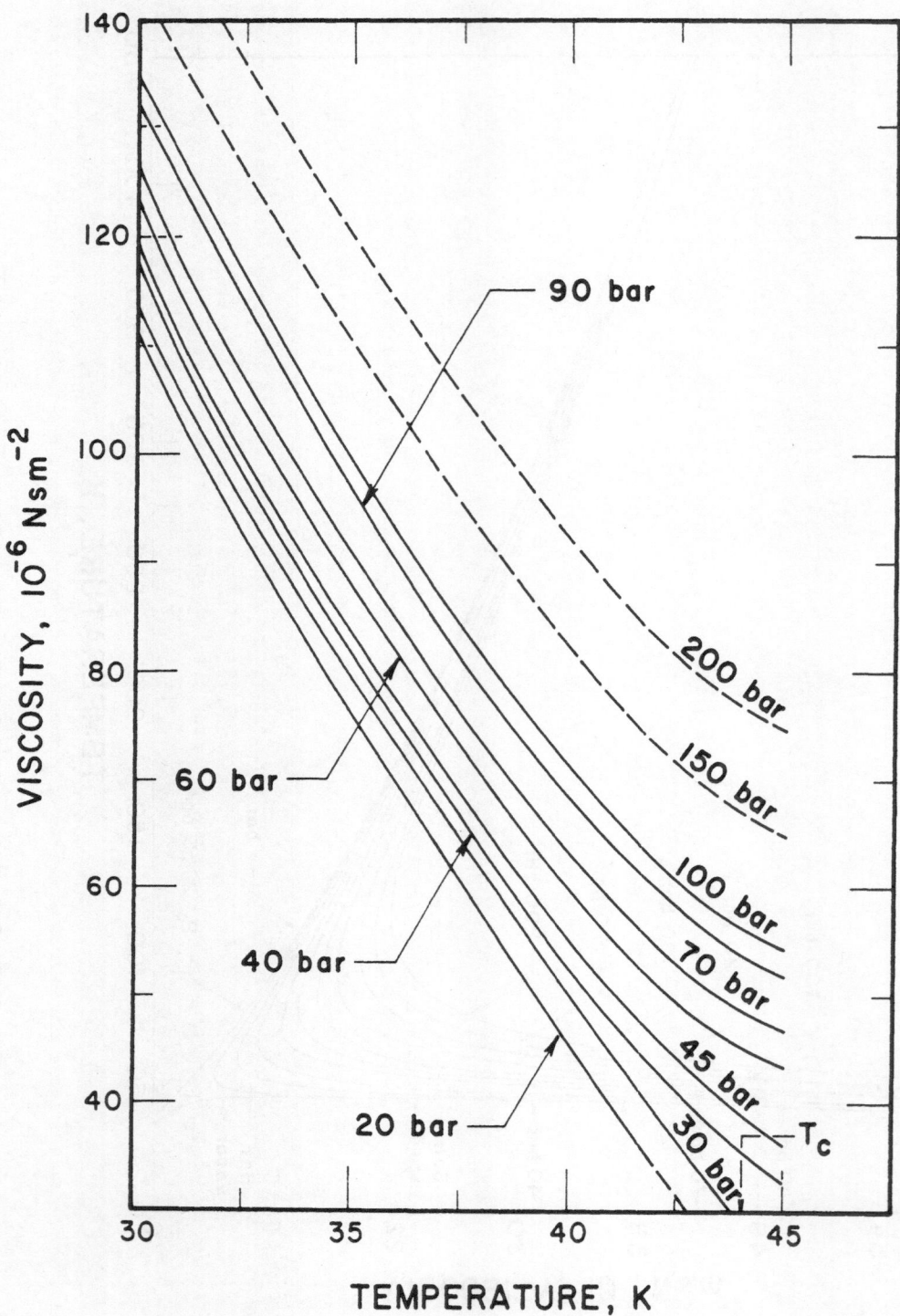

FIGURE 2B. VISCOSITY OF NEON [49, 82, 197].

NITROGEN

A large amount of data exists for the viscosity of nitrogen in all fluid states. The recommended values tabulated on the next two pages were taken from [203]. They cover a temperature range from 65 K to 1300 K and a pressure range from 1 bar to 1000 bar. Figure 1 shows these smoothed data plotted as isotherms against pressure. The saturated liquid line was found by extrapolation of the isotherms to the vapour pressure, the saturated vapour line was taken from [195]. Figure 2 shows the data as isobars plotted against temperature.

Additional works on the pressure dependence of the viscosity of nitrogen are listed in the summary table below. An extensive comparison between the recommended values and these other works is shown in the summary table below. In the dense gas region the agreement is generally quite satisfactory. The uncertainty in this region is estimated to be 2%. The agreement is considerably worse in the liquid region, where the viscosity values of van Itterbeck et al. [188] and Grevendonk et al. [67] are higher than the recommended values by 5 to 10%. Since the formulated data of [76] lie between the recommended values and those of the two cited authors, we estimate the accuracy of the recommended values in this work to be around ± 5%, with the possibility that these values may turn out to be a little too low. A detailed comparison of the other authors' data and the recommended values can also be seen in the departure plots given in Figures 3A and 3B.

ADDITIONAL REFERENCES ON THE VISCOSITY OF NITROGEN

Authors	Year	Ref. No.	Temperature K	Pressure	Method	Departure % (no. points)
Golubev and Kurin	1974	59	273-423	1-4000	Capillary tube	-0.775 ± 1.201 (25)
Latto and Saunders	1972	129	80-450	1-172	Capillary tube	-0.187 ± 1.035 (63)
Shepeleva and Golubev	1971	179	83-276	9-506	Capillary tube	0.78 ± 3.34 (16)
Grevendonk et al.	1969	67	66.5-123	6-195	Torsional cryst.	6.21 ± 4.28 (49)
Gracki et al.	1969	66	183-298	6-155	Capillary tube	0.024 ± 1.143 (18)
Chierici and Paratella	1969	33	323	6-913	Capillary tube	-0.168 ± 0.765 (12)
Keo and Kobayashi	1967	97	183-323	10-506	Capillary tube	0.016 ± 0.855 (35)
Van Itterbeck et al.	1966	88	70-90	0.5-99	Oscillating disk	4.26 ± 3.16 (34)
Makavetskas et al.	1963	136	285-813	16-599	Capillary tube	5.319 ± 3.44 (32)
Flynn et al.	1963	49	195-373	6-178	Capillary tube	0.34 ± 0.61 (34)
Goldman	1963	56	195-298	78-127	Capillary tube	1.52 ± 1.56 (16)
Michels and Gibson	1962	147	298-348	10-965	Capillary tube	0.28 ± 0.64 (42)
Baron et al.	1959	11	342-426	7-552	Capillary tube	1.426 ± 1.637 (40)
Ross and Brown	1957	171	223-298	34-689	Capillary tube	-1.47 ± 3.15 (40)

VISCOSITY OF NITROGEN

$[\mu,\ 10^{-6}\ N\ s\ m^{-2}]$

T, K	Pressure, bar												
	1	10	20	30	40	50	60	70	80	90	100	125	150
65	263.3	267.7	272.8	277.7	282.6	287.6	292.7	297.8	302.7	-	-	-	-
70	211.3	215.4	219.7	224.4	229.0	233.6	238.3	242.9	247.5	252.3	256.8	268.7	280.8
75	167.7	171.3	175.3	179.3	183.2	187.5	191.4	195.7	199.8	204.1	208.4	219.1	230.1
80	5.52	136.6	140.0	143.3	146.7	150.1	153.7	157.3	160.9	164.5	168.2	177.7	187.2
85	5.86	111.1	113.6	116.4	119.2	121.9	124.8	127.8	130.8	133.8	136.9	144.8	153.0
90	6.20	93.26	95.28	97.37	99.51	101.8	104.0	106.3	108.7	111.1	113.6	120.0	126.7
95	6.54	80.93	82.66	84.29	86.02	87.77	89.51	91.35	93.19	95.07	97.00	102.0	107.5
100	6.88	71.30	72.97	74.59	76.16	77.72	79.22	80.76	82.29	83.86	85.42	89.49	93.76
105	7.22	7.83	63.97	65.84	67.60	69.29	70.89	72.42	73.92	75.39	76.81	80.34	83.95
110	7.56	8.11	55.17	57.28	59.25	61.12	62.94	64.69	66.36	67.94	69.48	73.08	76.44
115	7.89	8.40	46.26	49.02	51.40	53.54	55.53	57.37	59.10	60.78	62.43	66.36	69.93
120	8.21	8.68	9.53	39.74	43.35	46.12	48.49	50.41	52.55	54.37	56.08	60.05	63.76
125	8.52	8.96	9.68	11.55	33.69	38.10	41.35	44.07	46.33	48.36	50.23	54.45	58.21
130	8.83	9.25	9.88	11.07	18.05	29.45	34.24	37.63	40.33	42.70	44.79	49.31	53.23
135	9.14	9.54	10.10	11.03	13.14	19.98	26.86	31.31	34.62	37.30	39.70	44.27	48.63
140	9.45	9.83	10.28	11.12	12.50	15.51	20.68	25.44	29.23	32.29	34.87	40.12	44.38
150	10.06	10.41	10.85	11.45	12.32	13.65	15.70	18.47	21.47	24.34	26.96	32.48	36.94
160	10.66	10.98	11.38	11.86	12.52	13.40	14.60	16.14	18.00	20.04	22.11	26.98	31.26
170	11.24	11.54	11.89	12.32	12.86	13.53	14.37	15.42	16.66	18.05	19.58	23.47	27.19
180	11.81	12.09	12.41	12.79	13.25	13.80	14.46	15.25	16.16	17.19	18.32	21.39	24.54
190	12.37	12.63	12.93	13.27	13.67	14.15	14.69	15.32	16.04	16.85	17.74	20.20	22.81
200	12.92	13.16	13.44	13.76	14.11	14.52	14.99	15.53	16.12	16.78	17.50	19.51	21.73
210	13.46	13.69	13.95	14.24	14.56	14.93	15.34	15.80	16.32	16.88	17.48	19.18	21.05
220	13.99	14.21	14.45	14.72	15.02	15.35	15.72	16.13	16.57	17.58	17.99	19.12	20.38
230	14.50	14.71	14.94	15.19	15.46	15.77	16.11	16.47	16.87	17.30	17.76	19.03	20.45
240	15.00	15.20	15.42	15.66	15.91	16.19	16.50	16.83	17.19	17.58	17.99	19.12	20.38
250	15.49	15.68	15.89	16.12	16.35	16.61	16.89	17.20	17.53	17.88	18.25	19.27	20.40
260	15.97	16.15	16.35	16.57	16.79	17.03	17.29	17.58	17.88	18.19	18.54	19.46	20.48
270	16.44	16.62	16.80	17.01	17.22	17.45	17.70	17.96	18.24	18.52	18.84	19.69	20.62
280	16.91	17.08	17.26	17.45	17.65	17.87	18.10	18.35	18.61	18.87	19.17	19.95	20.81
290	17.37	17.53	17.71	17.89	18.08	18.29	18.50	18.74	18.98	19.23	19.50	20.23	21.03
300	17.82	17.97	18.15	18.32	18.50	18.70	18.90	19.12	19.35	19.59	19.84	20.52	21.26
310	18.26	18.41	18.58	18.74	18.92	19.10	19.30	19.50	19.72	19.95	20.18	20.82	21.51
320	18.70	18.85	19.00	19.16	19.34	19.51	19.70	19.89	20.10	20.32	20.53	21.13	21.78
330	19.13	19.28	19.42	19.58	19.75	19.91	20.09	20.28	20.47	20.68	20.89	21.45	22.06
340	19.56	19.71	19.85	20.00	20.15	20.32	20.48	20.67	20.85	21.05	21.26	21.78	22.36
350	20.00	20.14	20.28	20.42	20.56	20.73	20.87	21.06	21.24	21.42	21.62	22.12	22.68
400	22.04	22.16	22.28	22.40	22.53	22.67	22.80	22.94	23.08	23.24	23.39	23.80	24.23
450	23.96	24.06	24.17	24.28	24.39	24.51	24.63	24.74	24.87	24.99	25.13	25.46	25.83
500	25.77	25.86	25.96	26.06	26.15	26.26	26.36	26.47	26.58	26.69	26.80	27.09	27.40
550	27.47	27.56	27.64	27.74	27.82	27.91	28.00	28.10	28.19	28.29	28.39	28.64	28.92
600	29.08	29.16	29.24	29.32	29.40	29.48	29.57	29.65	29.74	29.82	29.91	30.14	30.38
650	30.62	30.69	30.77	30.84	30.91	30.99	31.07	31.14	31.22	31.30	31.38	31.59	31.80
700	32.10	32.17	32.24	32.30	32.37	32.44	32.51	32.58	32.66	32.73	32.81	32.99	33.18
800	34.91	34.97	35.03	35.09	35.14	35.21	35.27	35.33	35.39	35.46	35.52	35.68	35.85
900	37.53	37.58	37.63	37.69	37.74	37.79	37.85	37.90	37.95	38.01	38.07	38.21	38.35
1000	39.99	40.04	40.08	40.13	40.18	40.23	40.27	40.32	40.37	40.42	40.47	40.60	40.73
1100	42.32	42.36	42.40	42.45	42.49	42.53	42.58	42.62	42.67	42.71	42.76	42.87	42.99
1200	44.53	44.57	44.61	44.65	44.69	44.73	44.76	44.81	44.85	44.89	44.93	45.03	45.14
1300	46.62	46.66	46.69	46.73	46.77	46.80	46.84	46.87	46.91	46.95	46.99	47.08	47.18

VISCOSITY OF NITROGEN (continued)

T,K	Pressure, bar											
	175	200	250	300	400	450	500	600	700	800	900	1000
65	-	-	-	-	-	-	-	-	-	-	-	-
70	292.7	304.7	328.8	353.3	-	-	-	-	-	-	-	-
75	241.1	252.3	274.7	297.9	344.5	367.8	391.4	-	-	-	-	-
80	197.1	207.1	227.9	249.1	292.8	315.1	337.6	-	-	-	-	-
85	161.4	170.2	188.4	207.5	247.5	268.5	289.6	-	-	-	-	-
90	133.8	141.2	156.8	173.3	209.2	228.3	247.8	-	-	-	-	-
95	113.1	119.1	132.1	146.2	177.4	194.5	212.1	-	-	-	-	-
100	98.25	103.0	113.5	125.3	152.0	166.8	182.6	-	-	-	-	-
105	87.63	91.50	99.92	109.4	131.7	144.5	158.1	-	-	-	-	-
110	79.72	83.05	90.02	97.70	116.1	126.8	138.6	-	-	-	-	-
115	73.25	76.40	82.55	89.02	104.1	113.1	123.0	-	-	-	-	-
120	67.30	70.56	76.55	82.32	95.03	102.5	110.8	-	-	-	-	-
125	61.67	65.07	71.27	76.86	88.00	94.27	101.2	-	-	-	-	-
130	56.77	60.07	66.31	72.04	82.41	87.82	93.70	-	-	-	-	-
135	52.29	55.63	61.72	67.48	77.50	82.64	87.74	-	-	-	-	-
140	48.11	51.54	57.65	63.25	73.11	77.94	82.67					
150	40.82	44.25	50.32	55.76	65.44	70.00	74.30	82.71	-	-	-	-
160	35.03	38.36	44.27	49.59	58.95	63.31	67.49	75.50	83.16	-	-	-
170	30.66	33.82	39.46	44.50	53.54	57.61	61.67	69.36	76.71	83.74	-	-
180	27.55	30.45	35.73	40.49	49.03	52.92	56.75	64.01	71.11	77.89	84.43	-
190	25.43	27.97	32.80	37.57	35.29	49.06	52.63	59.63	62.27	72.75	79.05	85.45
200	23.97	26.24	30.60	34.70	42.27	45.84	49.26	55.84	66.50	68.32	74.35	80.50
210	23.01	25.00	23.91	32.68	39.75	43.13	46.38	52.60	58.67	64.51	70.26	76.06
220	21.73	24.13	27.67	31.14	37.73	40.85	43.96	49.87	55.62	61.18	66.71	72.11
230	21.98	23.54	26.73	29.90	36.02	39.01	41.88	47.50	52.95	58.27	63.56	68.71
240	21.73	23.14	26.04	28.95	34.64	37.43	40.15	45.47	50.64	55.70	60.74	65.73
250	21.61	22.88	25.52	28.20	33.53	361.0	38.69	43.71	48.65	53.44	58.23	63.06
260	21.59	22.74	25.17	27.64	32.59	35.01	37.45	42.20	46.91	51.48	56.05	60.67
270	21.63	22.69	24.92	27.22	31.82	34.12	36.41	40.91	45.39	49.76	54.14	58.53
280	21.74	22.71	24.77	26.91	31.21	33.37	35.54	39.82	44.06	48.25	52.45	56.61
290	21.88	22.79	23.72	26.69	30.73	32.76	34.81	33.87	42.91	46.90	50.93	54.89
300	22.06	22.89	24.66	26.54	30.35	32.26	34.20	38.05	41.90	45.70	49.56	53.36
310	22.26	23.04	24.70	26.44	30.04	31.86	33.70	37.34	41.01	44.65	48.34	52.00
320	22.48	23.21	24.77	26.41	29.80	31.53	33.28	36.75	40.24	43.74	47.25	50.78
330	22.71	23.41	24.87	26.42	29.63	31.27	32.93	36.25	39.58	42.95	46.29	49.69
340	22.97	23.63	25.01	26.47	29.52	31.08	32.66	35.83	39.01	42.25	45.46	48.72
350	23.26	23.87	25.18	26.56	29.47	30.95	32.47	35.50	38.55	41.63	44.76	47.85
400	24.70	25.18	26.16	27.30	29.61	30.86	32.09	34.50	37.01	39.54	42.12	44.71
450	26.21	26.61	27.44	28.34	30.25	31.25	32.26	34.33	36.45	38.60	40.76	43.01
500	27.73	28.06	28.77	29.53	31.15	31.99	32.86	34.65	36.45	38.31	40.19	42.09
550	29.20	29.49	30.11	30.76	32.16	32.89	33.64	35.20	36.79	38.42	40.08	41.76
600	30.64	30.89	31.44	32.01	33.23	33.88	34.55	35.90	37.33	38.77	40.26	41.80
650	32.03	32.26	32.74	33.26	34.35	34.93	35.52	36.74	38.01	39.31	40.64	42.08
700	33.39	33.60	340.4	34.50	35.48	36.00	36.54	37.62	38.78	39.97	41.18	42.41
800	360.2	36.19	36.56	36.95	37.76	38.19	38.63	39.55	40.50	41.49	42.52	43.57
900	38.50	38.65	38.97	39.30	39.99	40.36	40.73	41.52	42.33	43.18	44.05	44.95
1000	40.85	40.99	41.27	41.56	42.16	42.47	42.81	43.49	44.19	44.94	45.69	46.47
1100	43.10	43.22	43.46	43.72	44.26	44.53	44.82	45.42	46.05	46.70	47.37	48.07
1200	45.24	45.35	45.57	45.80	46.28	46.53	46.79	47.32	47.88	48.45	49.06	49.68
1300	47.28	47.37	47.58	47.78	48.19	48.44	48.67	49.16	49.66	50.18	50.72	51.28

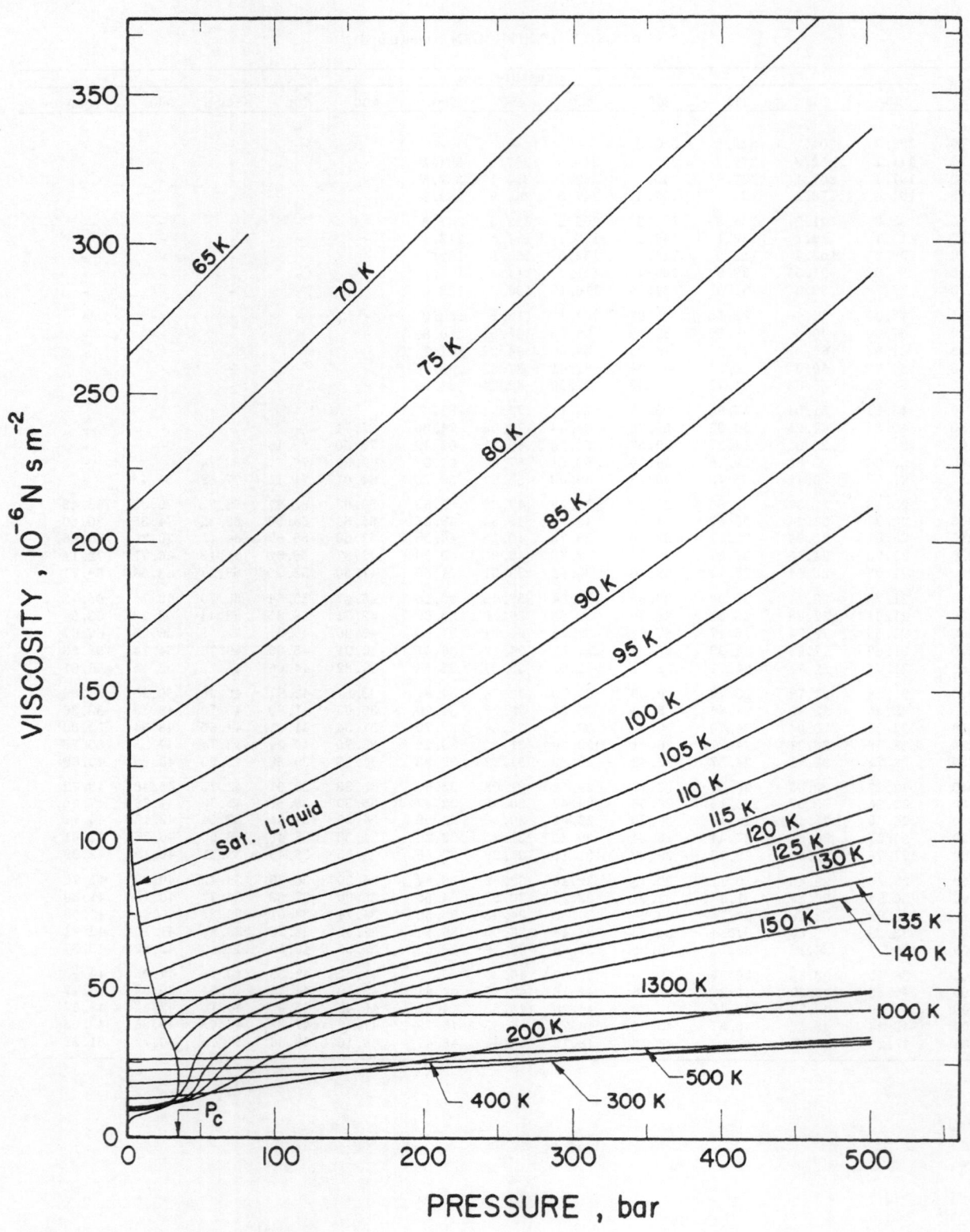

FIGURE 1A. VISCOSITY OF NITROGEN [203].

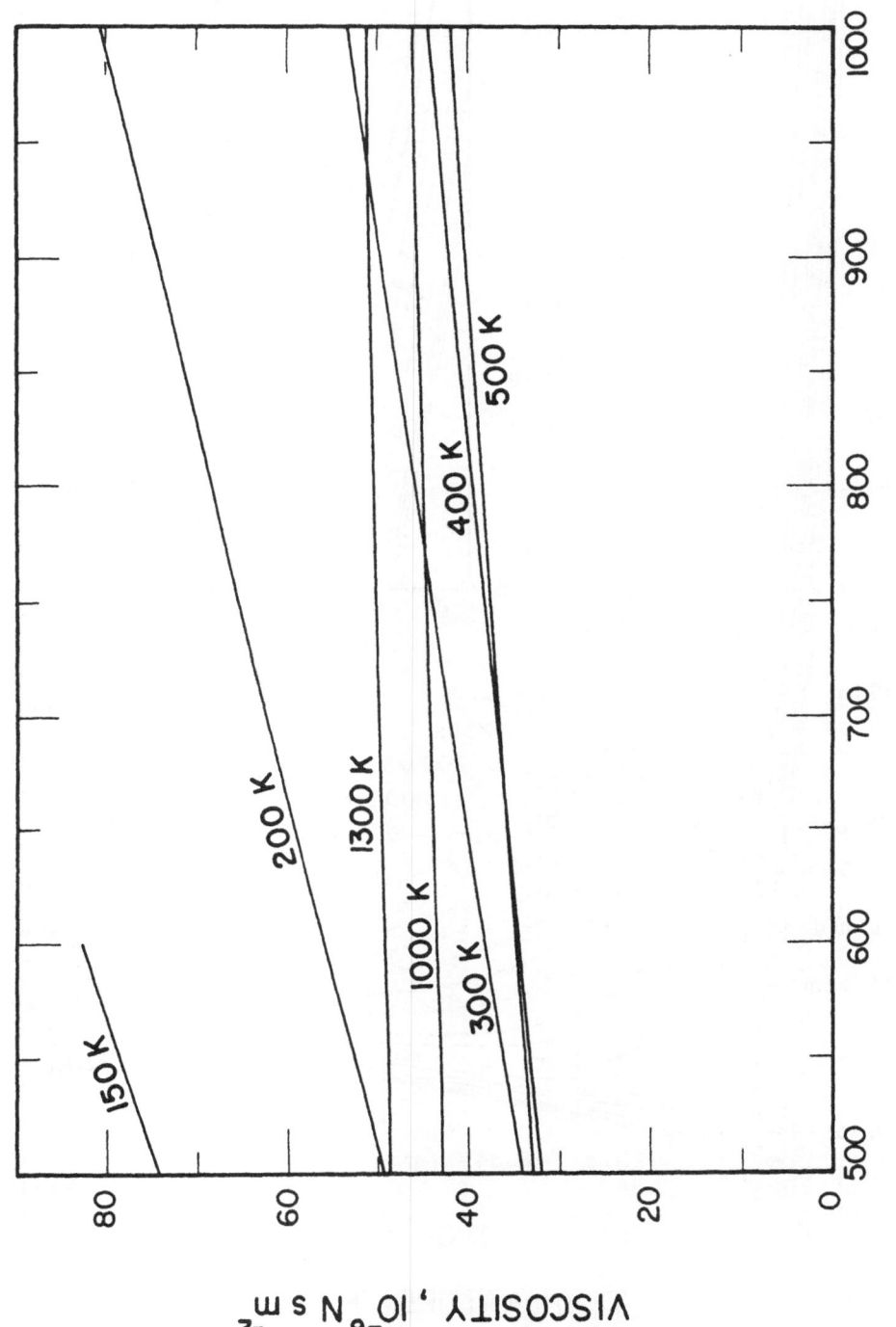

FIGURE 1B. VISCOSITY OF NITROGEN [203].

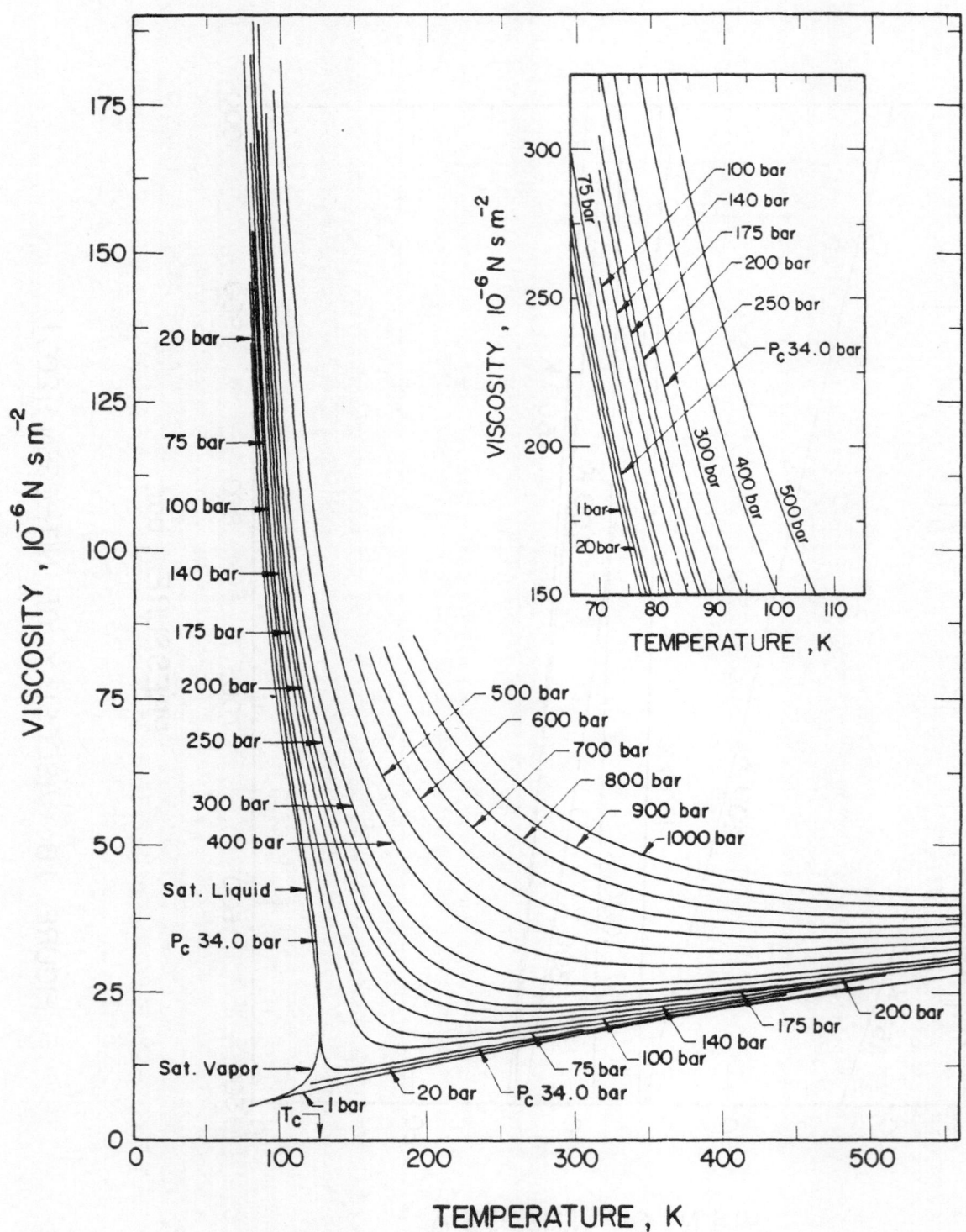

FIGURE 2A. VISCOSITY OF NITROGEN [203].

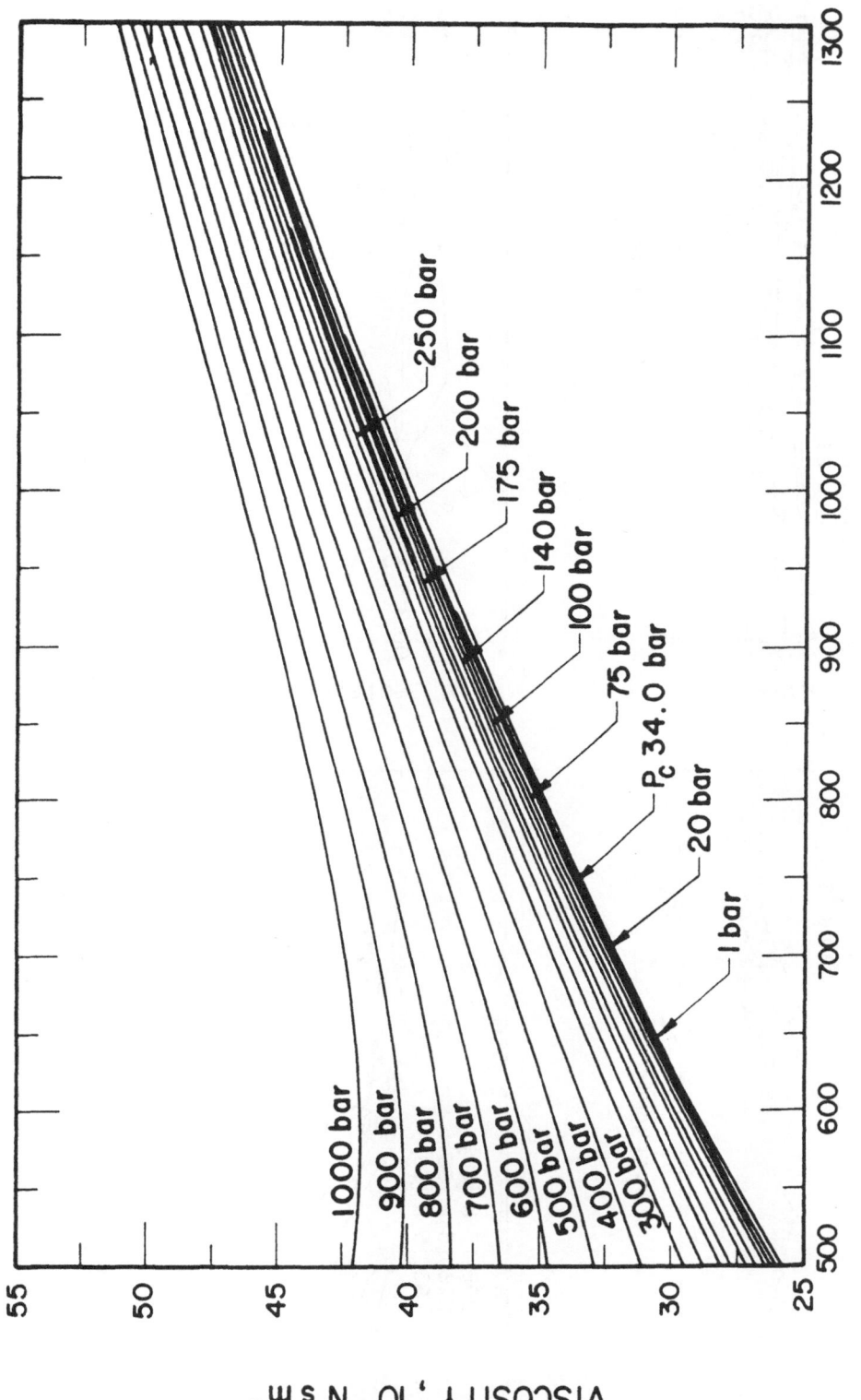

FIGURE 2B. VISCOSITY OF NITROGEN [203].

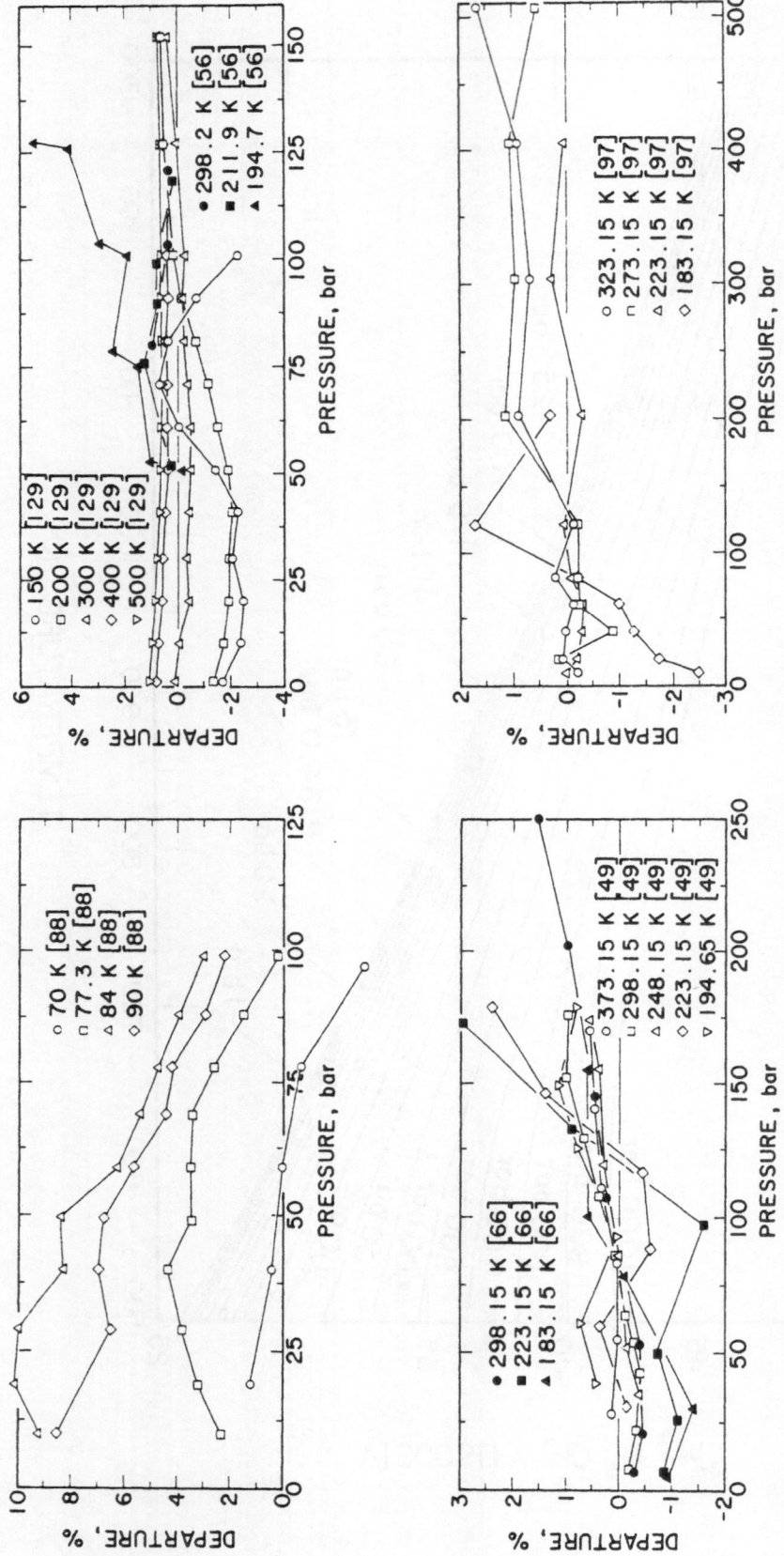

FIGURE 3A. DEPARTURE PLOT ON THE VISCOSITY OF NITROGEN.

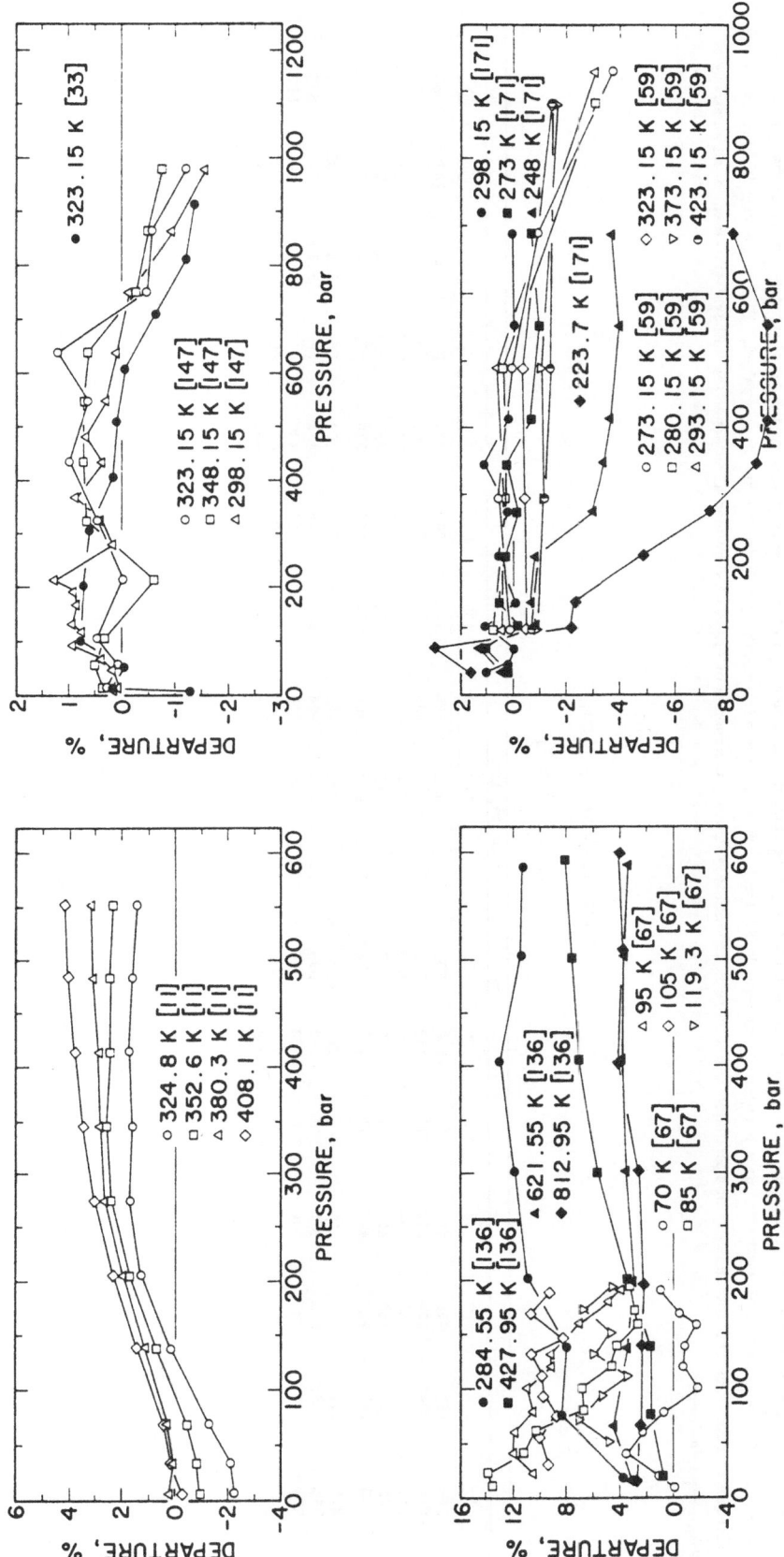

FIGURE 3B. DEPARTURE PLOT ON THE VISCOSITY OF NITROGEN.

n-NONANE

Only the data by Keramidi and Rastorguev [101], obtained by a capillary tube viscosimeter, are available in the literature on the pressure dependence of the viscosity of n-nonane. The data are plotted as isotherms as a function of pressure in Figure 1, where only minor smoothing adjustments had to be applied. Figure 2 shows isobars as a function of temperature. The recommended values tabulated below were read from this figure. In absence of further experimental evidence the authors' statement of an accuracy of ± 1.2% is reported here.

VISCOSITY OF n-NONANE

$[\mu, 10^{-6} \text{ N s m}^{-2}]$

T, K	Pressure, bar																
	1	22.8*	40	60	80	100	120	140	160	180	200	250	300	350	400	450	500
300	652	672	686	702	722	736	754	774	789	805	821	863	906	949	995	1036	1080
310	574	589	604	618	633	648	662	678	692	706	720	758	795	832	869	906	942
320	511	526	537	550	562	576	588	601	614	627	640	672	704	736	768	800	832
330	459	472	482	494	506	517	527	540	552	562	574	602	631	659	688	717	746
340	415	426	436	446	457	467	476	488	499	508	519	545	570	596	624	649	675
350	378	386	396	406	415	425	434	444	454	462	472	496	520	544	568	593	616
360	345	353	362	370	379	389	398	406	415	424	433	455	478	500	522	545	566
370	316	324	332	340	348	356	365	372	380	389	398	418	439	460	480	501	520
380	291	300	306	314	320	328	336	343	351	359	366	386	404	424	442	462	480
390	269	278	283	290	297	303	310	317	324	332	339	357	374	392	408	426	444
400	250	258	263	270	276	282	288	295	301	308	315	331	347	363	380	396	412
410	233	240	245	251	257	264	269	276	282	288	294	309	324	340	355	370	386
420	218	223	228	234	240	247	252	258	264	270	276	290	304	319	334	347	362
430	-	-	-	-	-	232	237	243	249	254	260	273	286	300	314	327	342
440	-	-	-	-	-	219	224	229	235	240	245	258	271	284	296	309	323
450	-	-	-	-	-	207	212	216	222	226	232	244	256	269	281	293	306
460	-	-	-	-	-	195	200	204	210	214	219	230	243	255	266	278	290
470	-	-	-	-	-	185	189	193	198	203	207	218	230	242	252	264	275

*Critical pressure.

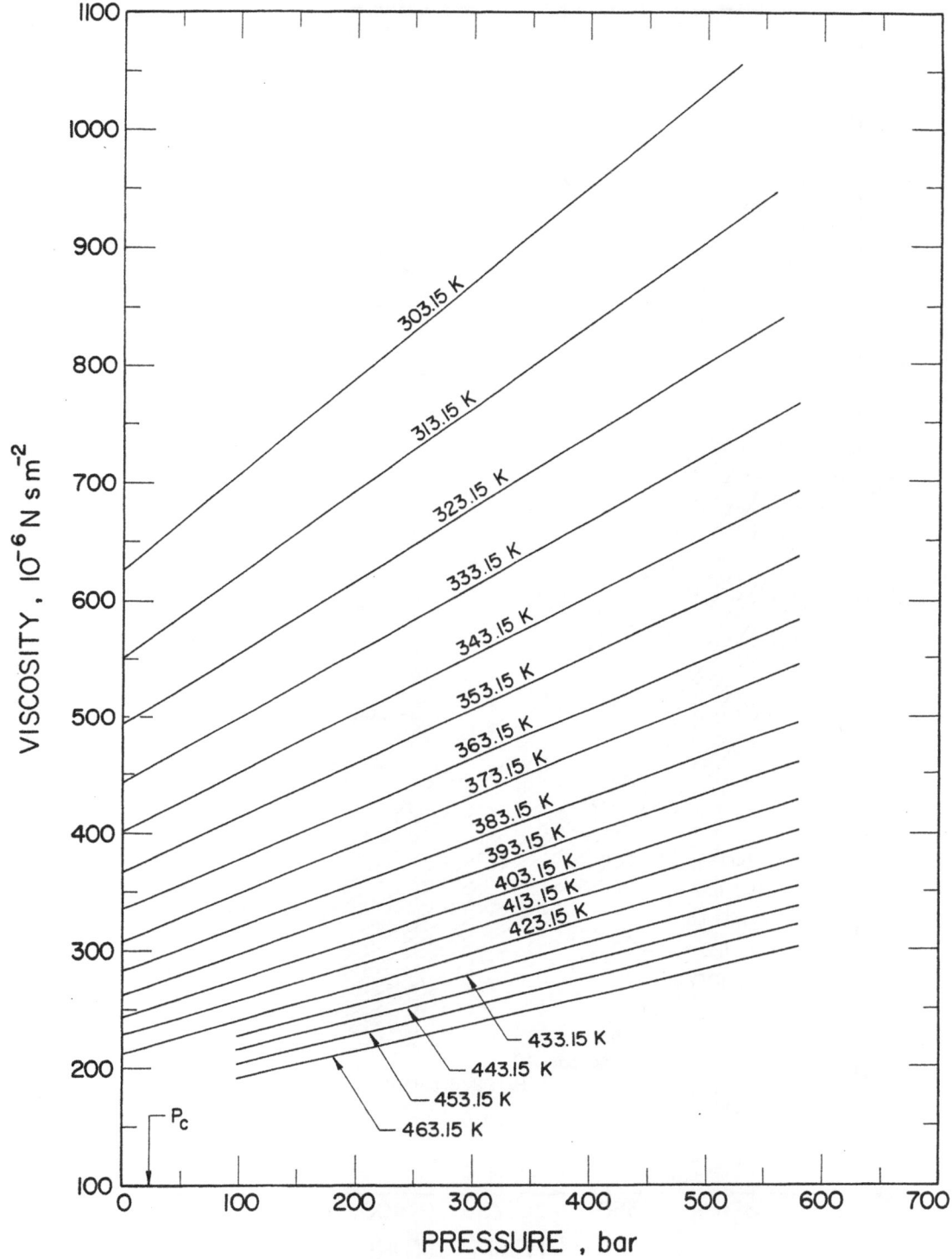

FIGURE I. VISCOSITY OF n - NONANE [101].

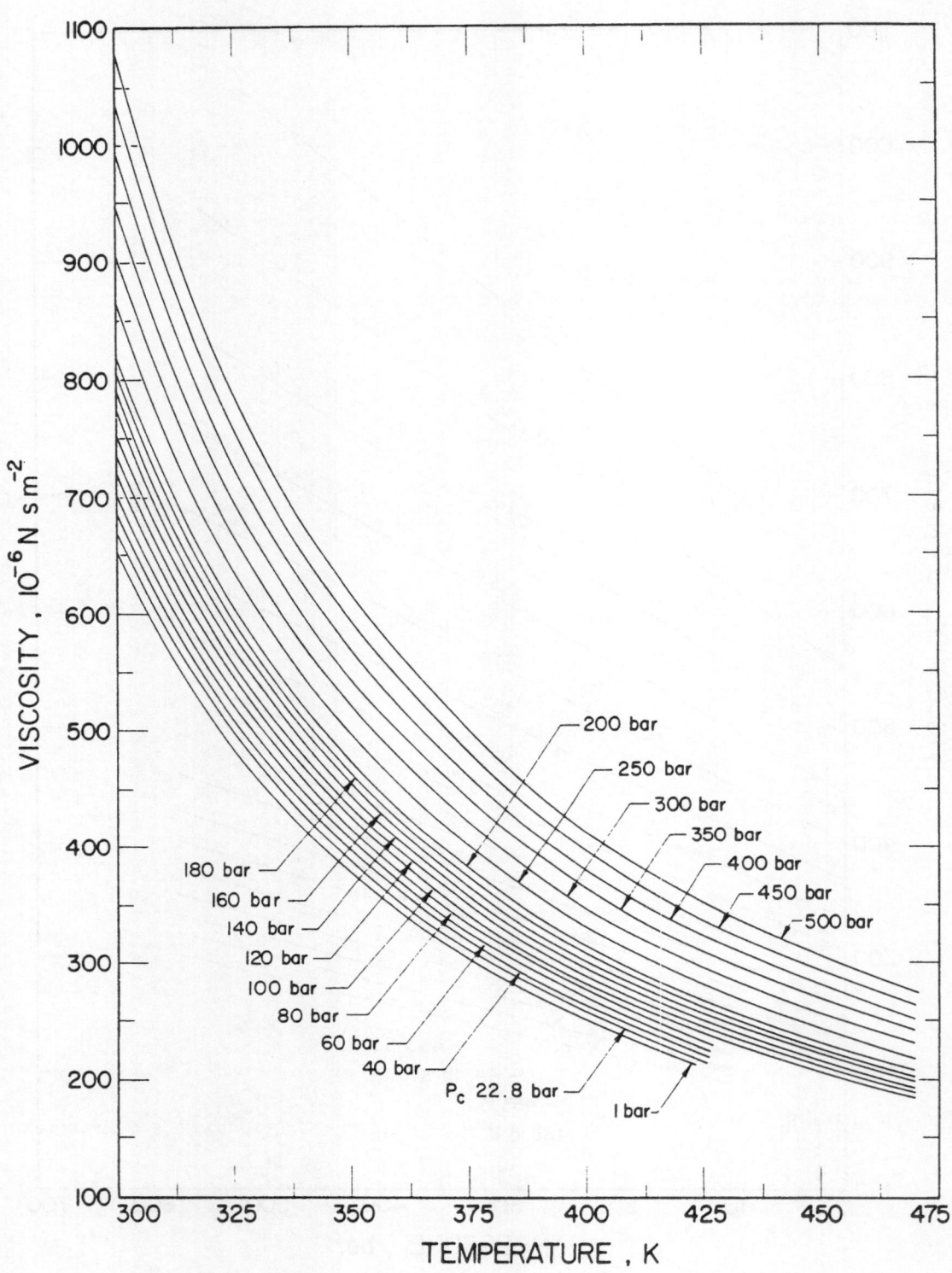

FIGURE 2. VISCOSITY OF n-NONANE [101].

i-OCTANE

Only one set of data by Agaev and Yusibova [5] was found in the literature for the pressure dependence of the viscosity of liquid iso-octane. The precision of the data is stated to be ± 1.5%. In Figure 1, these data are plotted as isotherms against pressure, where smoothing modifications up to 1.5% were applied. Using the vapor-pressure data of [127], the saturated liquid line was established by extrapolation of the measured isotherms to the vapor pressures and is shown in Figure 1. From Figure 1, isobars were plotted as a function of temperature to generate Figure 2. The recommended values presented below were read from this figure.

VISCOSITY OF i-OCTANE
$[\mu,\ 10^{-6}\ N\ s\ m^{-2}]$

T, K	\multicolumn{20}{c	}{Pressure, bar}
	1	5	10	15	20	25	30	40	50	60	80	100	150	200	250	300	350	400	450	500
290	525	528	532	535	537	541	545	551	557	562	576	590	622	655	687	725	760	792	830	870
300	470	472	475	477	480	485	490	495	500	504	512	525	562	592	625	657	687	720	750	787
320	372	374	375	378	382	389	395	397	400	405	415	425	455	482	510	535	560	585	610	642
340	305	307	310	312	314	316	317	323	330	334	342	350	375	395	420	445	462	487	505	530
360	257	260	262	265	267	269	270	275	280	283	290	300	317	337	355	375	397	420	435	450
380	-	220	222	225	227	229	230	235	240	243	250	260	275	290	310	325	345	365	380	397
400	-	188	190	191	192	195	197	201	205	208	215	225	237	255	270	285	302	320	335	352
420	-	160	162	165	167	169	170	174	177	180	187	195	210	225	240	255	270	285	297	315
440	-	135	137	140	142	145	147	151	155	157	162	170	185	200	215	230	245	257	270	282
460	-	-	117	119	125	126	127	131	135	138	145	150	165	180	195	207	220	232	245	257
480	-	-	-	101	107	109	110	112	115	118	125	132	147	160	175	187	200	212	222	235
500	-	-	-	82.5	87.5	90.0	92.5	96.0	100	103	110	115	132	142	157	170	180	192	202	215
520	-	-	-	-	67.5	70.0	72.5	77.5	82.5	86.5	95.0	102	117	130	142	152	165	175	185	197
540	-	-	-	-	-	-	52.5	51.0	70.0	74.0	82.5	90.0	105	117	127	140	150	160	170	180

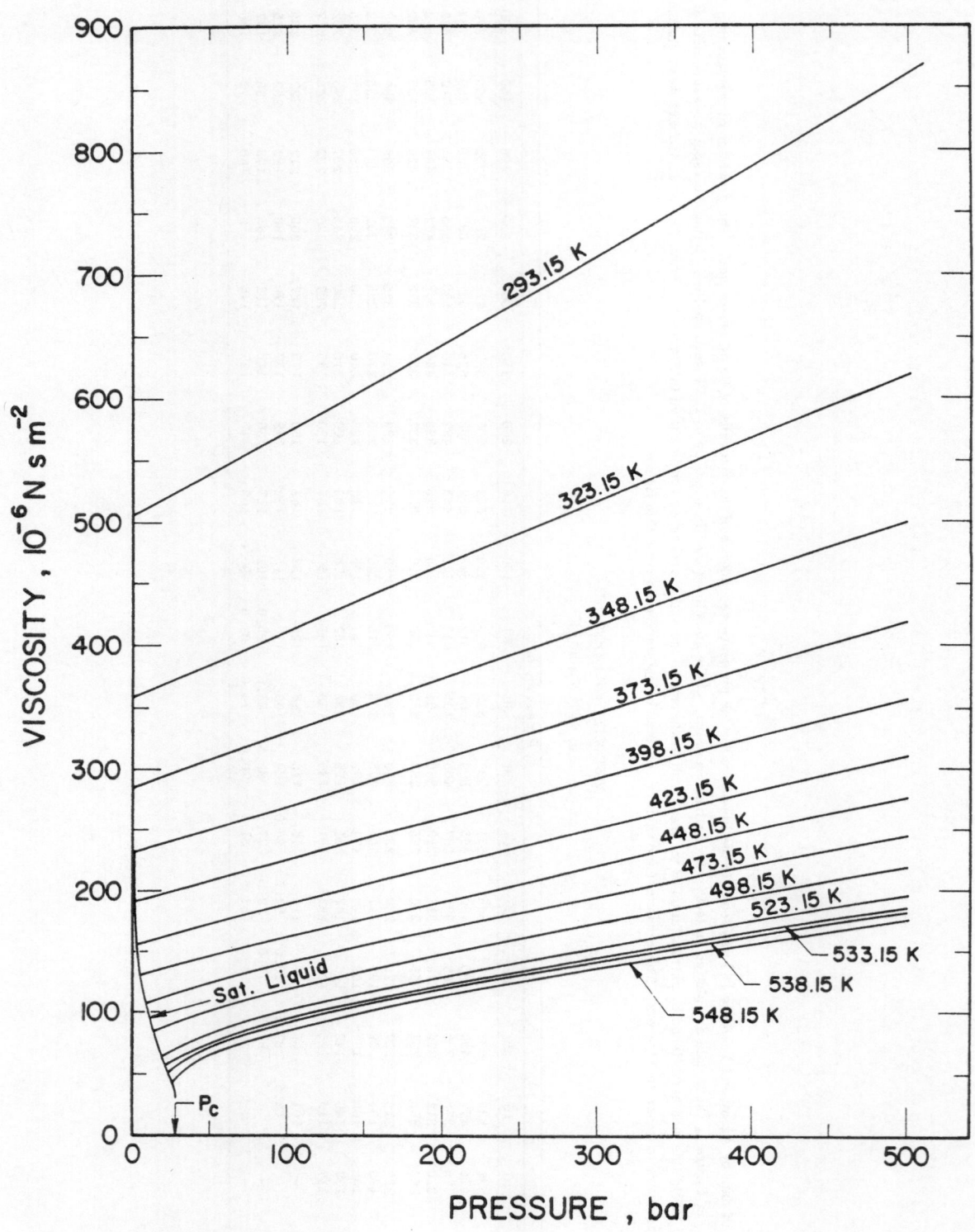

FIGURE 1. VISCOSITY OF i-OCTANE [5].

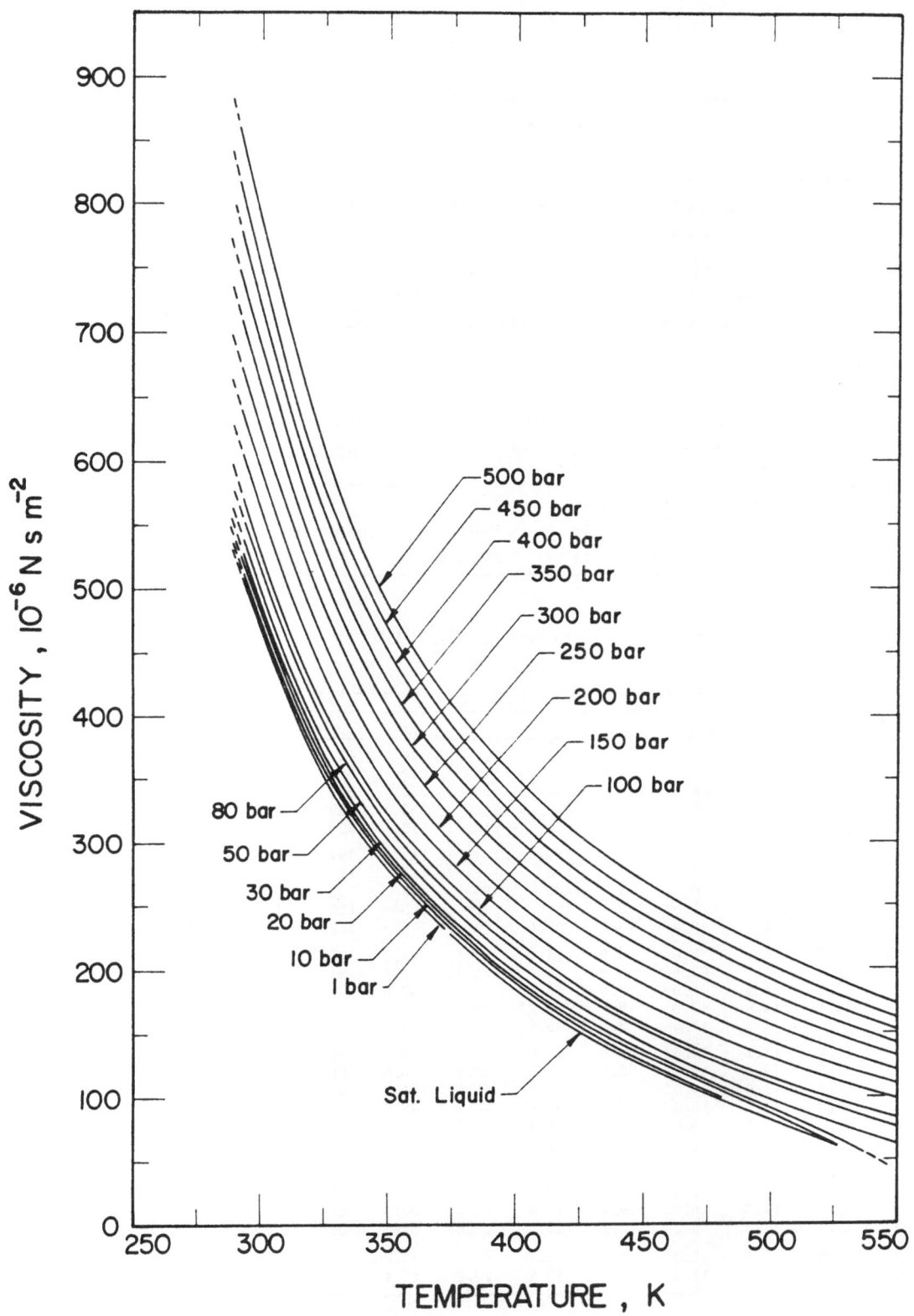

FIGURE 2. VISCOSITY OF i-OCTANE [5].

n-OCTANE

The data of Agaev and Golubev [3], obtained by a capillary tube viscosimeter were used to generate the recommended values presented below. The data are plotted as isotherms against pressure in Figures 1 where the smoothing modifications did not surpass 1%. Using the P-V-T data of [47, 143], the dilute gas data of [195], with a slight linear extrapolation, and the residual viscosity concept, an extrapolation from 569 K to 670 K up to 100 bar was made. The values are shown as isobars against temperature in Figures 2A and 2B, where the extrapolated values are indicated by dashed lines. The recommended values were read from the latter figures.

Additional works on the pressure dependence of the viscosity of n-octane are listed in the summary table below. A limited comparison was possible with the data of Bridgman [21], revealing deviations from the recommended values of 6%. The accuracy of the recommended values is estimated to be ± 5% in the region of experiments and ± 7 in the region of extrapolation.

ADDITIONAL REFERENCES ON THE VISCOSITY OF n-OCTANE

Authors	Year	Ref. No.	Temperature K	Pressure bar	Method
Brazier and Freeman	1969	20	273-333	1-4000	Rolling ball
Kozlov et al.	1966	119	293-553	Isochores	Falling cylinder
Bridgman	1926	21	303, 348	1-11768	Falling weight

VISCOSITY OF n-OCTANE

$[\mu, 10^{-6} \text{ N s m}^{-2}]$

T, K	Pressure, bar													
	1	10	20	24.9*	30	40	50	60	80	100	200	300	400	500
320	393	397	402	404	406	411	416	420	429	439	483	530	–	–
340	329	332	336	339	341	345	349	353	361	370	412	456	497	543
360	286	290	294	296	298	302	305.	309	317	325	364	406	447	491
380	230	233	237	239	241	244	248	252	259	266	301	338	375	413
400	7.40	198	202	204	205	209	213	216	223	230	263	297	330	363
420	7.76	170	173	175	177	180	184	187	194	200	232	263	293	323
440	8.11	146	149	151	153	156	160	163	169	176	205	233	260	286
460	8.47	124	128	130	132	135	139	142	148	154	181	205	230	254
480	8.82	105	109	111	113	116	120	123	129	135	160	182	204	226
500	9.20	87.0	91.0	93.0	95.0	99.0	102	106	112	118	143	165	185	205
520	9.50	10.8	76.0	78.0	80.0	84.0	87.4	91.0	97.3	103	129	149	169	187
540	9.90	11.0	60.0	63.0	66.0	70.0	74.6	78.0	84.8	90.8	115	135	154	172
560	10.2	11.3	14.9	48.0	52.0	58.0	62.8	67.8	73.2	79.1	102	122	140	158
565	10.3	11.4	14.6	42.0	48.0	55.0	60.0	64.0	70.5	76.3	100	119	137	155
575	10.5	11.5	14.2	17.6	38.5	49.0	54.4	58.7	65.0	70.8	95.0	114	132	149
580	10.6	11.6	14.1	16.6	32.8	46.0	51.6	56.0	62.4	68.2				
585	10.6	11.7	14.0	16.2	25.6	42.9	49.0	53.5	60.0	65.6				
590	10.7	11.8	14.0	16.0	20.8	39.7	46.5	51.2	57.6	63.2				
595	10.8	11.9	14.0	15.8	19.4	36.4	44.2	48.8	55.2	60.9				
600	10.9	12.0	14.0	15.6	18.8	33.1	41.8	46.6	53.1	58.7				
610	11.1	12.2	14.0	15.4	17.8	28.0	37.5	42.4	49.1	54.6				
620	11.3	12.4	14.1	15.7	17.1	24.8	33.8	38.8	45.7	51.1				
630	11.4	12.5	14.2	15.3	16.7	22.8	30.8	36.0	43.0	48.3				
640	11.6	12.7	14.3	15.2	16.4	21.4	28.5	33.6	40.6	45.9				
650	11.8	12.9	14.3	15.2	16.2	20.4	26.6	31.6	38.5	43.8				
660	12.0	13.0	14.4	15.3	16.2	19.7	25.1	29.9	36.7	42.0				
670	12.2	13.2	14.5	15.3	16.3	19.2	23.8	28.4	35.1	40.4				

*Critical pressure.

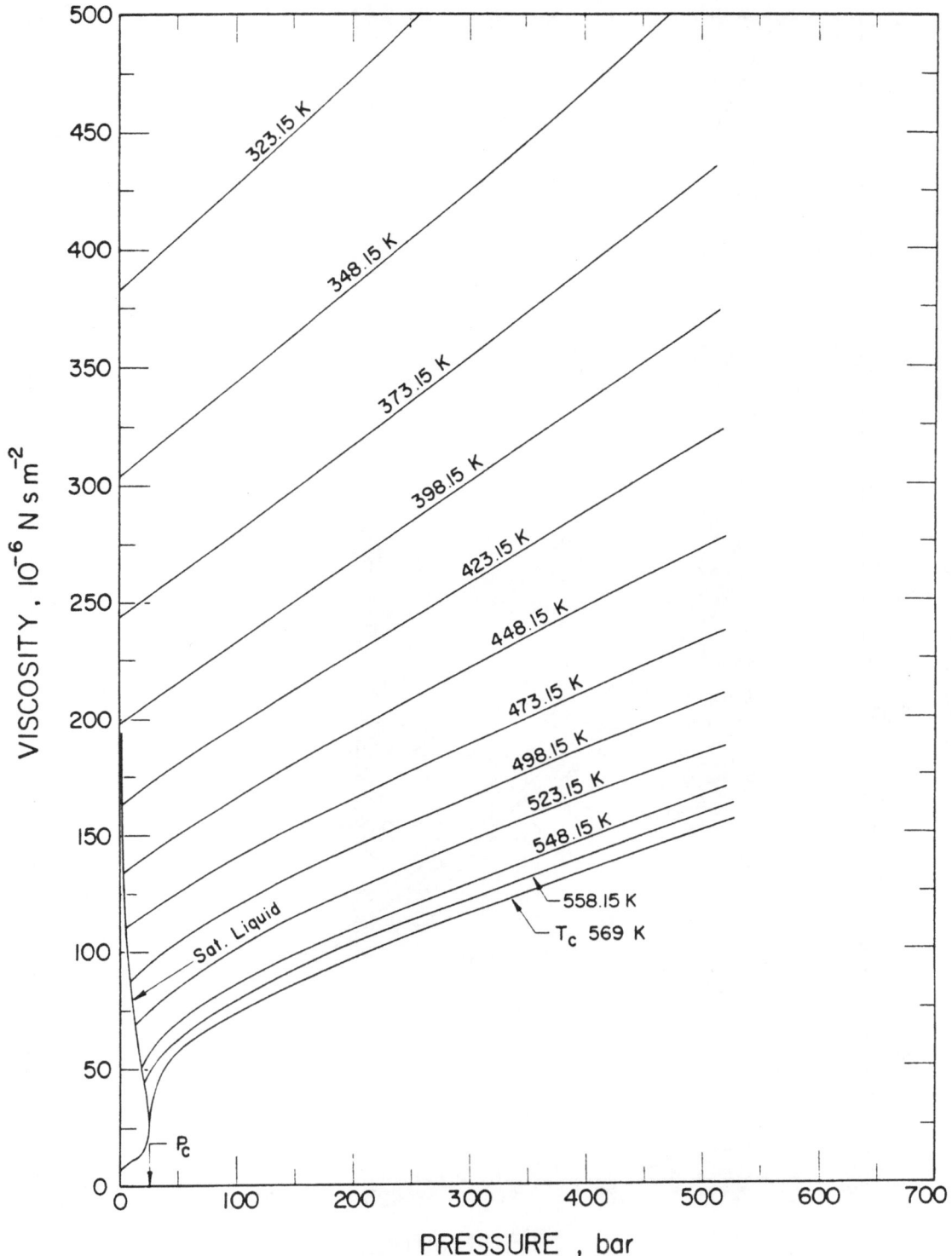

FIGURE 1. VISCOSITY OF n-OCTANE [3].

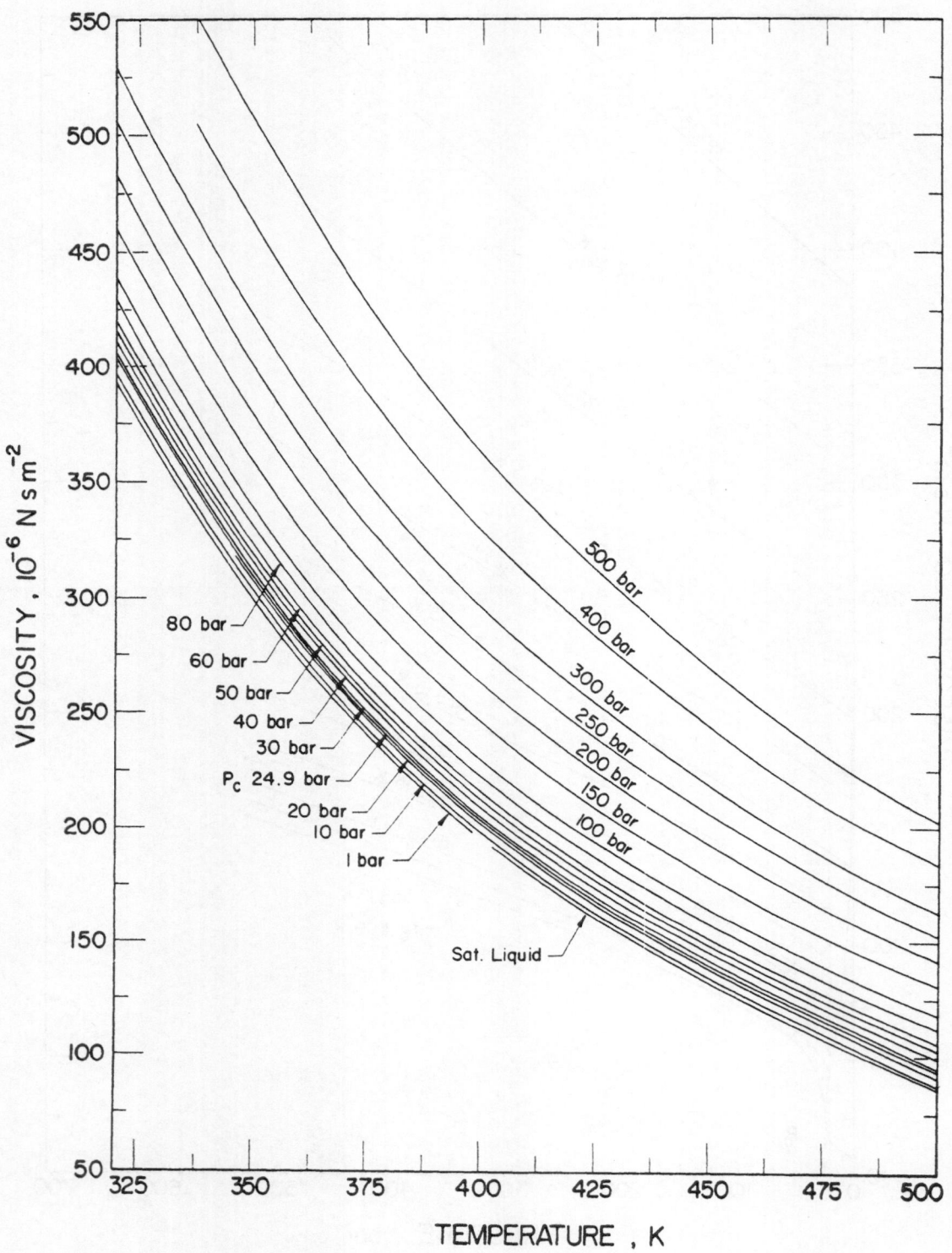

FIGURE 2A. VISCOSITY OF n-OCTANE [3].

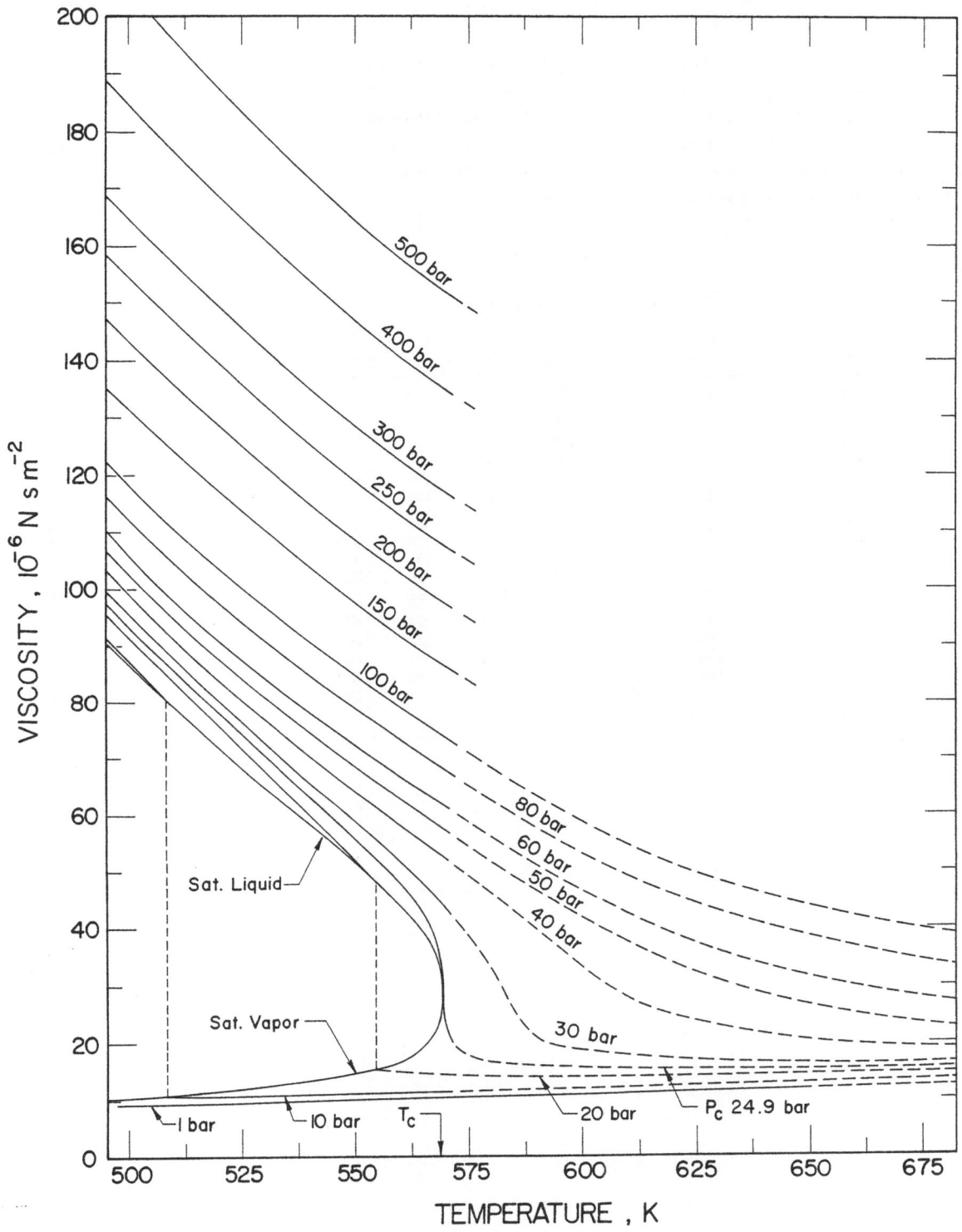

FIGURE 2B. VISCOSITY OF n-OCTANE [3].

n-OCTENE

The data of [157], obtained by a falling body method, were used to generate the recommended values presented below. They are plotted as isotherms against pressure in Figure 1, where some smoothing modifications were applied. The measured isotherms were extrapolated to the vapor pressure of n-octene which was estimated using the Harlacher-Braun-constants as described in [166]. From Figure 1 even isobars were plotted against temperature to generate Figure 2 including slight extrapolations. The recommended values were read from this figure. No comparison with further works was possible. The uncertainty in the recommended values is estimated to be ± 5%.

VISCOSITY OF n-OCTENE
$$[\mu, \; 10^{-6} \; N \; s \; m^{-2}]$$

T, K	Pressure, bar												
	1	30	50	70	100	150	200	250	300	350	400	450	500
300	438	452	464	473	488	510	532	554	574	600	618	638	662
305	416	428	440	450	464	486	508	530	552	574	592	614	636
310	394	406	418	428	442	464	486	508	530	550	570	590	612
320	356	370	380	390	402	424	446	470	490	570	530	550	570
330	322	334	346	356	366	388	410	434	452	472	492	512	532
340	294	304	314	324	334	356	378	398	418	438	456	474	494
350	268	280	288	300	308	328	348	366	386	406	422	440	458
360	248	260	266	278	284	302	320	338	356	376	392	408	424
370	232	242	248	258	266	282	298	316	330	348	364	382	396
380	217	226	232	240	248	265	280	296	310	324	340	358	372
400	−	200	206	212	220	234	248	262	276	288	300	316	330
420	−	179	186	190	196	209	220	232	242	254	265	276	288
440	−	161	167	170	175	186	194	204	213	222	232	240	249
460	−	144	148	152	157	166	173	182	190	198	207	214	222
480	−	128	132	136	144	150	159	169	177	184	193	200	208
490	−	126	129	134	141	148	157	167	174	181	191	198	206

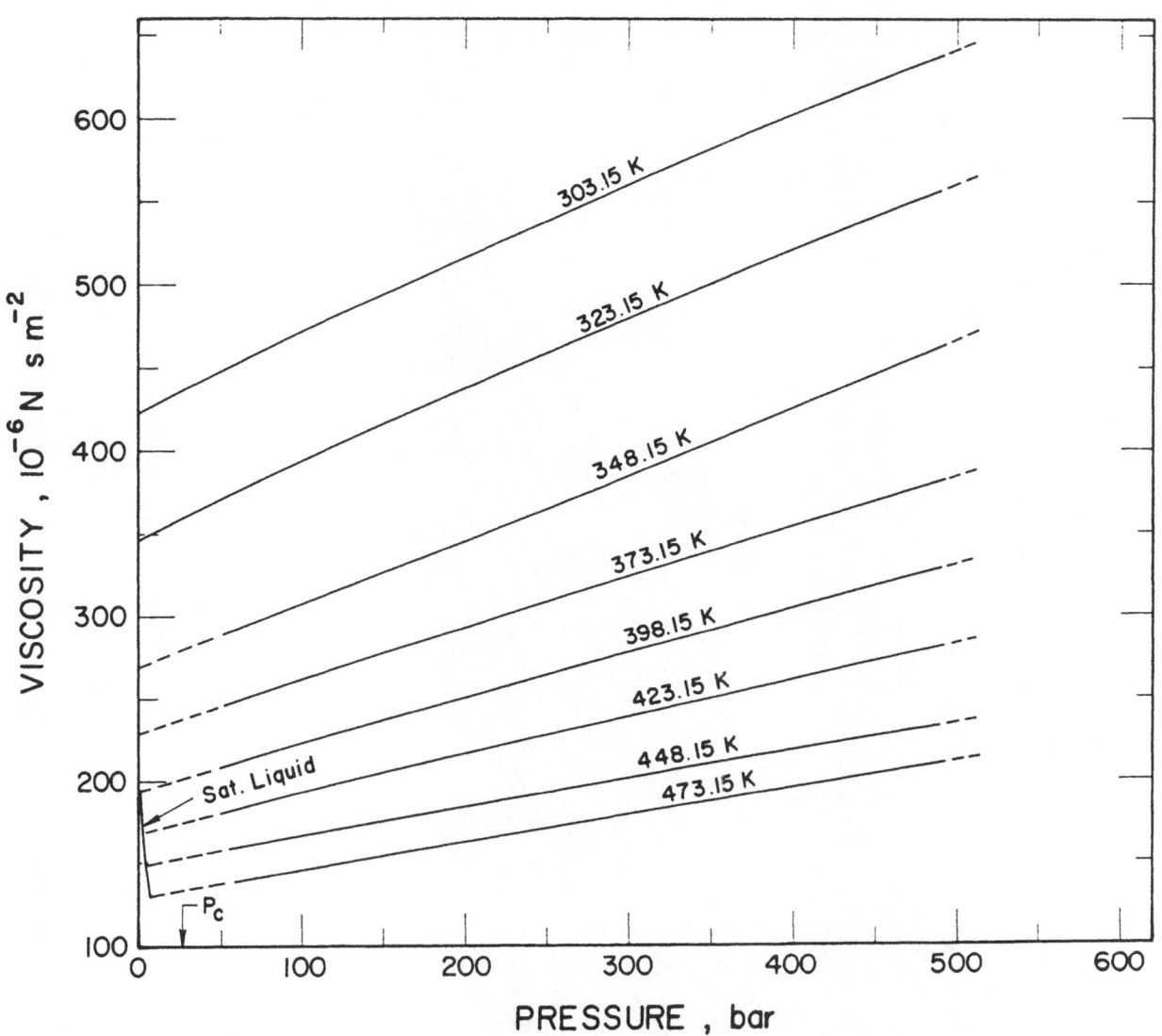

FIGURE 1. VISCOSITY OF n - OCTENE [157].

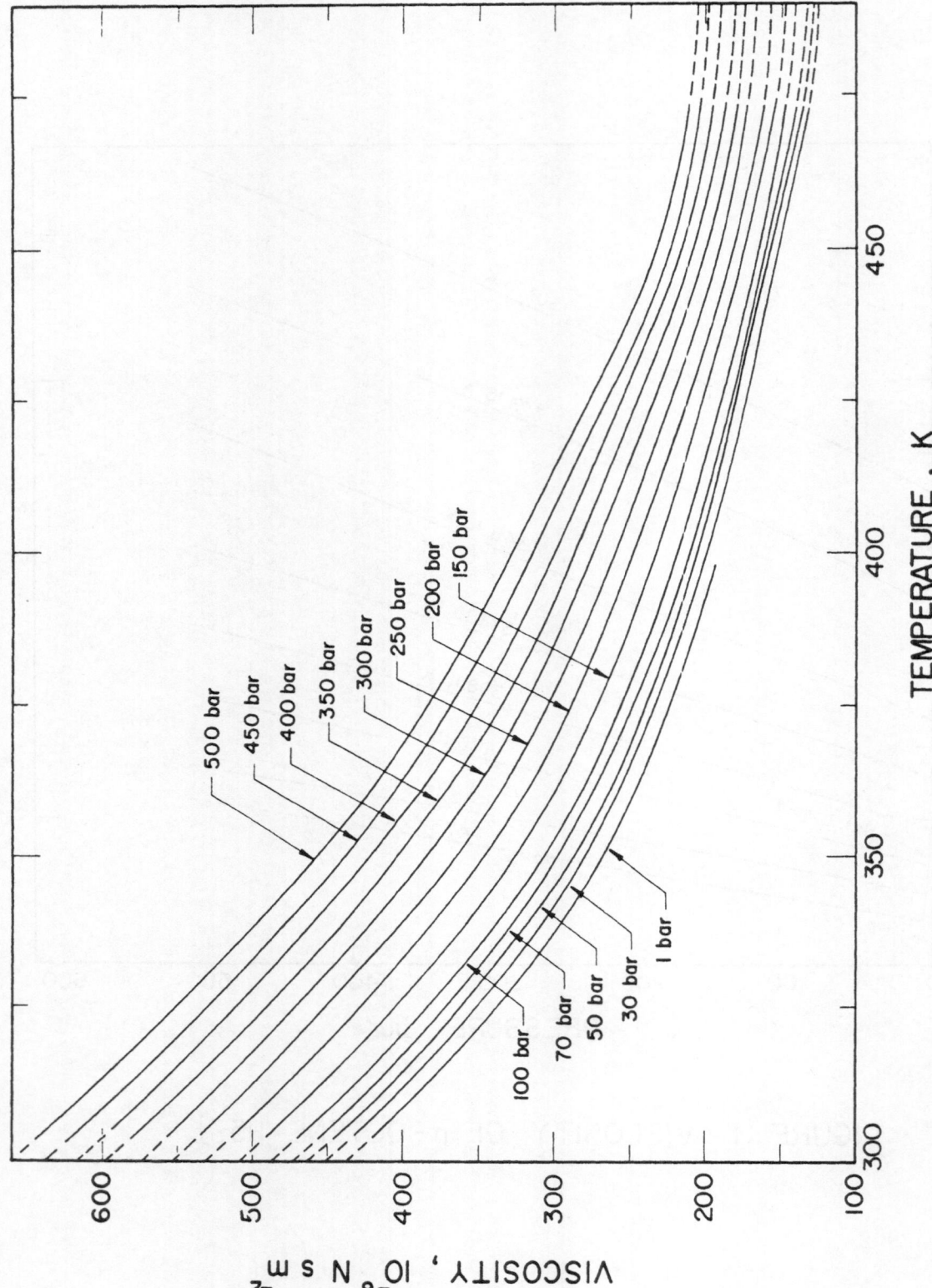

FIGURE 2. VISCOSITY OF n- OCTENE [157].

OXYGEN

A considerable amount of data exists for the viscosity of oxygen in all fluid states. The recommended values tabulated on the next two pages were taken from [203]. They cover a temperature range from 75 to 1300 K and a pressure range from 1 to 1000 bar, though some voids exist in this region of states. In an effort to fill in the gaps in data some extrapolation has been made using the given 1 bar data and the P-V-T data of [203] together with the residual viscosity concept. The extrapolated values are included in Figures 1 and 2, where a selection of the recommended values is plotted as isotherms against pressure and as isobars against temperature, respectively. The saturated liquid line was found by extrapolation of the isotherms to the saturation pressure. Along the 1000 K isotherm, the data at 400, 450, 500 and 600 bar were obviously incorrect in [203], and have been adequately corrected. The saturated vapour data were taken from [195].

Additional works on the pressure dependence of the viscosity of oxygen are listed in the summary table below. Considerable discrepancies are observed between the different sources. Extensive comparisons were made between these data and the recommended values and are shown in the summary table below as well as the departure plots in Figures 3A and 3B. Excellent agreement may be reported for the data of Kestin et al. [105] and Golubev [57] in the gaseous region. The agreement with the data of Grevendonk et al. [68], Kiyama et al. [114], van Itterbeek et al. [88] and de Bock et al. [38] is unsatisfactory. Large discrepancies occur for the data of Hellemans et al. [80]. The uncertainty in the recommended values in the gaseous region is estimated to be about ± 2%. For the liquid region, the discrepancies between the various authors are too large to report an uncertainty value confidently. It is tentatively estimated to be ± 5%. It can be observed that the recommended values are in good agreement with data recommended by [76], except perhaps in the low temperature liquid region.

ADDITIONAL REFERENCES ON THE VISCOSITY OF OXYGEN

Authors	Year	Ref. No.	Temperature K	Pressure bar	Method	Departure % (no. points)	
Golubev and Petrov	1970	57	288-271	1-150	Oscillating disk	-0.58 ± 0.78	(37)
Hellemans et al.	1970	80	96-152	2-97	Oscillating disk	-9.29 ± 14.74	(49)
Grevendonk et al.	1968	68	77-150	7-190	Torsional cryst.	3.49 ± 3.67	(102)
De Bock et al.	1967	38	77	1-138	Torsional cryst.	-9.15 ± 4.00	(17)
Van Itterbeek et al.	1966	88	70-90	95-96	Oscillating disk	4.93 ± 3.08	(24)
Boon and Thomaes	1963	18	Saturated	Liquid	Capillary tube	-	
Kestin and Leidenfrost	1959	105	293/298	0-53	Oscillating disk	0.66 ± 0.41	(15)
Glaser and Gebhardt	1959	55	287/361	10-190	Falling body	-	
Kiyama and Makita	1956	114	198-373	1-800	Capillary tube	4.44 ± 3.33	(50)
Kiyama and Makita	1952	116	298-473	1-97	Capillary tube	4.44 ± 3.33	(50)

VISCOSITY OF OXYGEN

$[\mu, 10^{-6} \text{ N s m}^{-2}]$

T, K	Pressure, bar												
	1	10	20	30	40	50	60	70	80	90	100	125	150
75	303.8	308.1	313.1	317.9	322.9	327.9	332.8	337.8	343.0	348.2	353.2	366.4	380.0
80	253.6	256.9	260.7	264.4	268.2	272.1	275.9	280.1	284.1	287.9	292.1	302.4	312.7
85	216.2	218.9	221.8	224.7	227.7	230.7	233.8	236.8	239.9	243.1	246.3	254.4	262.6
90	188.2	190.3	192.7	195.0	197.3	199.8	202.2	204.7	207.1	209.6	212.1	218.5	225.1
95	7.21	168.4	170.4	172.3	174.3	176.3	178.3	180.3	182.4	184.4	186.2	191.6	197.0
100	7.64	150.7	152.5	154.2	156.0	157.8	159.5	161.3	163.0	164.8	166.6	171.0	175.5
105	8.05	135.5	137.2	138.9	140.6	142.3	143.9	145.6	147.2	148.8	150.4	154.5	158.4
110	8.46	121.6	123.4	125.2	126.8	128.5	130.2	131.8	133.4	135.0	136.5	140.4	144.1
115	8.86	108.6	110.5	112.3	114.1	115.8	117.5	119.1	120.8	122.3	124.0	127.8	131.5
120	9.26	9.79	98.63	100.4	102.2	104.0	105.7	107.4	109.1	110.7	112.3	116.3	120.1
125	9.65	10.14	88.15	89.95	91.69	93.43	95.11	96.79	98.43	100.1	101.7	105.6	109.4
130	10.03	10.49	78.78	80.75	82.58	84.34	86.00	87.61	89.20	90.78	92.31	96.13	99.85
135	10.40	10.83	11.59	71.50	73.74	75.78	77.71	79.49	81.19	82.79	84.32	87.98	91.52
140	10.77	11.18	11.85	61.60	64.56	67.12	69.40	71.49	72.32	75.24	76.95	80.87	84.43
145	11.13	11.52	12.12	13.17	54.65	58.13	61.00	63.53	65.79	67.86	69.78	74.12	77.96
150	11.48	11.85	12.40	13.28	15.09	48.43	52.35	55.50	58.19	60.61	62.77	67.57	71.72
155	11.82	12.18	12.68	13.44	14.72	18.57	41.78	46.77	50.64	53.43	55.96	61.24	65.70
160	12.16	12.50	12.97	13.65	14.67	16.53	23.24	35.53	41.68	45.82	49.00	55.21	59.95
165	12.49	12.82	13.26	13.87	14.05	16.34	19.20	26.39	32.40	37.72	41.72	49.22	54.50
170	12.82	13.14	13.55	14.11	14.24	15.95	17.60	20.44	25.22	30.51	35.08	43.51	49.43
180	13.47	13.76	14.14	14.62	14.73	16.04	17.09	18.50	20.44	23.02	26.08	33.75	40.18
190	14.11	14.38	14.73	15.15	15.26	16.33	17.12	18.10	19.31	20.79	22.57	27.91	33.35
200	14.75	15.01	15.32	15.71	15.81	16.72	17.37	18.13	19.03	20.08	21.28	24.97	29.18
210	15.38	15.62	15.92	16.27	16.36	17.16	17.71	18.35	19.07	19.89	20.80	23.54	26.77
220	16.01	16.24	16.52	16.84	16.93	17.64	18.12	18.67	19.28	19.96	20.70	22.87	25.43
230	16.63	16.85	17.11	17.41	17.50	18.14	18.57	19.05	19.58	20.16	20.79	22.59	24.68
240	17.25	17.46	17.70	17.98	18.07	18.65	19.04	19.47	19.95	20.45	21.00	22.55	24.31
250	17.86	18.06	18.28	18.55	18.64	19.17	19.52	19.92	20.35	20.80	21.29	22.64	24.18
260	18.45	18.64	18.85	19.10	19.19	19.68	20.00	20.37	20.76	21.17	21.62	22.83	24.18
270	19.02	19.21	19.41	19.65	19.73	20.19	20.49	20.82	21.18	21.56	21.97	23.06	24.28
280	19.59	19.77	19.96	20.19	20.27	20.70	20.98	21.29	21.62	21.97	22.35	23.34	24.45
290	20.16	20.33	20.52	20.73	20.81	21.21	21.48	21.77	22.08	22.40	22.75	23.67	24.68
300	20.72	20.88	21.07	21.27	21.35	21.72	21.98	22.25	22.54	22.84	23.16	24.02	24.95
310	21.27	21.42	21.60	21.80	21.88	22.23	22.47	22.73	23.00	23.28	23.58	24.38	25.25
320	21.80	21.95	22.12	22.31	22.39	22.72	22.95	23.19	23.45	23.72	24.00	24.75	25.56
330	22.32	22.47	22.63	22.81	22.89	23.20	23.42	23.65	23.90	24.15	24.42	25.13	25.89
340	22.84	22.98	23.14	23.31	23.39	23.69	23.90	24.12	24.35	24.59	24.84	25.51	26.23
350	23.35	23.48	23.64	23.80	23.88	24.17	24.37	24.58	24.80	25.03	25.27	25.90	26.58
400	25.82	25.94	26.07	26.21	26.28	26.51	26.68	26.84	27.02	27.21	27.40	27.89	28.45
450	28.14	28.25	28.36	28.48	28.55	28.74	28.88	29.02	29.17	29.33	29.49	29.91	30.36
500	30.33	30.43	30.53	30.63	30.70	30.86	30.98	31.11	31.24	31.37	31.51	31.87	32.25
550	32.40	32.49	32.58	32.67	32.74	32.88	32.98	33.10	33.21	33.32	33.45	33.76	34.09
600	34.37	34.45	34.53	34.62	34.68	34.80	34.90	35.00	35.10	35.20	35.31	35.59	35.89
700	38.08	38.14	38.22	38.29	38.34	38.44	38.52	38.60	38.69	87.77	38.86	39.09	39.33
800	41.52	41.58	41.64	41.70	41.75	41.83	41.90	41.97	42.04	42.12	42.19	42.38	42.58
900	44.72	44.77	44.83	44.88	44.92	44.99	45.06	45.11	45.18	45.24	45.31	45.47	45.64
1000	47.70	47.75	47.79	47.84	47.88	47.95	48.00	48.05	48.11	48.16	48.22	48.37	48.52
1100	50.55	50.59	50.63	50.68	50.72	50.77	50.82	50.87	50.92	50.97	51.02	51.15	51.28
1200	53.25	53.30	53.33	53.37	53.41	53.46	53.51	53.55	53.59	53.64	53.69	53.80	53.92
1300	55.84	55.88	55.91	55.95	55.98	56.03	56.07	56.11	56.15	56.19	56.23	56.34	56.45

VISCOSITY OF OXYGEN (continued)

T, K	\multicolumn{13}{c}{Pressure, bar}												
	175	200	250	300	350	400	450	500	600	700	800	900	1000
75	393.9	407.7	436.8	466.8	497.9	529.9	-	-	-	-	-	-	-
80	323.6	334.5	357.4	381.2	406.1	432.0	458.5	486.2	-	-	-	-	-
85	271.1	279.8	297.6	316.5	336.5	357.2	378.7	400.8	-	-	-	-	-
90	231.8	238.6	252.9	268.1	283.8	300.4	317.8	336.0	-	-	-	-	-
95	202.4	208.0	219.5	231.6	243.3	257.6	271.7	286.6	-	-	-	-	-
100	180.0	184.6	194.2	204.1	214.5	225.4	236.8	248.9	-	-	-	-	-
105	162.4	166.5	174.6	183.0	191.7	200.7	210.3	220.2	-	-	-	-	-
110	147.8	151.6	158.9	166.3	173.8	181.6	189.6	198.1	-	-	-	-	-
115	135.2	138.8	145.7	152.5	159.4	166.4	173.4	180.7	-	-	-	-	-
120	123.7	127.3	134.1	140.8	147.2	153.6	160.1	166.6	-	-	-	-	-
125	113.1	116.8	123.6	130.2	136.5	143.7	148.8	154.8	-	-	-	-	-
130	103.5	107.1	114.0	120.5	126.8	132.9	138.8	144.7	-	-	-	-	-
135	94.98	98.41	105.1	111.7	117.9	124.0	129.9	135.6	-	-	-	-	-
140	87.75	90.98	97.35	103.7	109.8	115.9	121.7	127.3	-	-	-	-	-
145	81.44	84.65	90.68	96.62	102.5	108.4	114.2	119.8	-	-	-	-	-
150	75.43	78.84	84.93	90.54	96.12	101.7	107.3	112.8	-	-	-	-	-
155	69.65	73.22	79.57	85.20	90.48	95.78	101.3	106.6	119.1	-	-	-	-
160	64.09	67.81	74.41	80.22	85.47	90.52	95.63	100.8	112.2	122.7	-	-	-
165	58.90	62.81	69.53	75.48	80.70	85.75	90.63	95.40	106.0	116.3	-	-	-
170	54.15	58.21	65.01	70.99	76.20	81.27	86.06	90.61	100.2	110.0	119.8	-	-
180	45.41	49.77	57.01	63.00	68.28	73.34	77.98	82.22	90.80	99.40	108.5	117.5	126.2
190	38.34	42.71	50.16	56.23	61.59	65.53	70.50	75.50	83.20	91.10	98.90	107.4	115.8
200	33.38	37.35	44.49	50.55	55.93	60.70	65.14	69.30	77.10	84.70	91.30	98.80	106.3
210	30.20	33.61	40.07	45.84	51.06	55.72	60.05	64.07	71.90	78.70	85.20	91.90	98.60
220	28.22	31.10	36.79	42.11	47.08	51.57	55.83	59.72	67.04	73.70	80.20	86.50	92.50
230	27.00	29.45	34.40	39.23	43.82	48.11	52.19	56.01	63.05	69.60	75.80	81.80	87.60
240	26.28	28.37	32.71	37.04	41.26	45.28	49.14	52.81	59.62	66.13	72.10	77.70	83.30
250	25.87	27.68	31.51	35.40	39.24	42.99	46.62	50.10	56.65	62.92	68.47	74.20	79.40
260	25.67	27.26	30.66	34.17	37.67	41.13	44.52	47.80	54.07	60.11	65.54	71.00	76.00
270	25.60	27.02	30.05	33.23	36.44	39.64	42.77	45.86	51.83	57.65	62.90	68.00	72.90
280	25.65	26.92	29.66	32.55	35.48	38.44	41.36	44.25	49.93	55.48	60.53	65.48	70.31
290	25.77	26.94	29.43	32.07	34.77	37.50	40.24	42.94	48.31	53.58	58.42	63.27	67.89
300	25.95	27.02	29.31	31.73	34.23	36.77	39.33	41.86	46.92	51.95	56.59	61.28	65.73
310	26.18	27.17	29.28	31.51	33.83	36.20	38.59	40.96	45.74	50.51	54.99	59.47	63.80
320	26.43	27.34	29.30	31.37	33.54	35.75	37.98	40.22	44.73	49.27	53.59	57.85	62.06
330	26.70	27.55	29.38	31.31	33.33	35.41	37.50	39.62	43.86	48.19	52.33	56.43	60.51
340	26.99	27.80	29.50	31.32	33.21	35.16	37.13	39.13	43.14	47.26	51.21	55.17	59.10
350	27.30	28.06	29.67	31.37	33.16	34.99	36.86	38.73	42.55	46.45	50.24	54.00	57.81
400	29.01	29.60	30.85	32.09	33.52	34.93	36.38	37.84	40.85	43.93	46.99	50.10	53.23
450	30.82	31.31	32.32	33.39	34.50	35.64	36.81	38.01	40.46	42.98	45.49	48.08	50.69
500	32.65	33.06	33.92	34.82	35.75	36.71	37.69	38.69	40.75	42.88	45.02	47.21	49.43
550	34.44	34.79	35.54	36.32	37.12	37.95	38.80	39.69	41.44	43.27	45.12	47.02	48.94
600	36.19	36.50	37.16	37.85	38.55	39.29	40.03	40.79	42.35	43.96	45.59	47.26	48.95
700	39.58	39.83	40.37	40.92	41.49	42.08	42.68	43.29	44.54	45.83	47.14	48.49	49.85
800	42.79	43.01	43.45	43.92	44.40	44.89	45.39	45.90	46.94	48.02	49.12	50.24	51.38
900	45.82	46.01	46.39	46.79	47.20	47.62	48.05	48.49	49.39	50.32	51.25	52.21	53.19
1000	48.68	48.84	49.18	49.52	49.88	50.24	50.64	51.00	51.80	52.61	53.43	54.27	55.12
1100	51.42	51.57	51.86	52.17	52.49	52.82	53.16	53.49	54.19	54.91	55.64	56.39	57.15
1200	54.05	54.18	54.45	54.72	55.01	55.30	55.61	55.91	56.53	57.18	57.84	58.51	59.19
1300	56.56	56.68	56.92	57.17	57.43	57.70	57.96	58.25	58.82	59.40	60.00	60.61	61.22

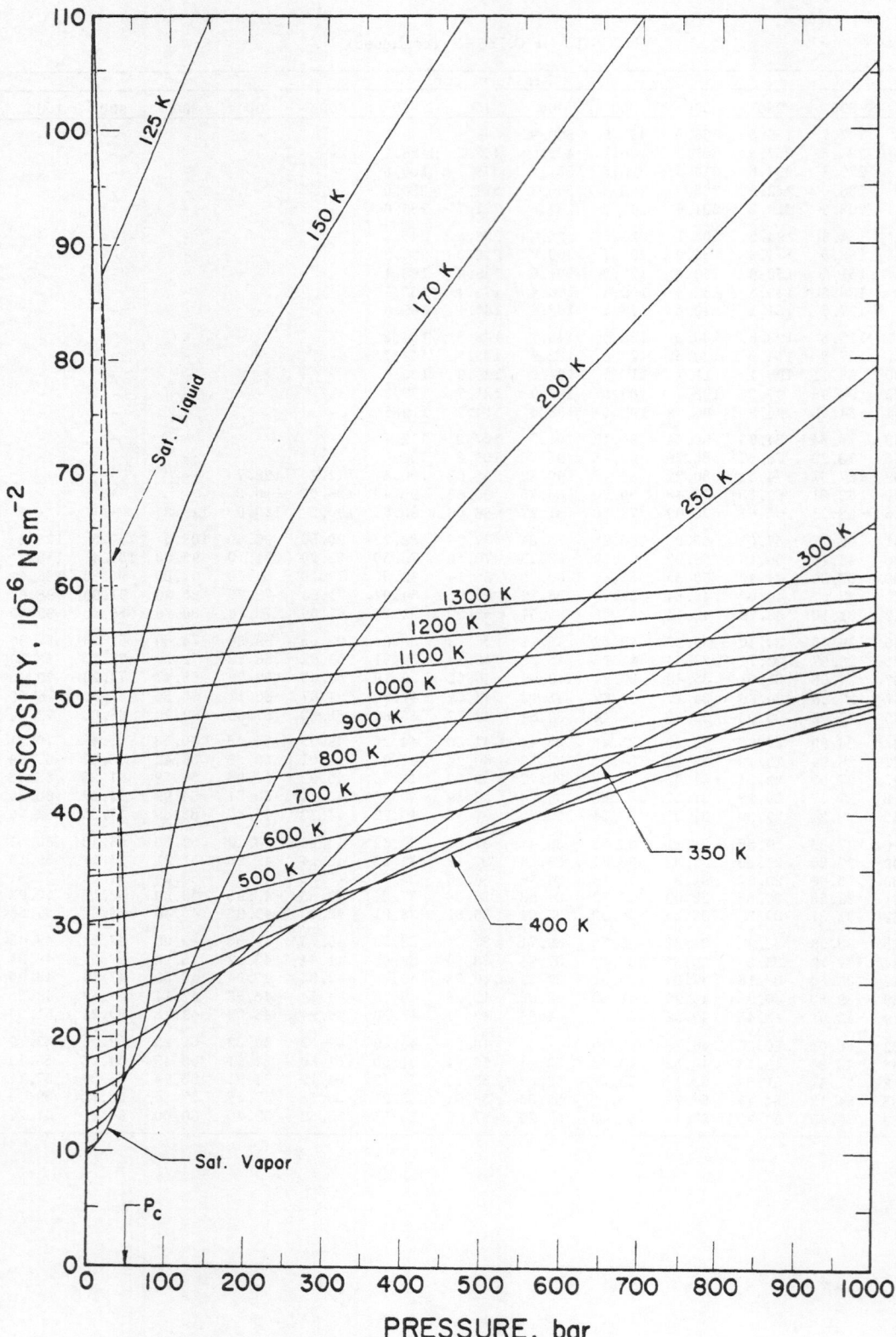

FIGURE 1. VISCOSITY OF OXYGEN [203].

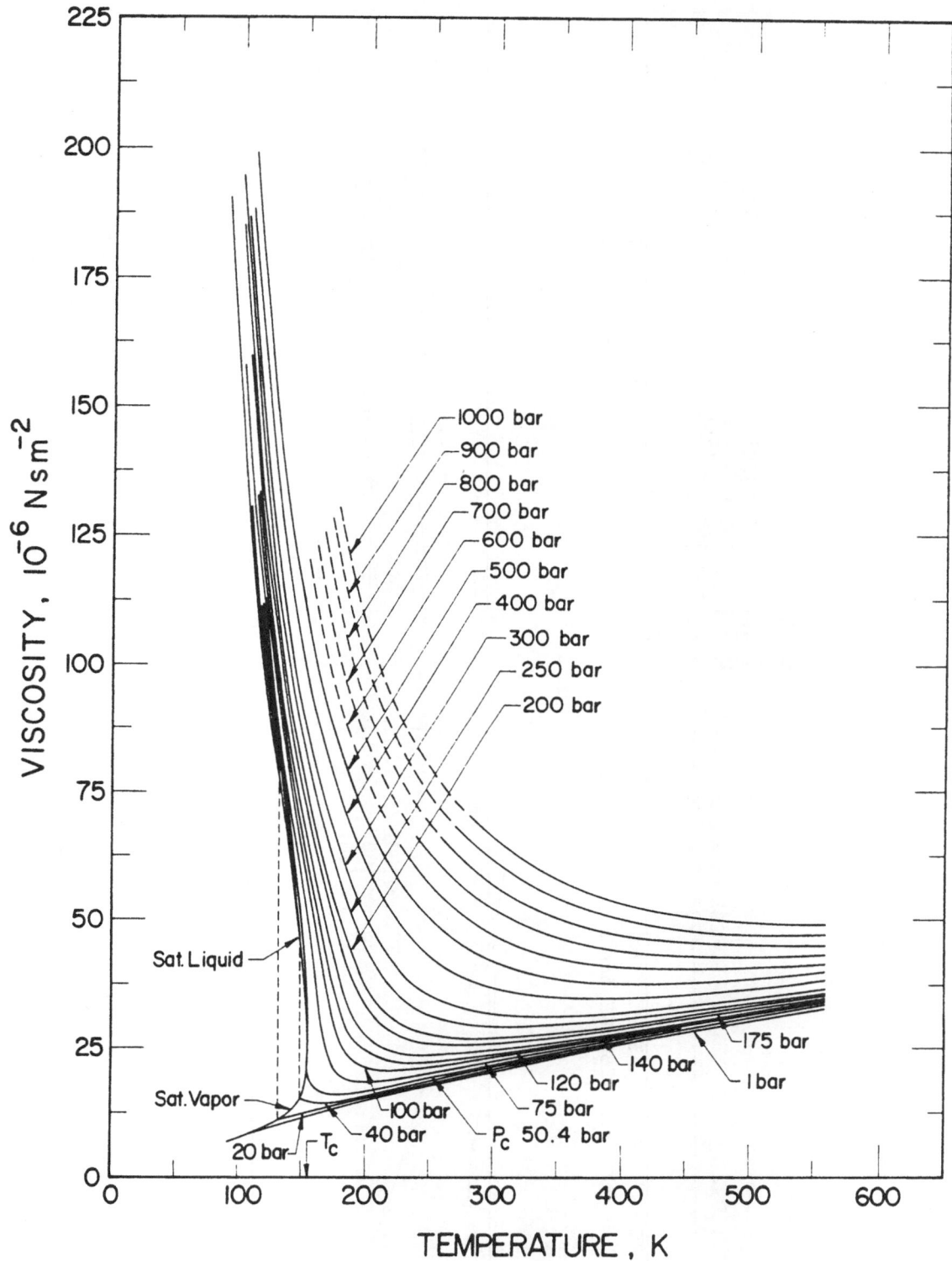

FIGURE 2. VISCOSITY OF OXYGEN [203].

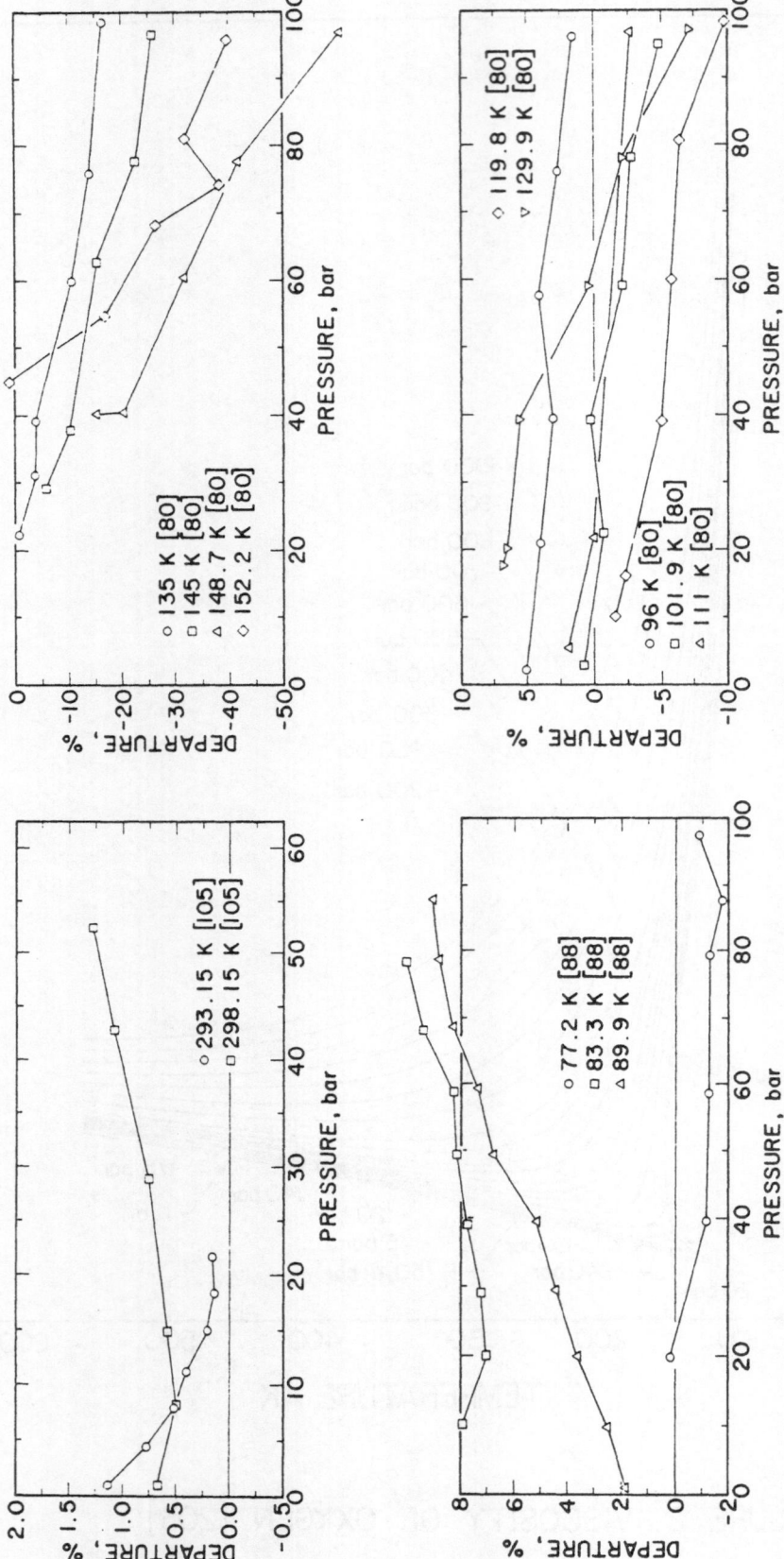

FIGURE 3A. DEPARTURE PLOT ON THE VISCOSITY OF OXYGEN.

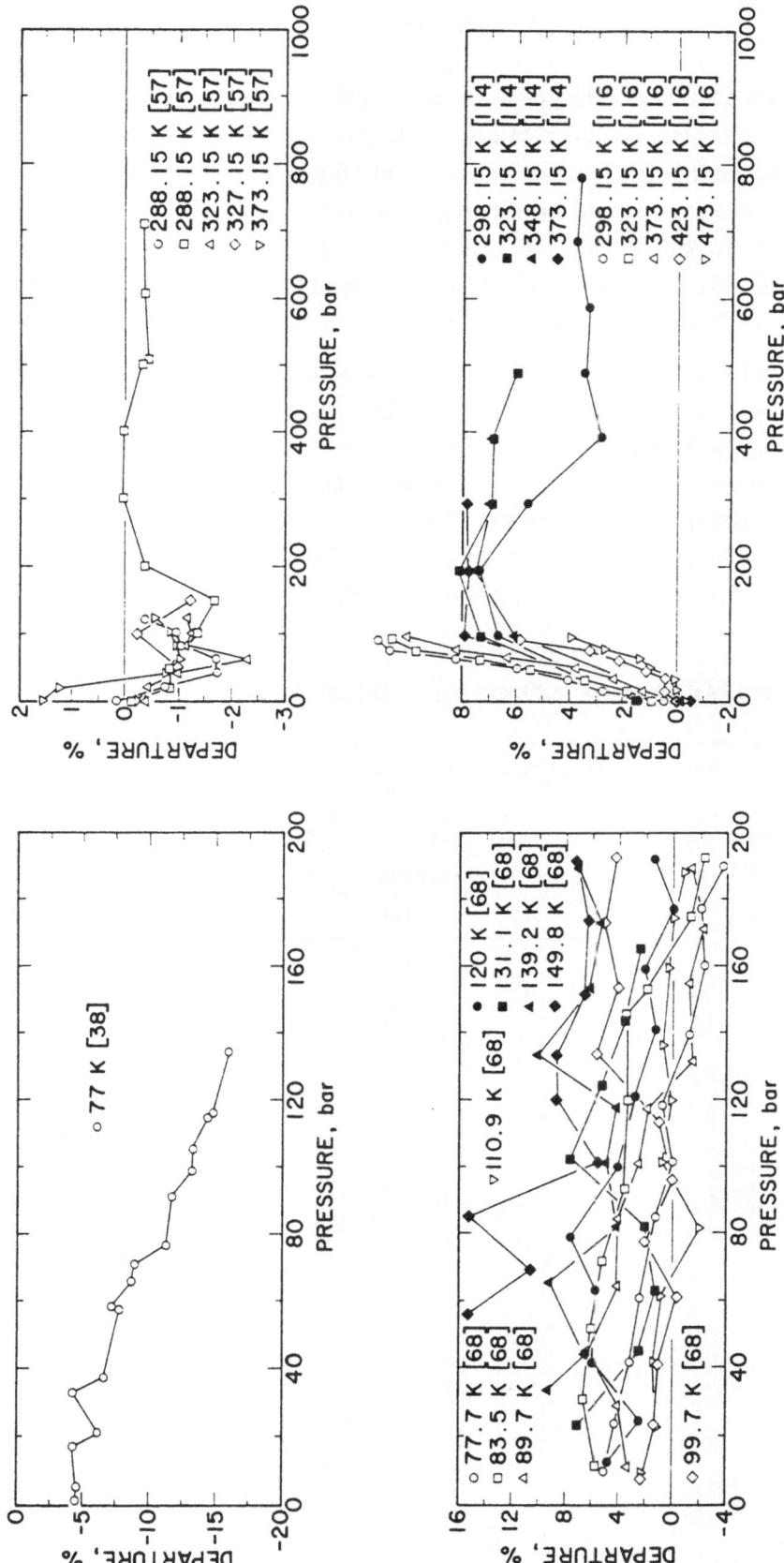

FIGURE 3B. DEPARTURE PLOT ON THE VISCOSITY OF OXYGEN.

i-PENTANE

The data of Yusibova and Agaev [209] obtained by a capillary tube viscosimeter were used to generate the recommended values tabulated on the next page. The data are plotted as isotherms against pressure in Figures 1A and 1B, where smoothing modifications of 1% maximum were applied. Using the
P-V-T data of [9] with some extrapolation, the dilute gas viscosities of [126], also, with some extrapolation, and the residual viscosity concept, an extrapolation from 523 K to 750 K up to 200 bar was made.
The values are shown as isobars as a function of temperature in Figures 2A and 2B. Dashed lines indicate
extrapolated values. The recommended values were read from the latter figures.

Additional works on the pressure dependence of the viscosity of i-pentane are listed in the summary
table below. A comparison with the data of Collings et al. [34] revealed deviations of 2%. The older data
of Bridgman [21] deviate by as much as 11%, while a comparison was not possible with the newer data of
Bridgman [22]. The uncertainty in the recommended values is estimated to be ± 5% in the measured region. Due to the extensive extrapolations of the P-V-T data and the dilute gas viscosities as well as due to
some irregularities found in the residual viscosity plot, the accuracy of recommended values deteriorate
somewhat in the extrapolated region and is estimated to be ± 10%.

ADDITIONAL REFERENCES ON THE VISCOSITY OF i-PENTANE

Authors	Year	Ref. No.	Temperature K	Pressure bar	Method	Departure % (no. points)
Collings and McLaughlin	1971	34	303-323	1-2942	Torsional crystal	2.1
Bridgman	1949	22	296	1-29420	Falling weight	–
Bridgman	1926	21	303-348	1-11768	Falling weight	11.0

VISCOSITY OF I-PENTANE
$[\mu, 10^{-6} \text{ N s m}^{-2}]$

T, K	\multicolumn{20}{c}{Pressure, bar}
	1	10	20	33.7*	40	45	50	60	70	80	90	100	120	140	160	200	300	400	500	600
280	258.5	261.0	263.0	268.0	270.0	271.0	272.5	275.0	278.0	284.0	284.0	287.0	293.5	300.0	306.0	319.0	350.0	380.0	411.0	443.0
300	212.5	215.0	218.0	221.0	223.0	225.0	226.0	228.5	231.0	234.0	237.0	239.5	244.5	250.0	255.0	266.0	293.0	319.5	347.0	373.0
320	7.66	177.0	180.0	183.0	184.5	186.0	187.0	190.0	192.0	195.0	197.0	199.5	204.0	209.0	213.5	223.0	246.5	270.0	293.5	316.0
340	8.15	147.0	149.5	153.0	154.5	156.0	157.0	159.0	161.0	164.0	166.0	168.5	172.5	177.0	181.5	190.0	211.5	232.5	253.0	274.0
360	8.62	122.5	125.0	128.0	130.0	131.0	132.5	134.5	137.0	139.0	141.0	143.5	147.5	152.0	156.0	164.0	183.5	202.5	221.5	240.0
380	9.10	102.0	104.4	107.8	109.5	110.8	112.0	114.0	116.0	118.4	120.4	122.5	126.4	130.4	134.4	142.4	160.4	177.9	195.3	213.4
400	9.60	–	87.3	90.8	92.6	94.0	95.4	98.0	100.6	102.6	104.8	106.8	110.4	114.0	118.2	125.8	143.1	159.6	176.0	192.4
420	10.05	–	72.0	76.4	78.4	80.0	81.5	83.3	86.0	89.4	90.8	93.0	96.8	100.5	104.8	111.7	128.4	144.0	159.2	174.5
440	10.53	–	–	61.3	63.6	66.0	68.0	70.4	73.4	75.9	78.5	81.0	85.1	89.2	93.2	100.0	116.0	131.0	145.2	159.3
450	10.77	–	–	50.8	55.5	58.4	61.1	64.4	67.6	70.4	73.0	75.6	80.0	84.0	88.0	93.7	110.7	125.2	139.2	152.7
455	10.90	–	–	43.2	51.1	54.6	57.5	61.4	64.8	67.7	70.4	73.0	77.5	81.5	85.6	92.3	108.2	122.6	136.3	149.6
460	11.00	–	–	–	46.7	50.8	54.0	58.4	62.0	65.2	67.9	70.6	75.1	79.2	83.2	90.0	105.8	120.2	133.6	146.7
470	11.25	11.6	12.8	16.8	–	42.0	46.6	52.4	56.6	60.2	63.2	65.8	70.4	74.8	78.6	85.6	101.4	105.4	128.5	141.2
480	11.50	11.8	13.0	16.0	20.4	30.8	38.9	46.3	51.2	55.6	58.7	61.4	66.4	70.6	74.4	81.6	97.4	111.2	124.0	136.1
490	11.73	12.0	13.2	16.0	18.7	23.3	30.5	40.0	46.0	50.9	54.4	57.2	62.4	66.8	70.7	78.0	93.6	107.3	119.8	131.6
500	11.97	12.3	13.4	16.1	18.2	20.5	24.0	33.6	91.0	46.4	50.2	53.3	58.7	63.2	67.2	74.7	90.4	103.7	116.0	127.6
510	12.20	12.5	13.6	16.1	17.8	19.2	21.3	28.0	36.2	42.0	46.0	49.5	55.2	59.9	64.0	71.6	87.2	100.4	112.6	123.8
520	12.45	12.8	13.8	16.2	17.6	18.8	20.5	25.4	32.0	37.8	42.4	46.0	51.5	56.8	61.0	68.7	84.4	97.5	109.5	120.4
530	12.70	13.0	14.0	16.2	17.4	18.5	20.0	24.0	29.1	34.4	38.9	42.8	48.8	54.0	58.4	66.1				
540	12.92	13.2	14.3	16.3	17.3	18.4	19.8	23.1	27.3	32.0	36.0	40.0	46.2	51.6	56.3	64.0				
560	13.40	13.7	14.7	16.4	17.2	18.1	19.3	22.0	25.0	28.6	32.3	35.8	42.2	47.8	52.7	60.4				
580	13.87	14.2	15.1	16.6	17.3	18.0	19.0	21.3	23.6	26.4	29.8	33.1	39.2	44.8	49.7	57.6				
600	14.35	14.7	15.6	16.8	17.4	18.1	18.9	20.8	22.7	25.2	28.1	31.1	36.8	42.3	47.3	55.3				
650	15.55	16.0	16.6	17.4	18.0	18.5	19.0	20.4	22.0	23.7	25.6	27.8	33.0	37.8	42.7	50.8				
700	16.75	17.2	17.6	18.3	18.9	19.3	19.6	20.7	22.0	23.2	24.7	26.4	30.6	35.0	39.4	47.6				
750	17.92	18.4	18.7	19.5	20.0	20.4	20.7	21.4	22.4	23.5	24.7	26.1	29.2	33.2	37.3	45.2				

*Critical pressure.

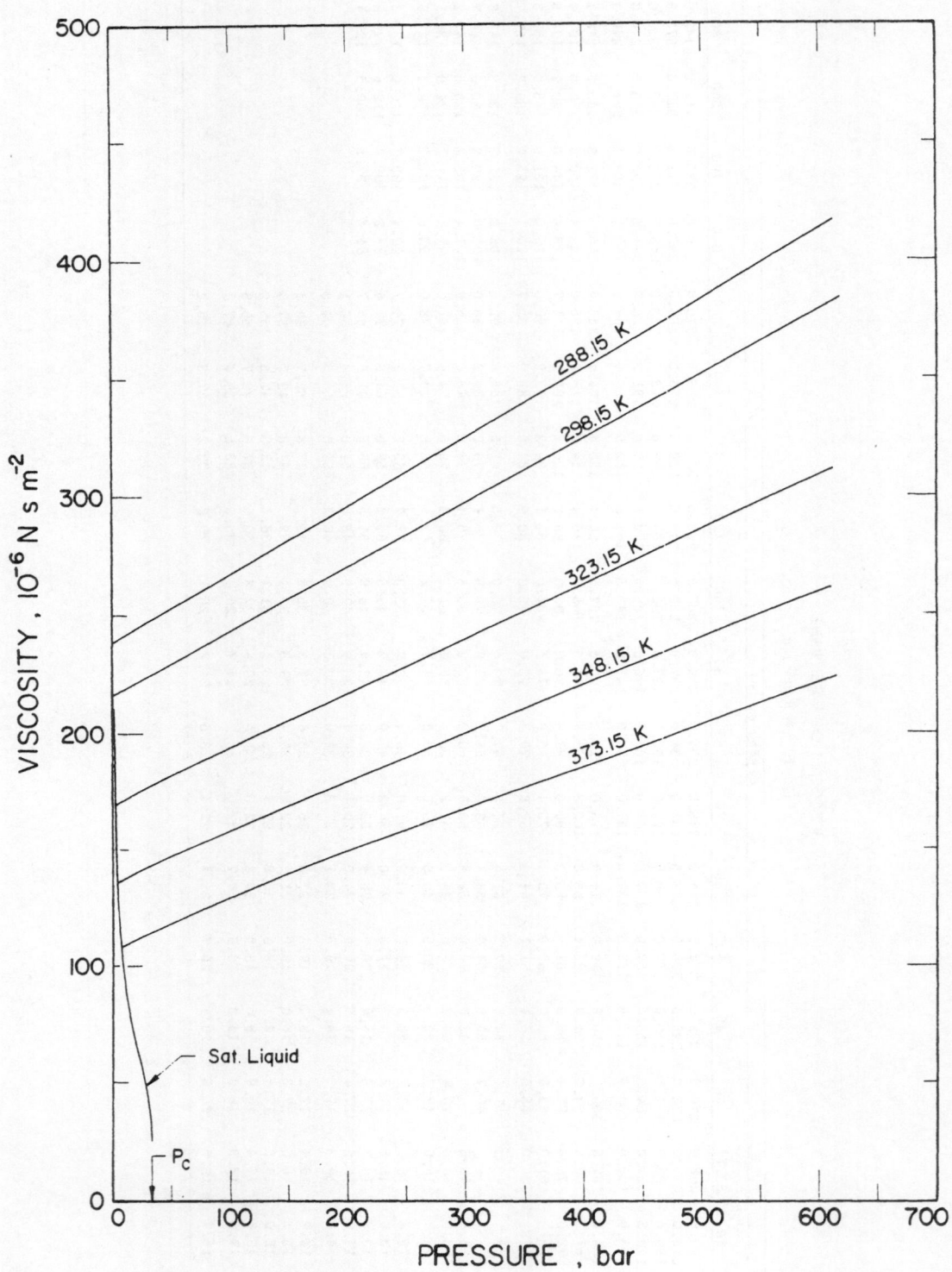

FIGURE 1A. VISCOSITY OF i-PENTANE [209].

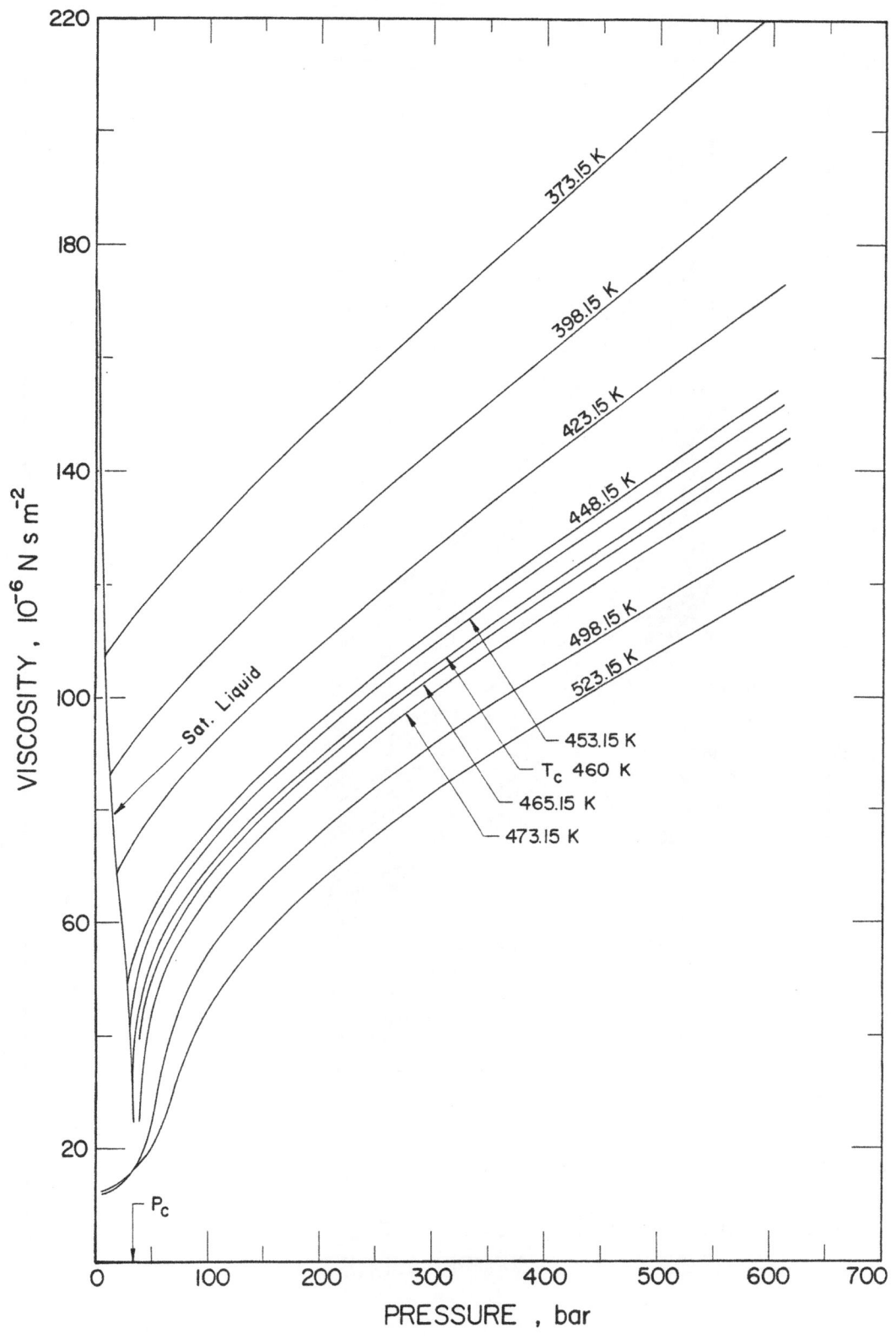

FIGURE IB. VISCOSITY OF i-PENTANE [209].

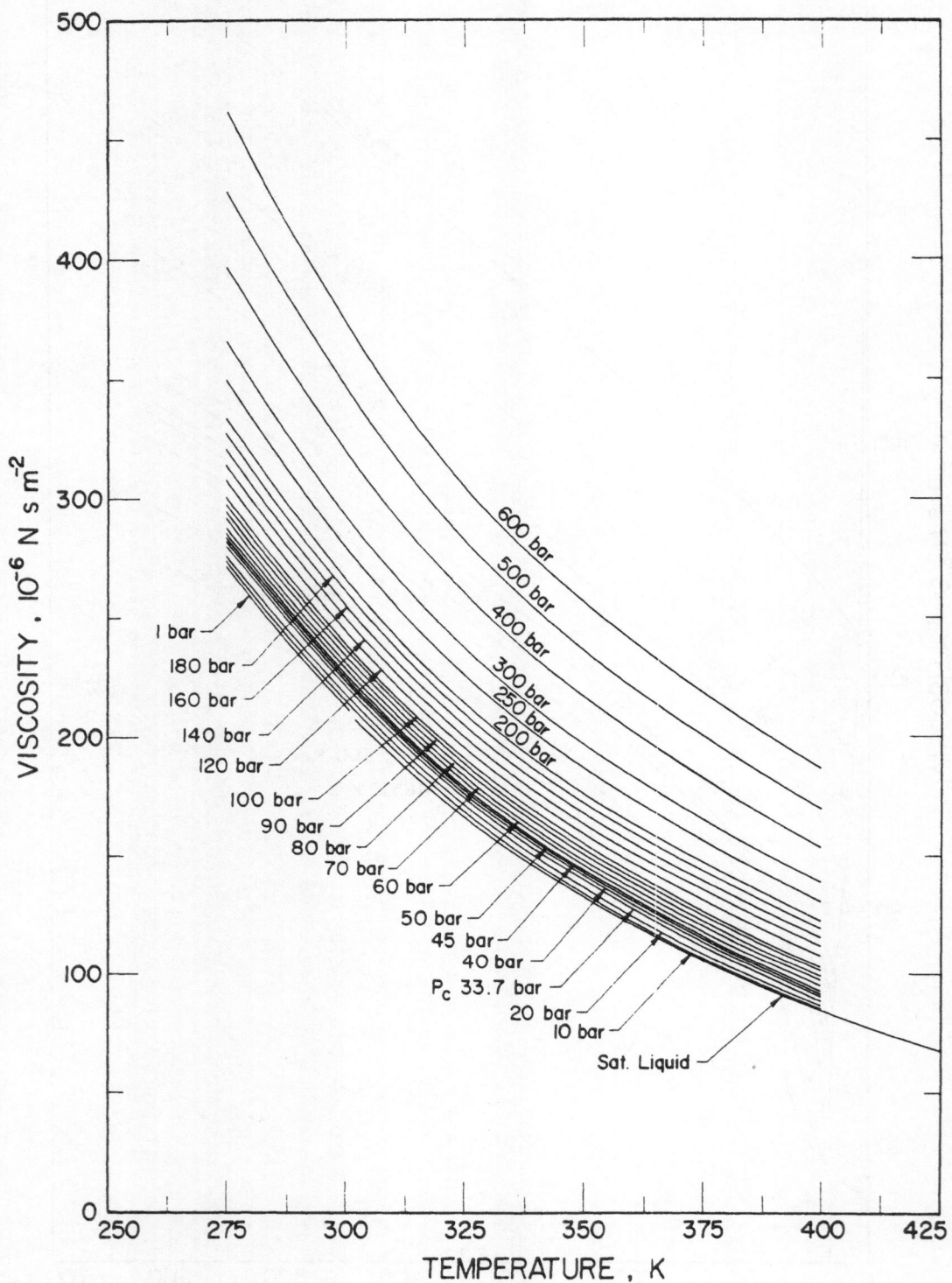

FIGURE 2A. VISCOSITY OF i-PENTANE [209].

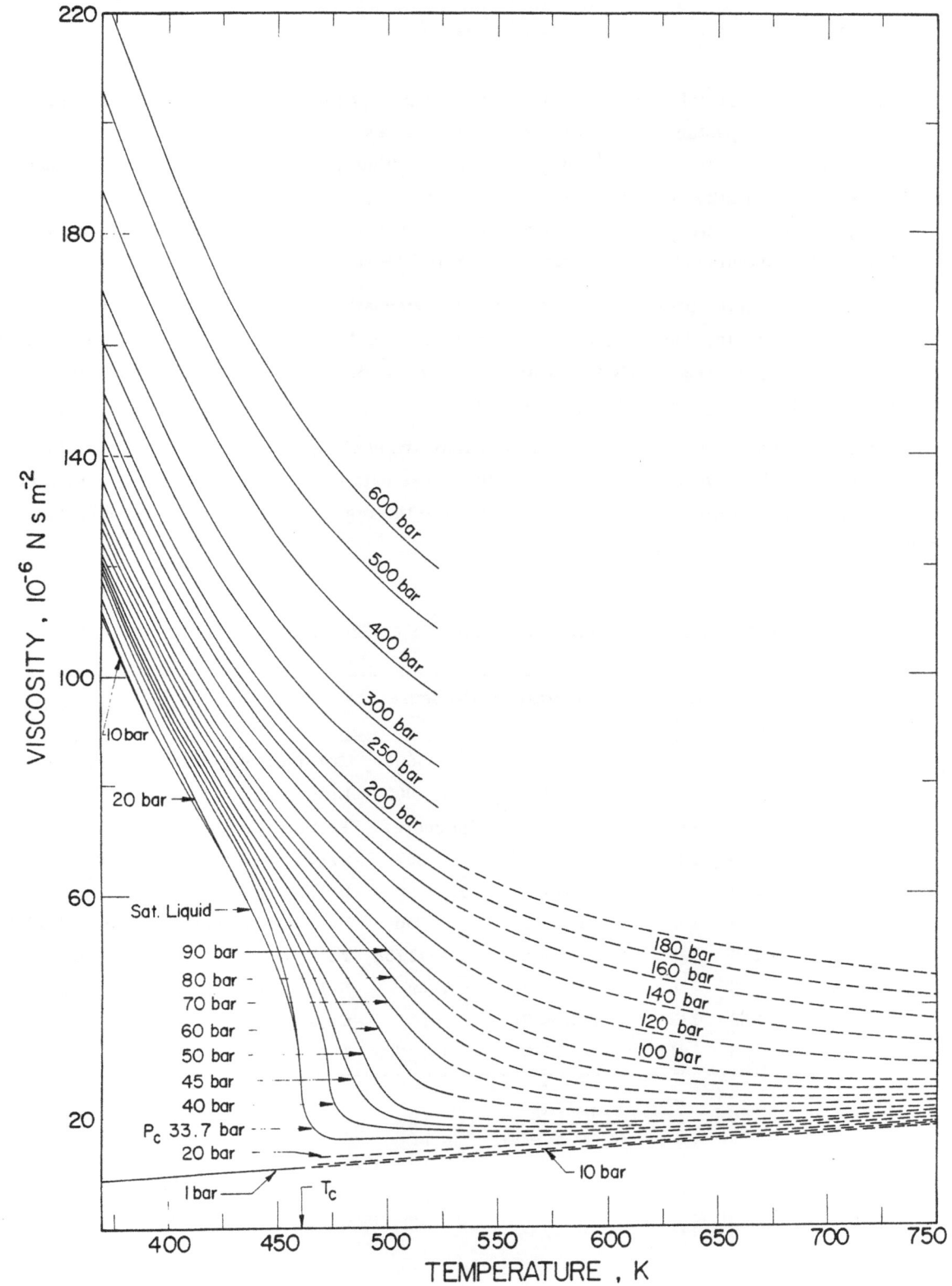

FIGURE 2B. VISCOSITY OF i-PENTANE [209].

n-PENTANE

The data of Agaev and Golubev [1] obtained by the capillary tube method were used to generate the recommended values for n-pentane which are tabulated on the next page and presented in Figure 1 as isotherms as a function of pressure. Extrapolating linearly the dilute gas values, using the P-V-T data of [143] and the residual viscosity concept, it was possible to extrapolate the data from 548 K to 950 K up to 150 bar. In Figure 2 isobars are plotted as a function of temperature. The extrapolated values are shown as dashed lines. The recommended values were read from Figure 2.

Additional works on the pressure dependence of the viscosity of n-pentane are listed in the summary table below. As seen from the departure plots in Figure 3, the data of Lee et al. [131] also obtained by a capillary tube method, agree reasonably well with the recommended values. The same may be stated about the data of Collings et al. [34] and of Bridgeman [21].

However, the agreement with the older data of Hubbard et al. and Reamer et al. is quite unsatisfactory as shown by the high deviations. It appears that these data have systematic errors. The uncertainty in the recommended values is estimated to be ± 5% in the region of experiments and ± 7% in the extrapolated region.

ADDITIONAL REFERENCES ON THE VISCOSITY OF n-PENTANE

Authors	Year	Ref. No.	Temperature K	Pressure bar	Method	Departure % (no. points)
Collings and McLaughlin	1971	34	303-323	1-6865	Torsional crystal	1.5
Brazier and Freeman	1969	20	303	1-4000	Rolling ball	-
Kozlov et al.	1966	119	293-453	Isochores	Falling cylinder	-
Lee and Ellington	1965	131	311-444	14-207	Capillary tube	1.69 ± 1.95 (32)
Babb and Scott	1964	10	303	2000-10000	Rolling ball	-
Reamer et al.	1959	165	311-411	7-350	Rotating cylinder	12.05 ± 3.76 (18)
Bridgman	1949	22	303-348	1-11768	Falling weight	4.0
Hubbard and Brown	1943	86	298-523	7-69	Rolling ball	12.26 ± 9.57 (41)
Khalilov	1939	113	293-423	SL	Capillary tube	-
Bridgman	1926	21	303-348	1-11768	Falling weight	4.0

VISCOSITY OF n-PENTANE
[μ, 10^{-6} N s m^{-2}]

T, K	\multicolumn{18}{c}{Pressure, bar}
	1	20	30	33.8*	40	45	50	60	70	80	100	110	130	150	200	300	400	500
320	7.2	184.4	186.7	187.6	189.0	190.2	191.3	193.4	195.5	197.6	201.8	203.8	207.7	211.6	222.6	244.5	266.5	288.0
340	7.7	156.1	158.9	160.0	161.7	163.1	164.5	166.7	168.9	171.0	175.4	177.2	180.9	184.5	194.5	212.8	231.8	249.6
360	8.3	131.8	134.4	135.3	136.9	138.2	139.5	141.8	144.0	146.3	150.8	152.7	156.4	160.1	169.6	187.9	205.3	220.0
380	8.7	111.5	114.0	114.9	116.4	117.7	118.9	121.1	123.3	125.6	130.0	131.9	135.7	139.5	148.5	165.7	183.3	197.4
400	9.2	94.4	96.5	97.3	98.7	99.9	101.0	103.8	106.1	108.3	112.4	114.4	118.4	122.0	131.0	147.3	163.9	179.1
420	9.7	77.3	80.3	81.3	82.9	84.3	85.6	87.9	90.4	92.5	96.9	99.0	103.0	106.7	115.5	132.1	147.4	162.5
440	10.1	77.9	64.0	66.0	68.1	69.8	71.5	74.0	76.7	79.4	83.8	86.2	90.4	94.1	103.2	119.2	134.0	148.4
490	11.2	12.2	14.1	16.5	21.8	28.8	35.0	43.0	48.1	52.2	58.6	61.0	65.8	70.3	79.5	95.2	109.2	122.2
495	11.3	12.3	14.0	16.0	20.2	25.4	31.0	39.4	45.3	49.6	56.4	58.9	63.7	68.3	77.5	93.1	107.0	120.0
500	11.4	12.4	14.0	15.6	19.1	23.2	27.8	35.7	42.5	47.0	54.3	56.9	61.7	66.3	75.5	91.0	105.0	117.9
510	11.6	12.5	14.0	15.1	17.8	20.4	23.5	30.5	37.4	42.4	50.0	52.9	57.9	62.5	71.7	87.1	101.0	113.8
520	11.8	12.7	14.1	14.9	16.9	18.9	21.2	27.0	33.1	38.2	46.0	49.1	54.2	58.9	68.1	83.3	97.1	109.9
530	12.0	12.8	14.1	14.9	16.5	18.2	20.1	24.7	29.6	34.5	42.4	45.5	50.7	55.3	64.6	79.6	93.5	106.1
540	12.2	13.0	14.2	14.8	16.3	17.7	19.3	23.0	27.2	31.5	39.0	42.3	47.4	52.0	61.4	76.3	90.0	102.5
550	12.4	13.2	14.3	14.9	16.2	17.5	18.7	22.0	25.5	29.3	36.2	39.1	44.4	48.7	58.0	73.0	86.5	99.0
560	12.6	13.4	14.5	14.9	16.2	17.3	18.4	21.5	24.6	27.7	33.9	36.3	41.1	45.9				
580	12.9	13.7	14.8	15.2	16.2	17.0	17.8	20.3	22.9	25.4	30.5	32.6	36.9	41.1				
600	13.4	14.1	15.0	15.4	16.2	16.9	17.6	19.7	21.8	24.0	28.2	30.1	34.0	37.8				
620	13.7	14.5	15.4	15.8	16.4	17.0	17.5	19.3	21.1	23.0	26.6	28.4	31.9	35.4				
640	14.1	14.8	15.6	16.0	16.6	17.1	17.6	19.2	20.8	22.3	25.5	27.1	30.4	33.6				
660	14.5	15.2	16.1	16.4	16.9	17.4	17.8	19.2	20.5	21.9	24.6	26.1	29.1	32.1				
680	14.9	15.5	16.3	16.6	17.1	17.5	17.9	19.1	20.3	21.6	24.0	25.4	28.2	31.0				
700	15.2	15.9	16.6	16.9	17.4	17.7	18.1	19.2	20.3	21.4	23.6	24.9	27.4	30.0				
750	16.0	16.7	17.3	17.5	17.9	18.2	18.5	19.4	20.4	21.3	23.2	24.2	26.3	28.4				
800	16.8	17.5	18.1	18.3	18.6	18.8	19.0	19.8	20.6	21.4	23.0	23.9	25.7	27.5				
850	17.6	18.3	18.8	19.0	19.2	19.3	19.5	20.2	20.9	21.7	23.1	23.9	25.4	27.0				
900	18.4	19.0	19.4	19.6	19.8	20.0	20.2	20.8	21.4	22.1	23.3	27.0	25.5	26.9				

*Critical pressure.

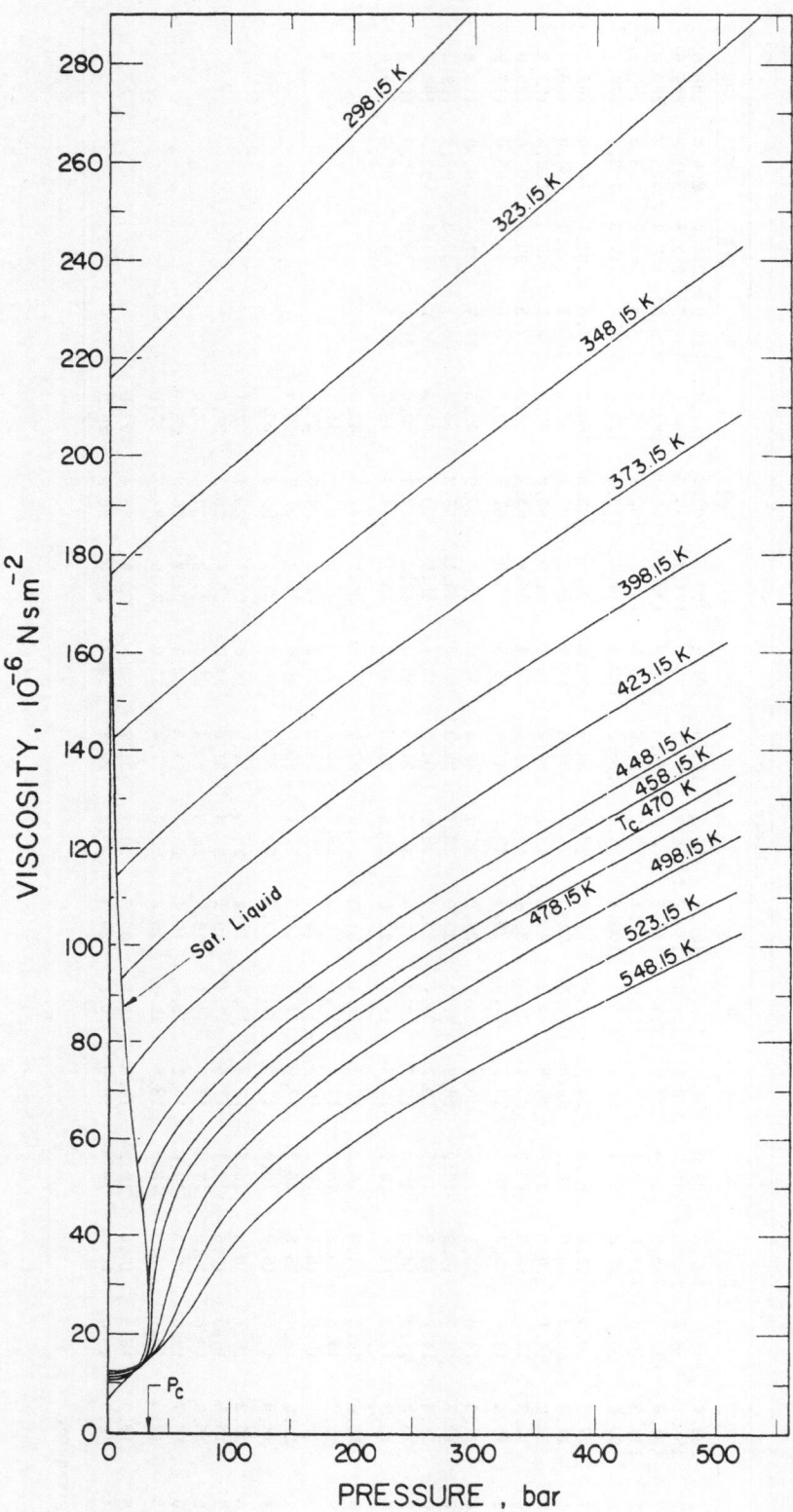

FIGURE 1. VISCOSITY OF n-PENTANE [1].

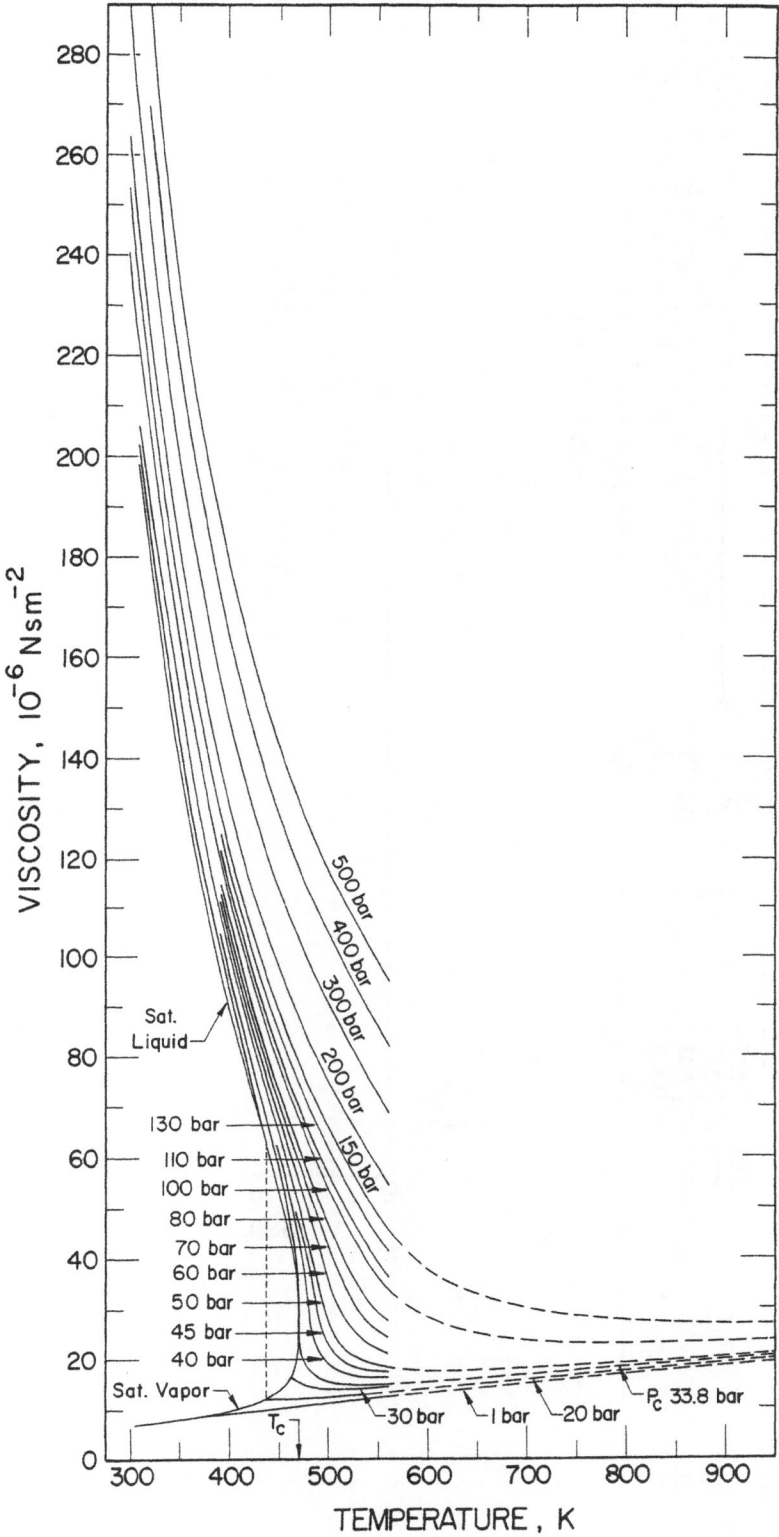

FIGURE 2. VISCOSITY OF n-PENTANE [1].

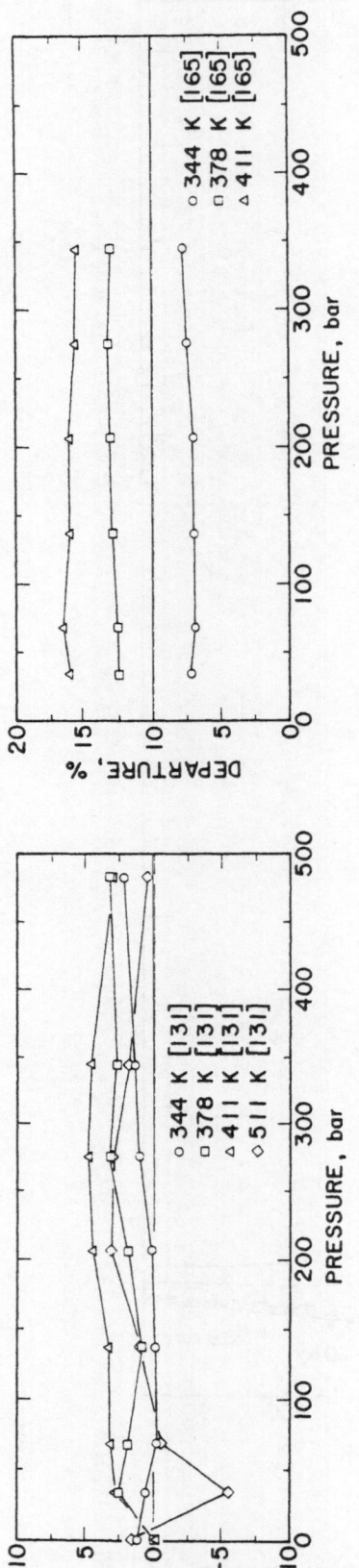

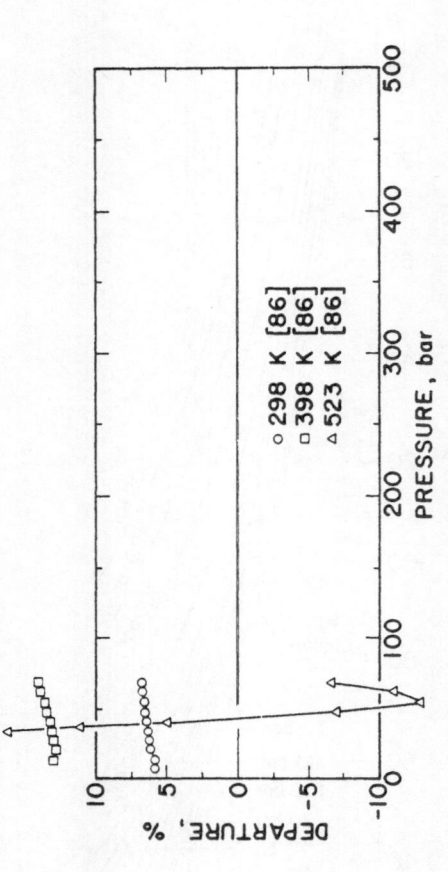

FIGURE 3. DEPARTURE PLOT ON THE VISCOSITY OF *n*-PENTANE.

PROPANE

A considerable volume of data exists for the viscosity of propane in all fluid states. The data of Huang et al. [85] and that of Carmichael et al. [24] were used to generate the recommended values, presented on the next page. These two sets of data were internally consistant and the smoothing corrections were only 0. 8%. Using P-V-T data for propane [120, 143] it was possible to extrapolate the measured data from 478 K to 750 K for pressures up to 250 bars. Figure 1 shows the measured isotherms as a function of pressure. Figures 2A and 2B show the interpolated isobars. Figure 2A refers specifically to the liquid region showing the good consistency between the two works selected to generate the recommended values. In Figure 2B the extrapolated curves are shown as dashed lines. The precision of the measured data is claimed to be quite high, though smoothing the data of [24] introduced a maximum modification of 1.7%.

Additional works on high pressure viscosity of propane are listed in the summary table below. The data of Giddings et al. [54] Starling et al. [185] and of Golubev et al. [57] show only minor deviations from the recommended values as seen from the departure plots of Figure 3. The data of Baron et al. [11] show somewhat higher deviations, while the data of Sage et al. [175] showing very large deviations are probably in error. For the data of Eakin et al. [43] Smith et al. [182] Bicher et al. [15] and Comings et al. [36] only the maximum percent deviations with respect to the recommended values are given in the summary table below. The earlier data of the latter three works do not appear to have an accuracy comparable to that of the recommended values. The uncertainty in the recommended values up to 478 K is estimated to be about ± 2%, while that of the extrapolated values is considered to be about ± 7%.

ADDITIONAL REFERENCES ON THE VISCOSITY OF PROPANE

Authors	Year	Ref. No.	Temperature K	Pressure bar	Method	Departure % (no. points)
Giddings et al.	1966	54	278-378	1-552	Capillary tube	0. 41 ± 1. 20 (48)
Babb and Scott	1964	10	303	2000-10000	Rolling ball	-
Starling et al.	1962	184	363-374	33-52	Capillary tube	-
Starling et al.	1960	185	298-411	7-552	Capillary tube	-0.2 ± 81 (50)
Swift et al.	1960	190	243-370	SL	Falling body	-
Baron et al.	1959	11	325-408	7-552	Capillary tube	1. 15 ± 2. 47 (28)
Eakin and Ellington	1959	43	298	7-621	Capillary tube	1. 2
Swift et al.	1959	189	90-363	SL	Falling body	-
Golubev and Petrov	1953	57	298-523	1-811	Capillary tube	-0.99 ± 1. 80 (36)
Comings et al.	1944	36	303-378	1-42	Capillary tube	4. 6
Bicher and Katz	1943	15	298-498	28-345	Rolling ball	8. 2
Smith and Brown	1943	182	295-463	7-345	Rolling ball	13. 0
Sage and Lacey	1938	175	311-378	1-138	Rolling ball	2. 60 ± 6. 45 (24)

VISCOSITY OF PROPANE
[μ, 10^{-6} N s m^{-2}]

T,K	Pressure, bar																	
	1	20	30	42.5*	50	55	60	70	80	90	100	110	130	150	200	250	300	350
175	419.5	424.5	427.2	430.5	432.5	434.2	435.9	439.3	442.7	446.1	449.5	453.0	460.0	467.0	484.5	503.5	521.0	538.5
180	388.0	392.6	395.1	398.2	400.0	401.5	403.0	406.0	409.0	412.0	415.0	418.2	424.6	431.0	447.0	464.0	480.0	496.0
190	333.0	337.1	339.2	341.9	343.5	344.8	346.1	348.7	351.3	353.9	356.5	359.2	364.6	370.0	384.0	398.0	411.5	425.0
200	289.5	293.4	295.4	298.0	299.5	300.7	301.9	304.3	306.7	309.1	311.5	314.0	319.0	324.0	336.5	348.5	360.5	372.5
210	255.5	259.2	261.1	263.6	265.0	266.2	267.3	269.6	271.9	274.2	276.5	278.8	283.4	288.0	300.0	311.0	322.0	333.0
220	227.0	230.7	232.6	235.1	236.5	237.6	238.6	240.7	242.8	244.9	247.0	249.2	253.6	258.0	268.5	279.0	289.0	300.0
230	201.5	205.2	207.1	209.6	211.0	212.0	213.0	215.0	217.0	219.0	221.0	223.0	227.0	231.0	241.0	250.5	260.0	270.0
240	6.9	183.5	185.3	187.6	189.0	190.0	190.9	192.8	194.7	196.6	198.5	200.4	204.2	208.0	217.0	226.0	234.5	244.0
250	7.1	164.5	166.3	168.6	170.0	170.9	171.8	173.6	175.4	177.2	179.0	180.7	184.1	187.5	196.0	205.0	213.0	221.5
260	7.4	147.5	149.3	151.6	153.0	153.9	154.7	156.4	158.1	159.8	161.5	163.2	166.6	170.0	178.5	187.0	194.5	202.5
270	7.6	133.5	135.2	137.3	138.5	139.4	140.3	142.1	143.9	145.7	147.5	149.1	152.3	155.5	163.5	171.0	178.5	185.5
280	7.8	120.5	122.2	124.3	125.5	126.4	127.2	128.9	130.6	132.3	134.0	135.7	139.1	142.5	150.0	157.5	165.0	171.5
290	8.1	109.0	110.5	112.4	113.5	114.4	115.2	116.9	118.6	120.3	122.0	123.6	126.8	130.0	138.0	145.0	152.5	159.5
300	8.3	96.6	98.6	101.2	102.8	103.8	104.8	106.8	108.8	110.2	111.6	113.2	116.4	119.5	127.1	134.4	141.7	148.1
320	8.8	76.4	78.9	82.0	83.7	84.7	85.7	87.7	89.6	91.4	93.1	94.9	98.4	101.2	109.0	116.1	123.1	129.1
340	9.3	9.9	60.9	65.2	67.5	68.4	69.3	71.9	74.2	76.0	78.1	80.0	83.4	86.4	94.2	101.0	107.6	113.5
360	9.8	10.5	11.7	48.7	51.9	53.8	55.6	57.9	59.8	62.1	64.3	66.0	69.9	73.2	81.2	88.1	94.5	100.3
400	10.7	11.5	12.2	13.6	15.1	16.6	19.4	25.1	31.7	36.1	40.0	43.1	48.0	52.4	60.5	67.4	73.6	79.5
405	10.9	11.7	12.3	13.5	14.8	16.0	18.0	23.1	29.0	33.6	37.7	40.8	45.9	50.2	58.5	65.5	71.7	77.4
410	11.0	11.8	12.4	13.4	14.6	15.6	17.2	21.4	26.8	31.5	35.4	38.7	43.8	48.1	56.6	63.5	69.8	75.4
420	11.2	12.0	12.6	13.4	14.5	15.2	16.3	19.1	23.4	27.6	31.3	34.8	39.9	44.1	52.9	59.9	66.2	71.8
430	11.4	12.2	12.7	13.4	14.4	14.9	15.8	17.9	21.2	24.7	28.1	31.3	36.3	40.5	49.4	56.6	62.9	68.4
440	11.7	12.4	12.9	13.5	14.4	14.8	15.5	17.3	19.8	22.6	25.6	28.4	33.1	37.2	46.4	53.5	59.8	65.3
450	11.9	12.6	13.1	13.6	14.4	14.9	15.4	16.8	19.0	21.2	23.9	26.0	30.6	34.5	43.4	50.6	56.8	62.4
460	12.2	12.8	13.3	13.7	14.5	15.0	15.4	16.6	18.4	20.3	22.5	24.4	28.7	32.4	40.9	47.9	54.0	59.7
470	12.4	13.0	13.5	13.9	14.5	15.1	15.4	16.5	18.1	19.6	21.5	23.2	27.2	30.7	38.8	45.4	51.5	57.1
480	12.6	13.2	13.7	14.0	14.6	15.1	15.5	16.5	17.9	19.2	20.7	22.4	25.9	29.2	36.8	43.2	49.0	54.6
500	13.1	13.6	14.0	14.3	14.9	15.3	15.6	16.6	17.6	18.6	19.6	21.0	23.8	26.8	33.5	39.5	44.8	50.0
520	13.5	14.2	14.5	14.9	15.2	15.6	15.9	16.7	17.4	18.2	18.9	20.1	22.5	24.9	31.0	36.6		
540	14.0	14.5	14.8	15.2	15.4	15.7	16.1	16.7	17.4	18.0	18.7	19.7	21.7	23.7	29.2	34.5		
560	14.4	14.9	15.2	15.5	15.7	16.0	16.3	16.9	17.4	18.0	18.6	19.5	21.2	22.9	27.7	32.8		
580	14.8	15.3	15.6	15.9	16.1	16.6	16.6	17.1	17.6	18.1	18.6	19.4	20.9	22.4	26.7	31.4		
600	15.2	15.7	15.9	16.2	16.4	16.6	16.9	17.3	17.8	18.2	18.7	19.4	20.7	22.1	25.9	30.2		
650	16.2	16.6	16.8	17.0	17.2	17.4	17.6	18.0	18.3	18.7	19.1	19.6	20.7	21.8	25.0	28.3		
700	17.1	17.4	17.6	17.9	18.0	18.2	18.4	18.7	19.1	19.4	19.8	20.2	21.1	21.9	24.4	27.3		
750	18.0	18.3	18.5	18.7	18.8	19.0	19.1	19.4	19.7	20.0	20.3	20.7	21.4	22.1		27.0		

*Critical pressure.

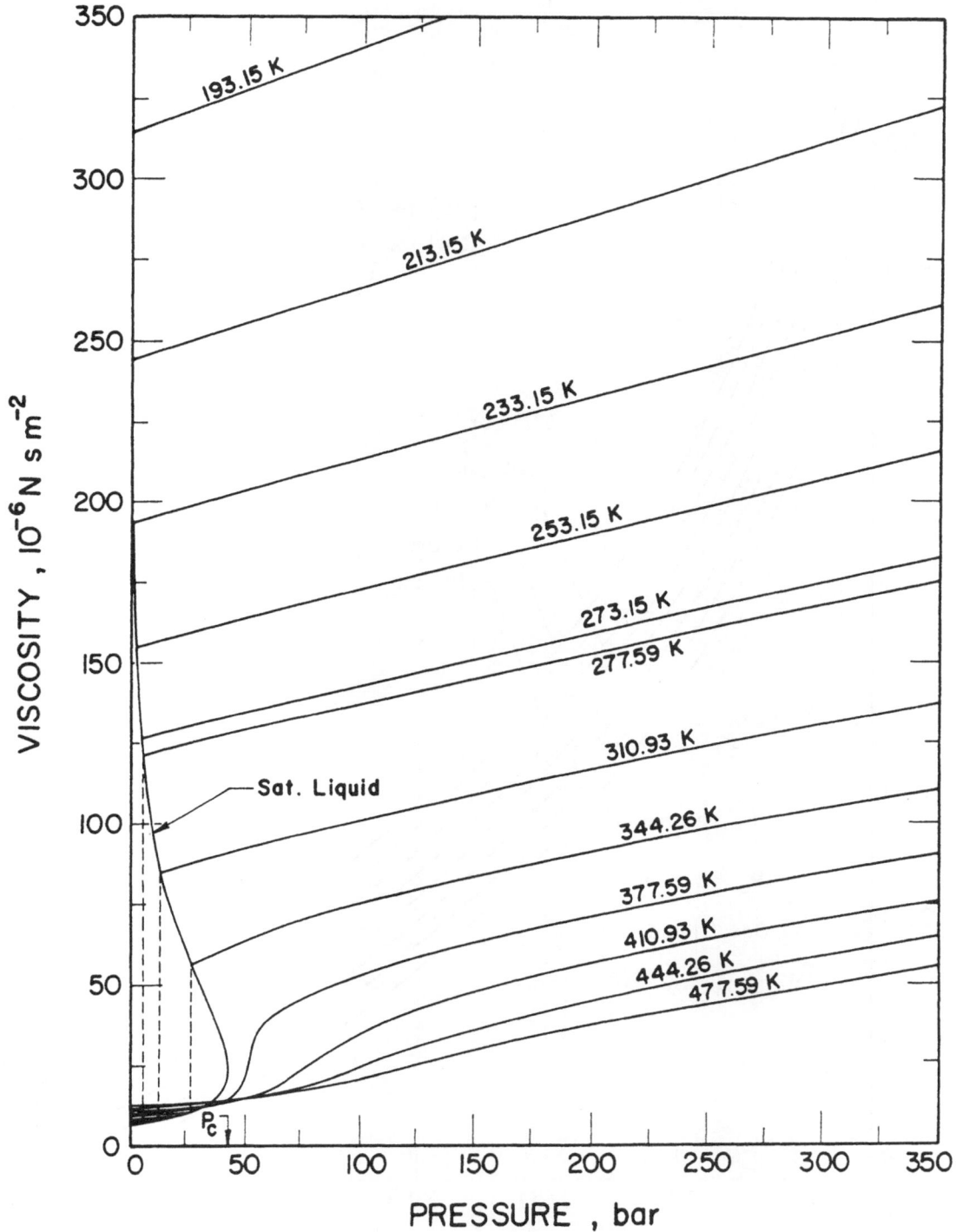

FIGURE 1. VISCOSITY OF PROPANE [24, 85].

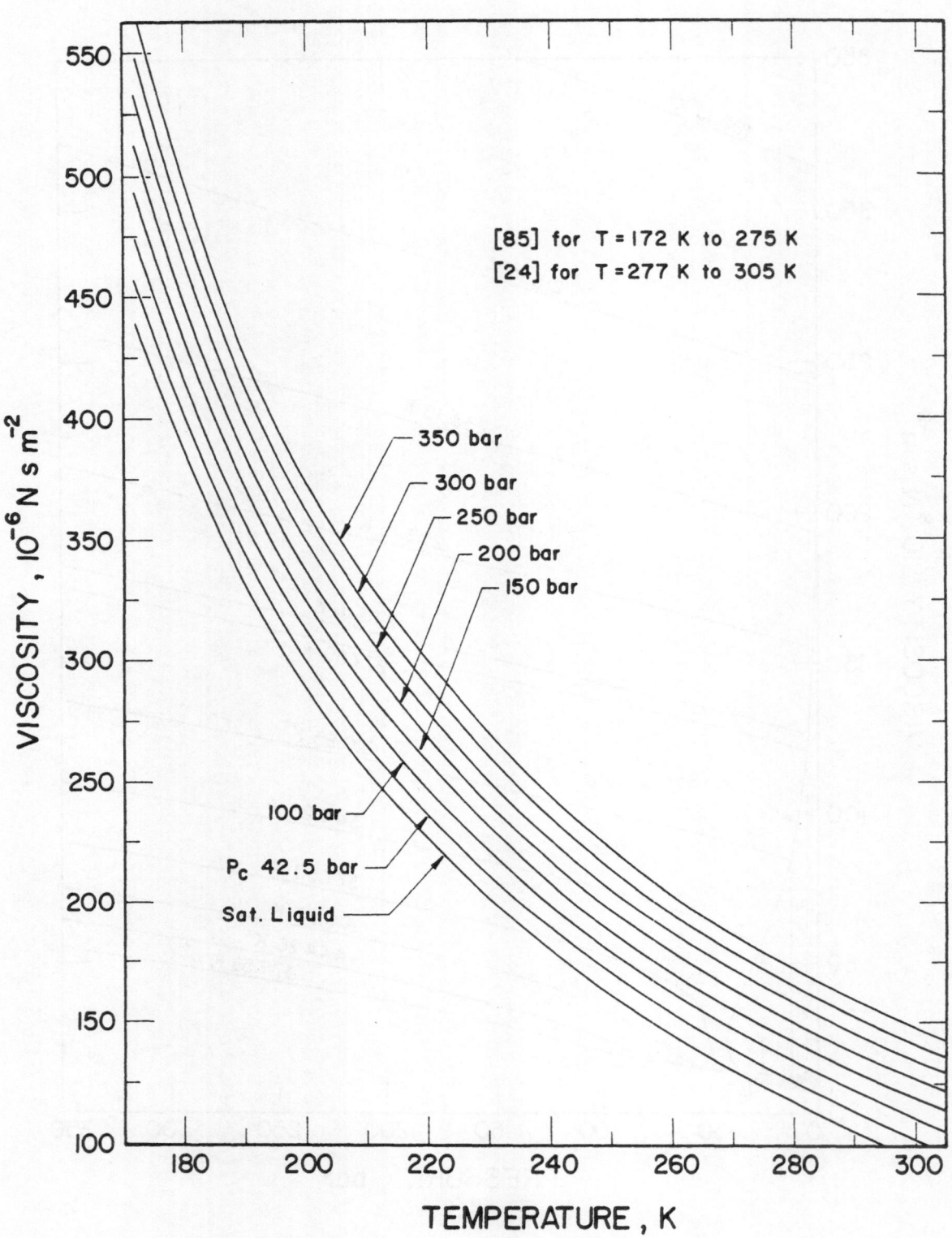

FIGURE 2A. VISCOSITY OF PROPANE [24, 85].

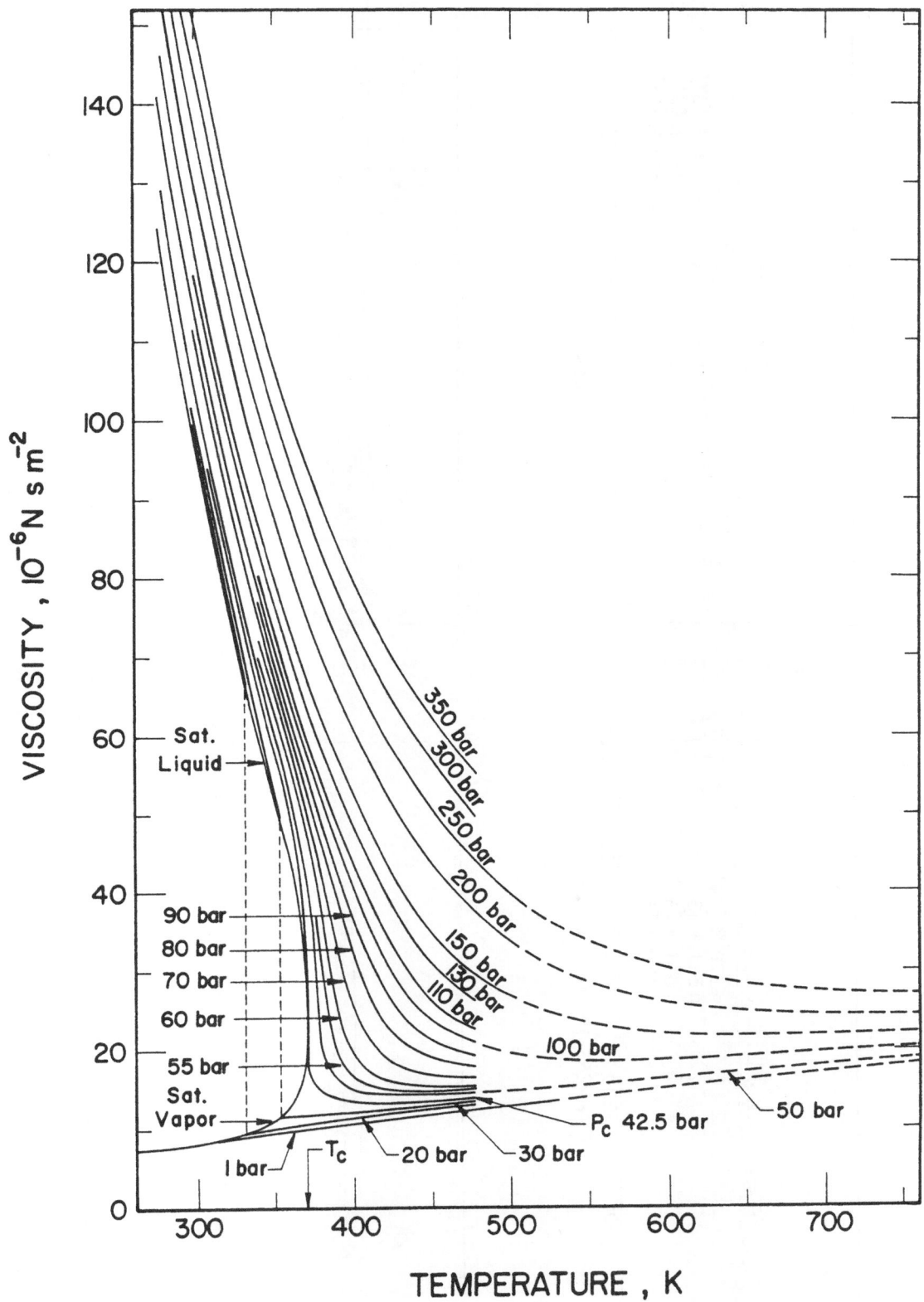

FIGURE 2B. VISCOSITY OF PROPANE [24].

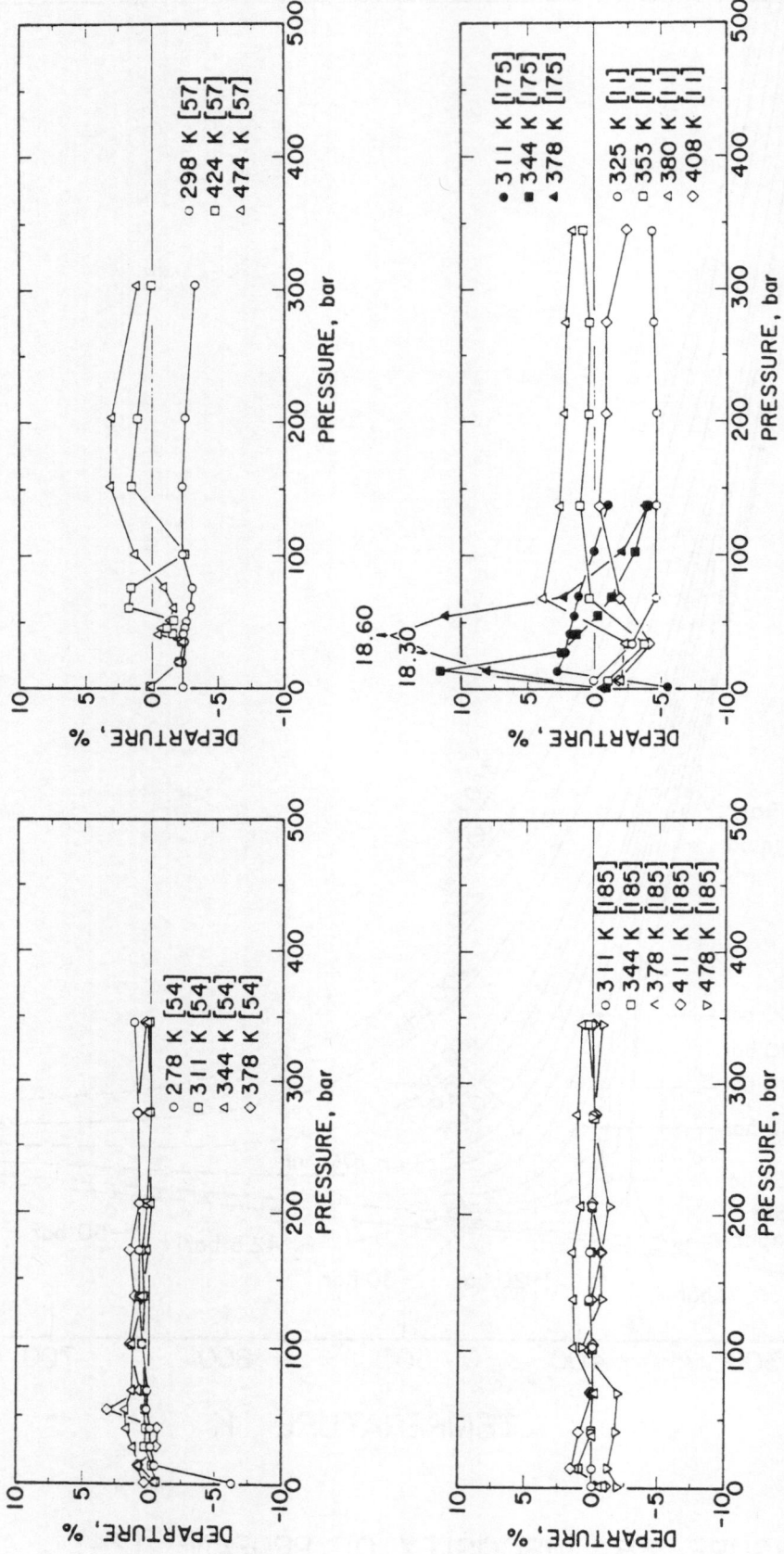

FIGURE 3. DEPARTURE PLOT ON THE VISCOSITY OF PROPANE.

i-PROPANOL

The data of [57] were used to generate the recommended values presented below. They are plotted as isotherms against pressure in Figure 1, and as isobars against temperature in Figure 2. The recommended values were read from this figure. Since no further detailed study on the viscosity of i-propanol has been made, a statement of accuracy for the recommended values cannot be given.

VISCOSITY OF i-PROPANOL

$[\mu,\ 10^{-6}\ N\ s\ m^{-2}]$

T, K	Pressure, bar											
	1	50	60	75	100	200	300	400	500	600	700	800
430	11.3	186	190	195	204	234	262	288	313	338	362	383
440	11.6	164	168	174	183	210	236	259	283	306	328	348
450	11.8	145	149	154	162	190	213	234	256	277	297	316
460	12.1	128	131	136	154	170	192	211	232	252	270	288
470	12.3	112	116	120	128	153	173	186	207	223	240	257
480	12.6	98	102	102	114	138	158	175	194	210	227	243
490	12.8	85	89	93	101	126	145	162	179	196	201	225
500	13.1	68	75	81	90	115	134	150	167	183	197	211
510	13.4	20	42	67	80	105	124	140	156	171	186	199
520	13.7	17	27	54	69	96	114	131	146	161	175	198
530	13.9	16	23	46	60	87	106	122	137	152	165	178

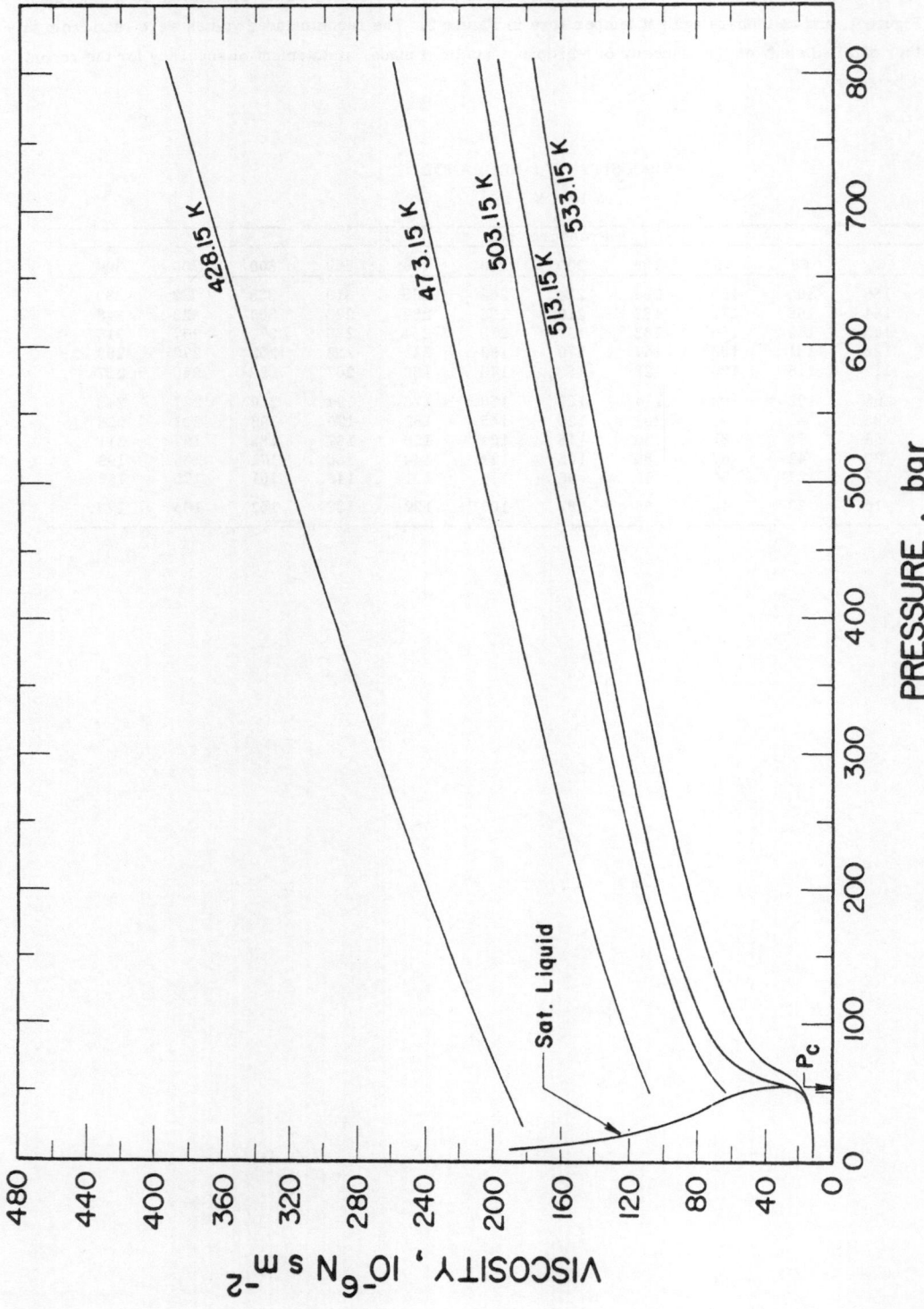

FIGURE 1. VISCOSITY OF i-PROPANOL [57].

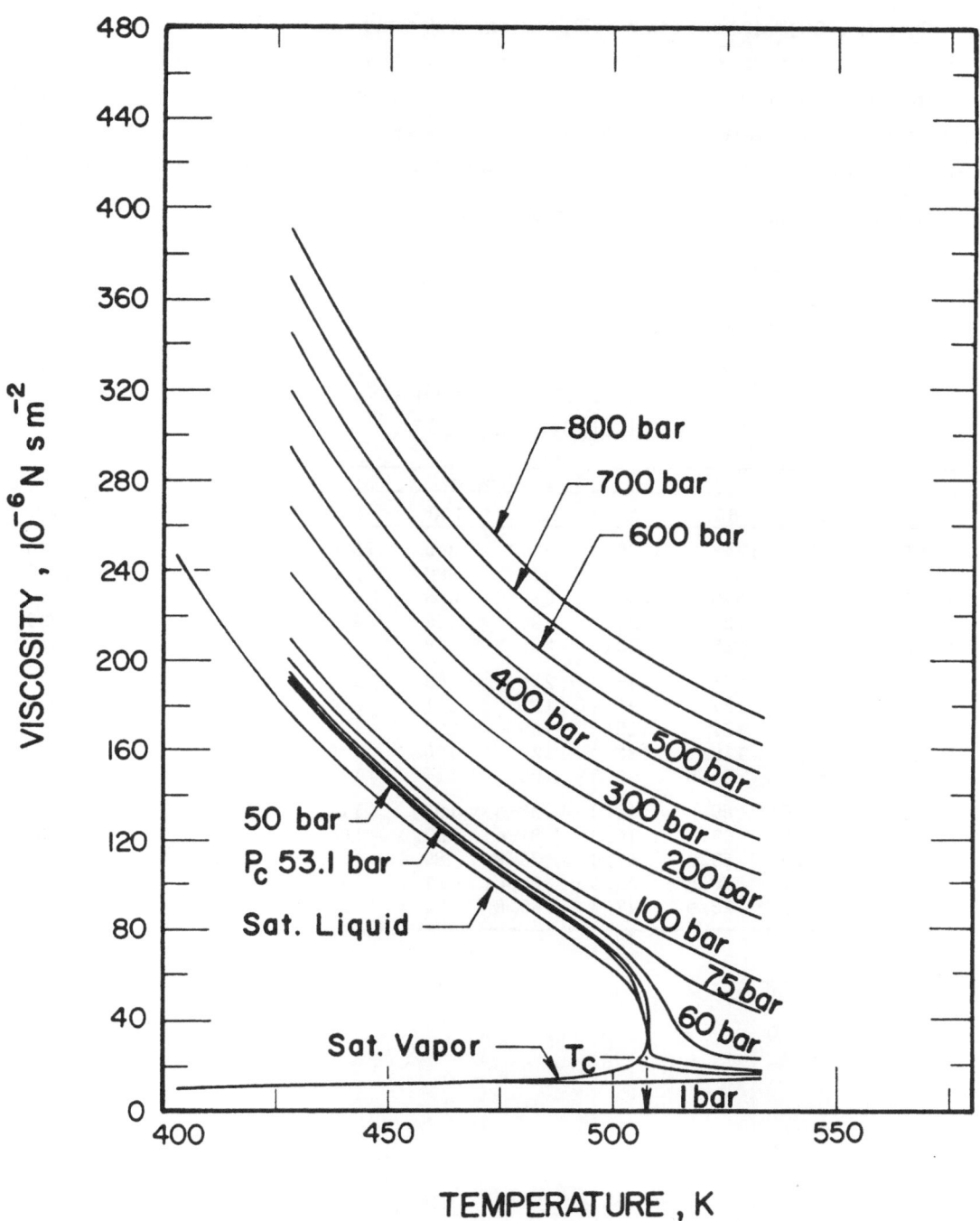

FIGURE 2. VISCOSITY OF i- PROPANOL [57].

n-PROPANOL

The values of [203] were used to develop the set of recommended values for the viscosity of n-propanol. The data are plotted as isotherms against pressure in Figure 1 where minor smoothing modifications were applied. The isotherms were extrapolated to the experimental vapour pressures of [127], thus giving the saturated liquid line, which is shown by a dashed line in the figure. Figure 2 shows isobars against temperature. From this figure, the recommended values presented below were read.

Since no detailed study was made on the original data of n-propanol, no estimate of accuracy is given.

VISCOSITY OF n-PROPANOL
$[\mu, 10^{-6} \text{ N s m}^{-2}]$

T,K	Pressure, bar									
	1	50	60	75	100	150	200	300	400	500
320	1190	1240	1248	1260	1285	1345	1390	1510	1625	1770
340	8.9	827	833	843	866	896	930	993	1060	1108
360	9.4	578	585	596	610	635	660	702	757	797
380	9.9	417	421	426	436	456	477	515	555	590
400	10.6	315	318	322	331	348	365	398	427	458
420	11.0	238	241	246	254	269	284	311	338	360
440	11.6	185	187	191	197	211	224	250	272	290
460	12.1	145	151	155	160	172	183	205	225	241
480	12.7	115	119	123	129	140	150	170	188	203
500	13.2	87.0	91.5	96.3	103	114	124	142	160	174
510	13.5	75.5	80.0	86.0	93.3	105	114	132	149	162
520	13.7	64.3	69.7	76.3	84.4	97.0	106	123	139	152
530	14.0	52.0	59.6	67.9	76.8	89.8	99.2	115	130	142
540	14.1	17.4	46.0	59.0	69.0	82.6	92.1	108	122	134
550	14.4	17.1	23.0	47.5	60.5	74.6	84.1	100	114	127

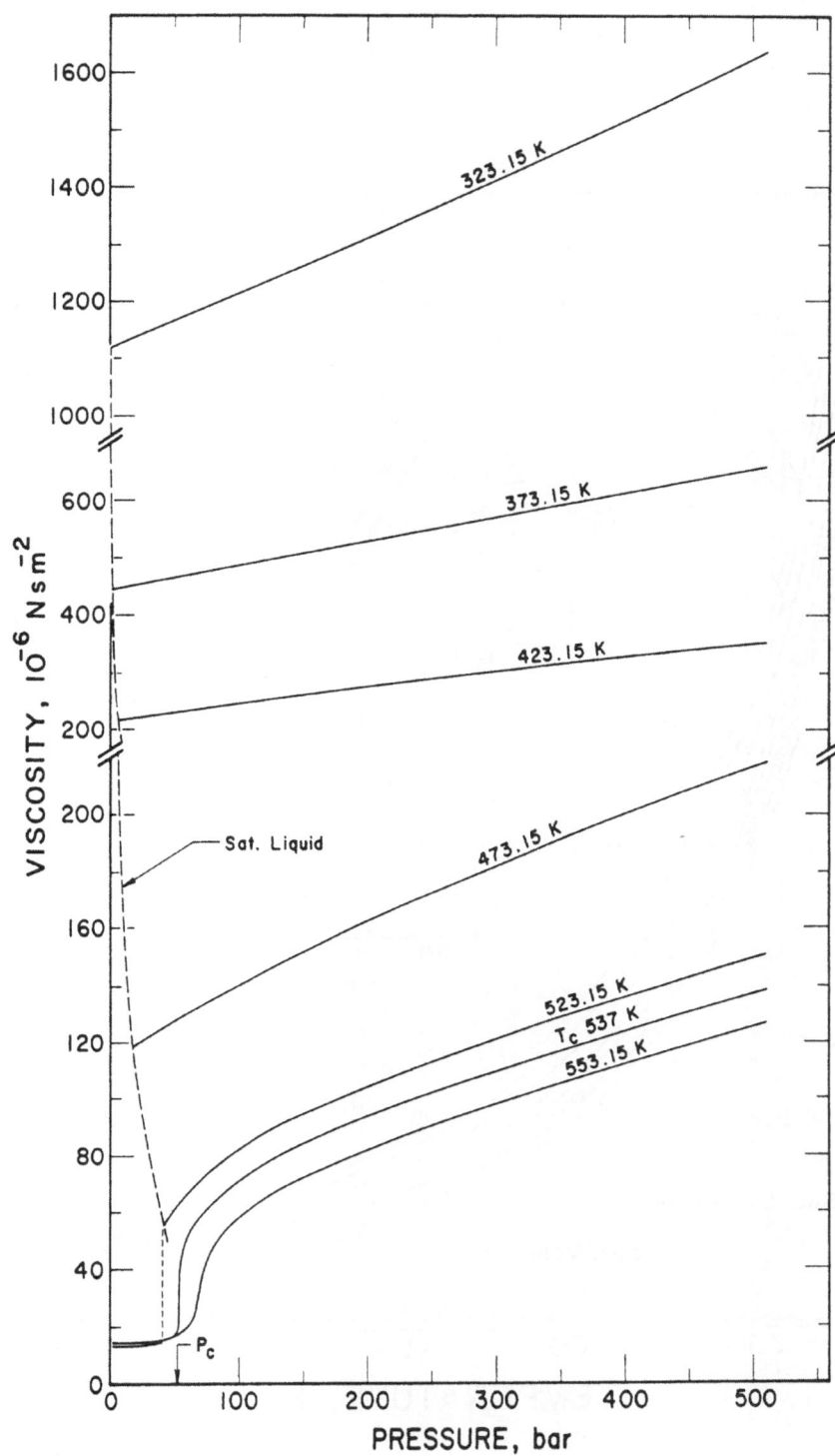

FIGURE I. VISCOSITY OF n-PROPANOL [203].

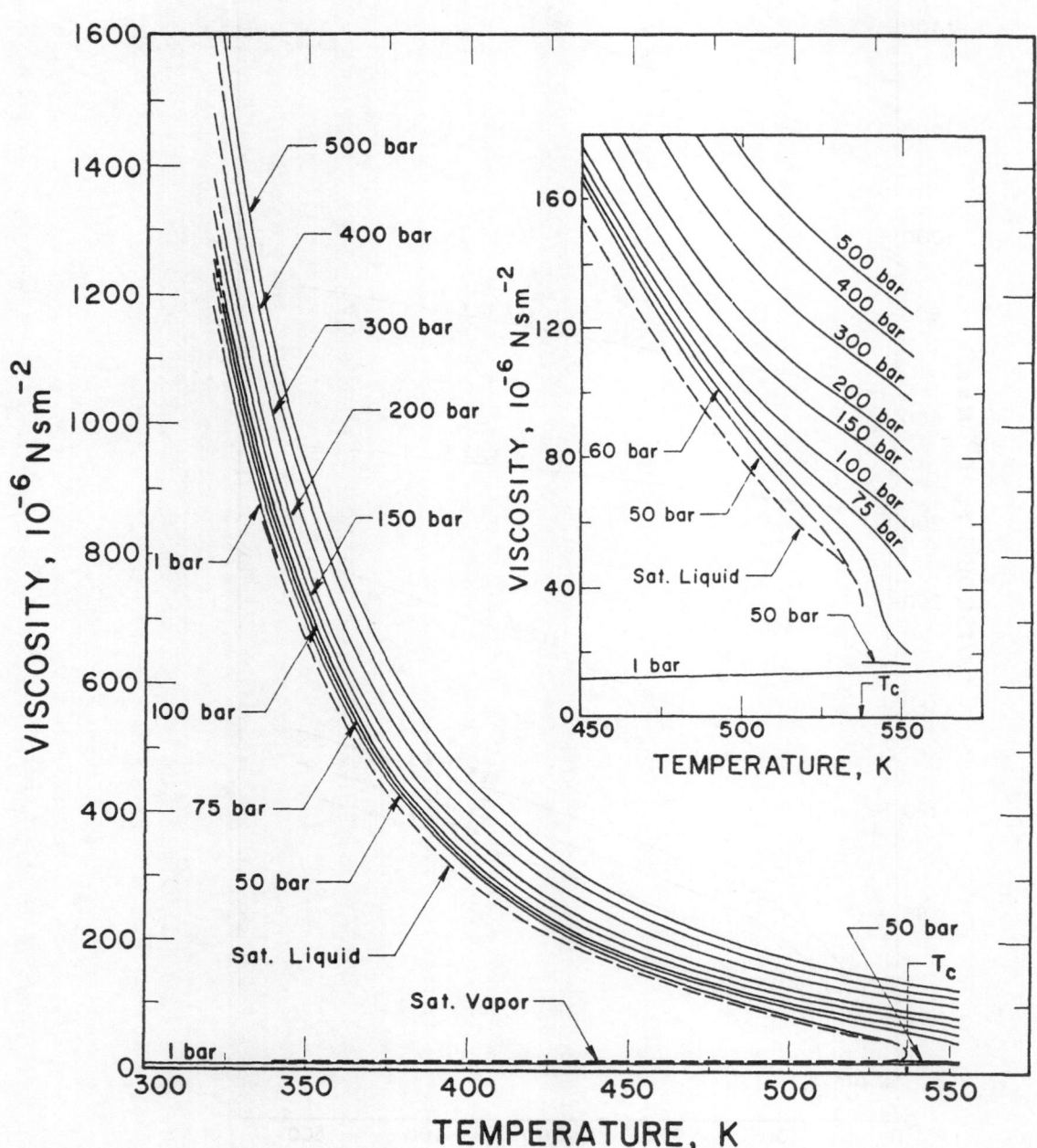

FIGURE 2. VISCOSITY OF n-PROPANOL [203].

PROPYLACETATE

The data by Guselnov and Kadschorov [70] were used to generate the recommended values presented below. These data were obtained with a capillary tube viscosi-meter, the uncertainty is stated to be 1.1%. The data are plotted as isotherms against pressure in Figure 1, and as isobars against temperature in Figure 2.

VISCOSITY OF PROPYLACETATE
$[\mu,\ 10^{-6}\ \text{N s m}^{-2}]$

T, K	Pressure, bar																		
	1	5	10	20	30	40	50	60	70	80	90	100	150	200	250	300	350	400	450
290	580	582	585	587	593	599	605	670	616	621	627	632	662	687	715	740	772	800	825
300	527	530	532	535	541	547	552	558	563	569	574	580	605	630	655	680	710	735	755
320	422	425	427	430	434	438	442	448	453	459	464	470	490	515	535	557	580	600	620
340	335	337	338	340	344	348	353	356	360	364	368	372	390	407	425	442	655	472	490
360	267	269	270	272	275	277	280	283	286	289	292	295	370	325	340	352	365	380	395
380	–	–	–	–	–	–	232	834	236	238	240	242	257	267	282	292	302	377	325
400	–	–	–	–	–	–	195	197	199	201	203	205	217	230	242	252	260	272	282
420	–	–	–	–	–	–	165	167	169	171	173	175	185	197	207	220	227	240	247
440	–	–	–	–	–	–	137	139	141	143	145	147	160	170	180	190	200	210	217
460	–	–	–	–	–	–	112	114	116	118	120	122	135	145	155	165	172	182	190
480	–	–	–	–	–	–	92.5	95.0	97.5	100	102.5	105	112	125	132	142	150	160	167
500	–	–	–	–	–	–	75.0	97.5	80.0	82.5	85.0	87.5	97.5	107	115	125	135	142	150

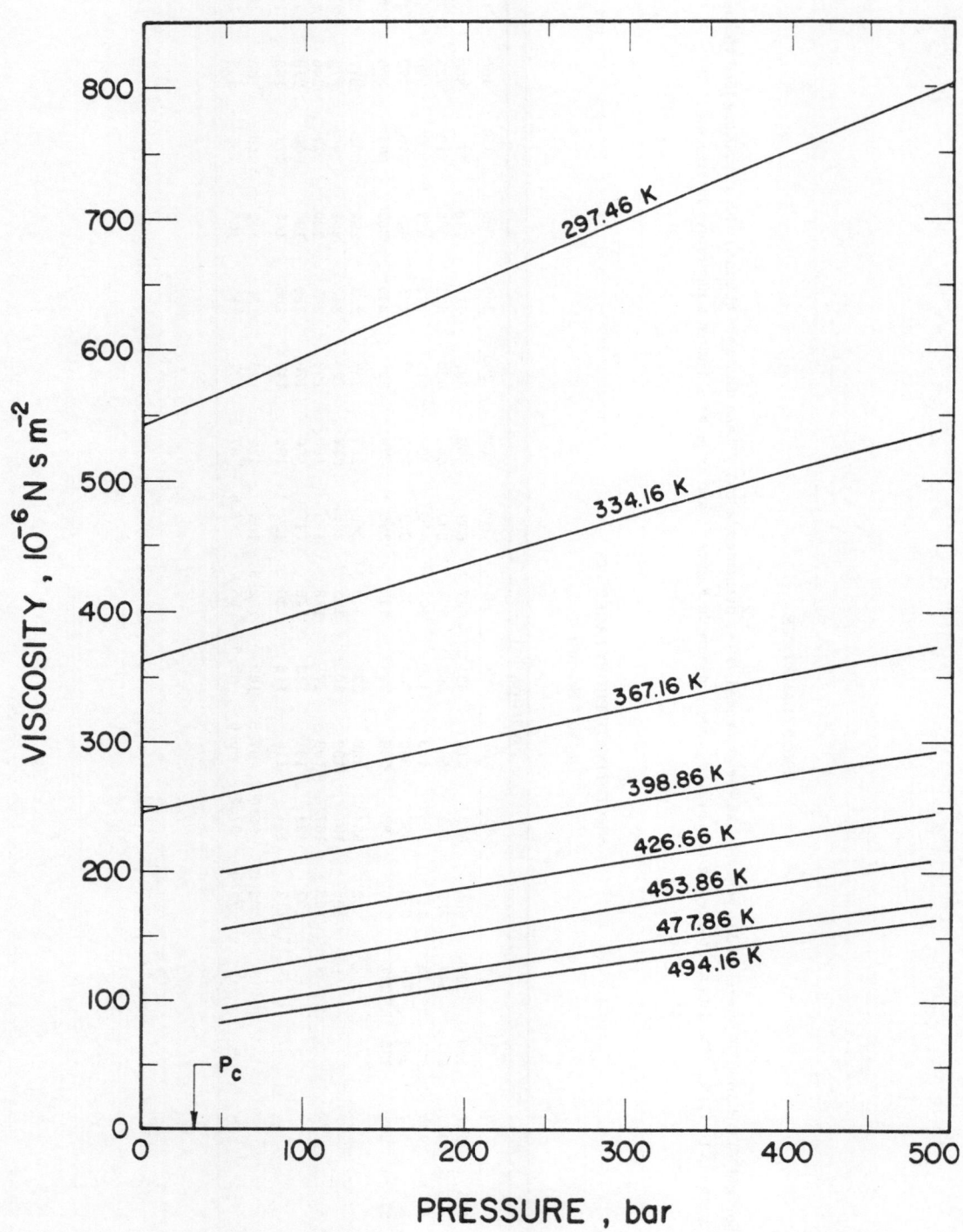

FIGURE 1. VISCOSITY OF PROPYLACETATE [70].

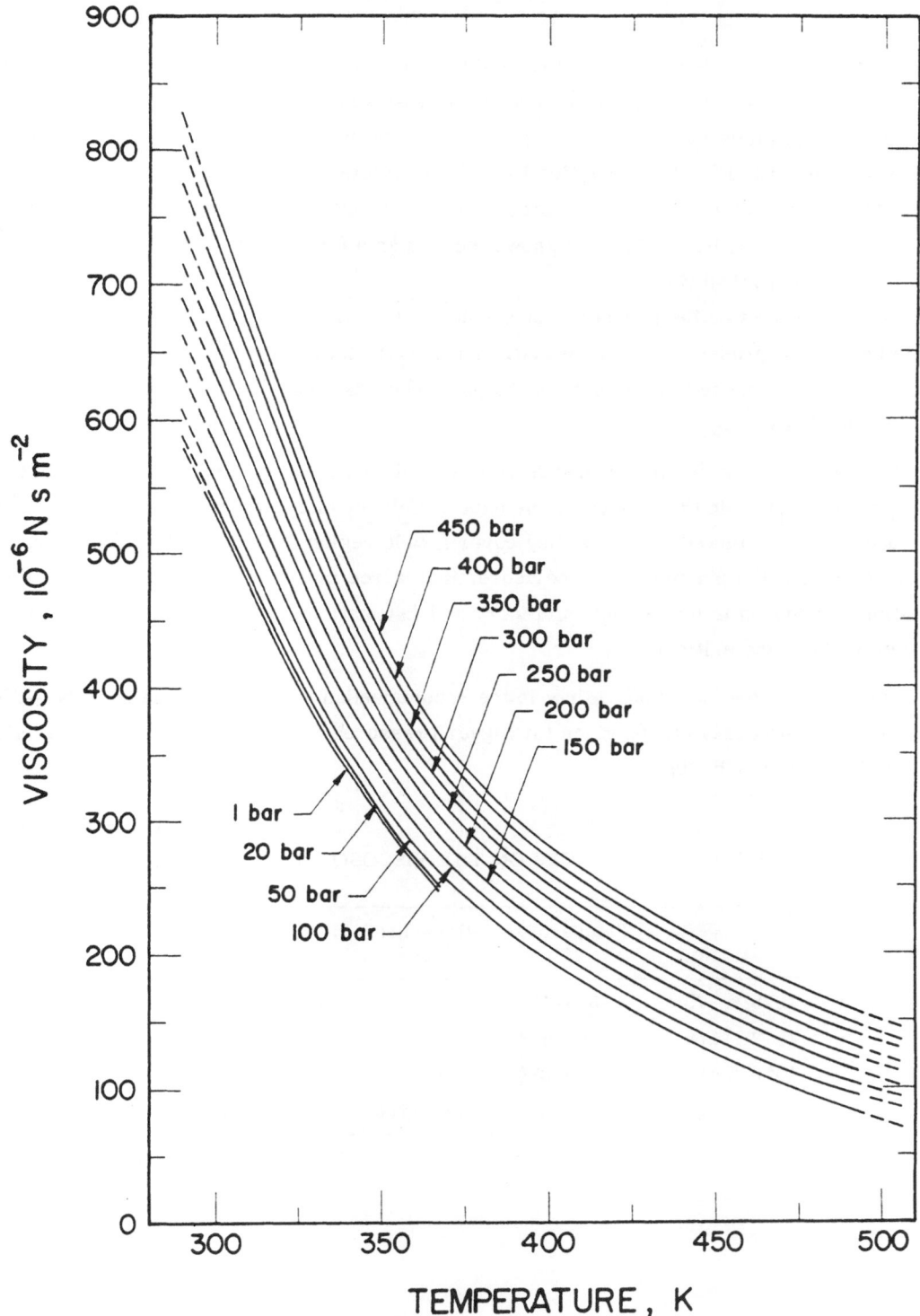

FIGURE 2. VISCOSITY OF PROPYLACETATE [70].

PROPYLENE

The data of Golubev and Petrov [57] were selected to generate the recommended values tabulated on the next page. The authors used a capillary tube viscosimeter and reported values from 291 K to 523 K up to 811 bar. The data are plotted as isotherms against pressure in Figure 1. The smoothing modifications amounted to a maximum of 1.1%. Extrapolating linearly the dilute gas viscosities, using the P-V-T data of [125, 128, 143] and the residual viscosity concept, it was possible to extrapolate up to a temperature of 650 K and a pressure of 900 bar. Figure 2 shows isobars as a function of temperature. The extrapolated values are shown in dashed lines.

Four additional works on the pressure dependence of the viscosity of propylene are listed in the summary table below. At pressures up to the critical (46 bar), there is very good agreement between all reported data. For higher pressures the data of Haepp [75] deviate considerably from the recommended values, as shown in Figure 3.

Further experiments on the high pressure viscosity of gaseous propylene are necessary to put confidence on either set of data. In this analysis, the data of Golubev and Petrov are selected as the basis of the recommended values, primarily because they cover a wide region of fluid states. However, the data of Haepp have a high standard of precision, furthermore, they agree much better with the corresponding states correlations discussed in the section on Theory and Estimation, than do the values recommended here for pressures above the critical.

In view of these factors, the uncertainty in the recommended values is estimated to be ± 2% up to 46 bars. No estimation of accuracy can be given for higher pressures at the present time, as deviations up to 50% are found with the data of Haepp.

ADDITIONAL REFERENCES ON THE VISCOSITY OF PROPYLENE

Authors	Year	Ref. No.	Temperature K	Pressure bar	Method	Departure % (no. points)
Haepp	1975	75	304–485	1–100	Oscillating disk	−5.86 ± 12.3 (34)
Naziev et al.	1972	156	300–474	1–9	Capillary tube	3.1
Neduzhii and Khmara	1970	158	210–310	1–8	Capillary tube	2.0
Babb and Scott	1964	10	303	2000–12000	Rolling ball	–

VISCOSITY OF PROPYLENE

[μ, 10^{-6} N s m^{-2}]

Pressure, bar

T,K	1	20	40	46.0*	60	80	100	120	140	160	180	200	250	300	400	500	600	700	800	900
290	8.3	107.0	110.8	112.3	114.8	118.5	121.9	125.2	128.6	131.4	134.5	138.0	145.3	152.0	164.9	178.0	190.0	201.8	213.7	225.6
300	8.6	96.4	100.3	101.6	104.4	108.0	111.2	114.4	117.6	120.8	123.6	126.4	133.4	140.0	152.7	165.0	177.0	189.3	200.9	211.7
310	8.9	86.6	90.5	91.7	94.4	98.0	101.2	104.2	107.5	110.8	113.5	116.0	122.8	129.2	141.6	154.0	165.5	177.6	189.0	199.0
320	9.2	77.5	81.3	82.5	85.2	88.8	92.0	95.2	98.4	101.5	104.4	106.8	113.6	120.0	132.0	143.8	155.0	166.8	177.7	187.6
330	9.5	-	73.0	74.0	76.6	80.2	83.6	86.8	90.0	93.2	96.0	98.6	105.3	111.6	123.5	134.9	145.7	156.8	167.3	177.1
340	9.8	-	65.3	66.4	68.8	72.2	76.0	79.5	82.7	85.5	88.5	91.3	98.0	104.4	116.0	127.0	137.5	147.8	158.0	167.6
350	10.0	-	57.2	58.3	61.0	65.1	69.2	72.7	76.0	78.8	82.0	84.8	91.5	97.7	109.2	120.0	130.0	139.7	149.5	159.2
360	10.3	11.7	-	47.5	53.4	58.4	62.8	66.4	70.0	72.8	76.2	79.0	85.7	92.0	103.2	113.6	123.6	132.7	142.0	151.4
380	10.9	11.7	13.7	15.0	38.0	46.2	51.6	55.7	59.7	63.1	66.6	69.3	76.3	82.4	93.3	103.4	112.8	121.3	130.4	139.0
390	11.2	12.0	13.6	14.5	30.1	40.0	46.6	51.2	55.4	59.0	62.5	65.2	72.1	78.2	89.0	99.0	108.0	116.6	125.4	133.8
400	11.5	12.2	13.6	14.4	23.3	34.5	42.1	47.0	51.4	55.2	58.8	61.4	68.4	74.4	85.2	94.9	104.0	112.2	120.9	129.1
410	11.8	12.5	13.8	14.5	19.9	30.0	38.0	43.2	47.8	51.6	55.2	58.0	64.9	71.0	81.6	91.2	100.0	108.2	116.8	124.6
420	12.0	12.7	14.0	14.6	18.2	26.6	34.5	39.8	44.5	48.2	51.8	54.8	61.6	67.7	78.2	87.7	96.4	104.5	112.7	120.4
430	12.3	13.0	14.1	14.8	17.3	24.0	31.5	36.7	41.4	45.2	48.6	51.7	58.8	64.8	75.1	84.4	93.0	101.0	108.8	116.5
440	12.6	13.2	14.3	14.8	16.8	22.3	28.8	33.9	38.4	42.2	45.6	48.8	55.8	61.7	72.0	81.3	89.9	97.6	105.2	112.8
450	12.8	13.5	14.4	15.0	16.6	21.0	26.4	31.2	35.7	39.3	42.8	46.0	53.0	59.0	69.2	78.4	86.8	94.4	101.7	109.2
460	13.1	13.7	14.7	15.2	16.4	20.0	24.4	29.1	33.3	36.8	40.4	43.6	50.4	56.4	66.4	75.6	84.0	91.6	98.6	106.0
470	13.3	13.9	14.8	15.3	16.5	19.4	23.0	27.2	31.2	34.6	38.0	41.2	48.0	54.0	64.0	73.2	81.4	88.8	95.7	103.1
480	13.6	14.2	15.0	15.4	16.6	19.0	22.0	25.6	29.4	32.8	36.1	39.1	45.8	51.6	61.6	70.7	78.8	86.2	93.1	100.2
490	13.8	14.4	15.2	15.6	16.7	18.8	21.4	24.6	28.0	31.4	34.4	37.3	44.0	49.6	59.6	68.4	76.5	84.0	90.6	97.5
500	14.1	14.7	15.4	15.8	16.8	18.7	21.1	23.8	27.0	30.2	33.0	35.8	42.2	48.0	57.6	66.4	74.2	81.6	88.2	95.0
520	14.6	15.1	15.8	16.2	17.1	18.7	20.5	22.9	25.5	28.3	30.9	33.3	39.4	44.8	54.0	62.4	70.0	77.2	84.0	90.3
540	15.1	15.6	16.4	16.6	17.4	18.7	20.2	22.3	24.4	26.9	29.2	31.4	37.0	42.3	50.9	58.8	66.1	73.3	79.8	85.9
560	15.5	16.1	16.7	16.9	17.6	18.8	20.0	22.0	23.7	25.9	28.0	30.0	35.1	40.0	48.1	55.6	62.4	69.4	75.9	81.7
580	16.0	16.6	17.1	17.3	18.0	19.0	20.0	21.6	23.3	25.1	27.0	28.8	33.4	38.1	45.6	52.5	58.8	65.7	72.0	77.7
600	16.5	17.0	17.5	17.7	18.3	19.4	20.1	21.6	23.0	24.6	26.3	28.0	32.0	36.4	43.2	49.6	55.6	62.1	68.4	74.0
620	16.9	17.4	17.9	18.1	18.6	19.6	20.2	21.6	22.9	24.3	25.8	27.3	31.0	34.8	41.0	47.0	52.4	58.6	64.8	70.4
640	17.2	17.8	18.3	18.4	18.9	19.7	20.4	21.6	22.9	24.1	25.5	26.8	30.0	33.4	39.1	44.5	49.3	55.3	61.2	66.9
650	17.4	18.0	18.5	18.6	19.1	19.9	20.6	21.6	22.9	24.1	25.4	26.6	29.6	32.8	38.2	43.3	47.9	53.6	59.4	65.2

*Critical pressure.

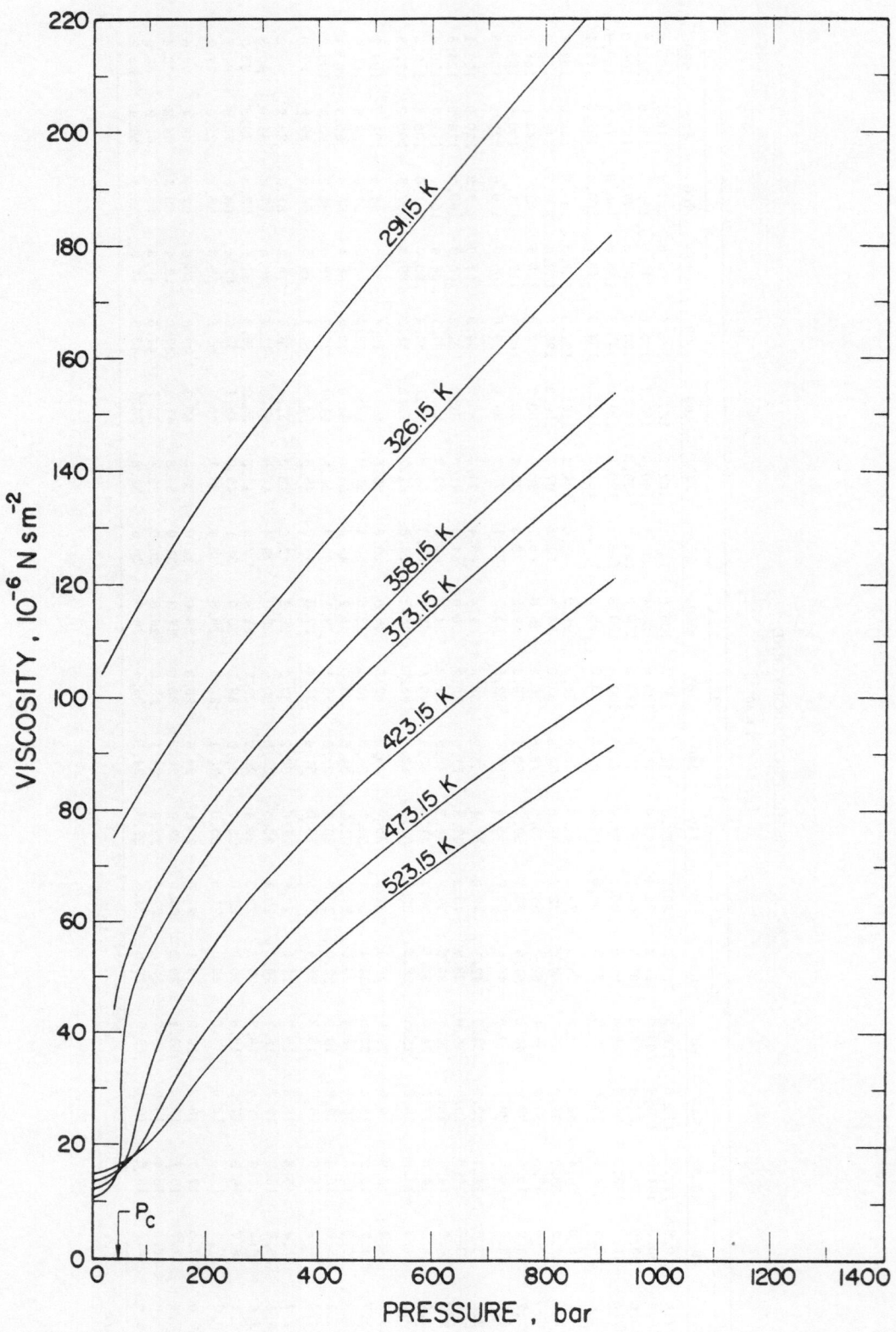

FIGURE 1. VISCOSITY OF PROPYLENE [57].

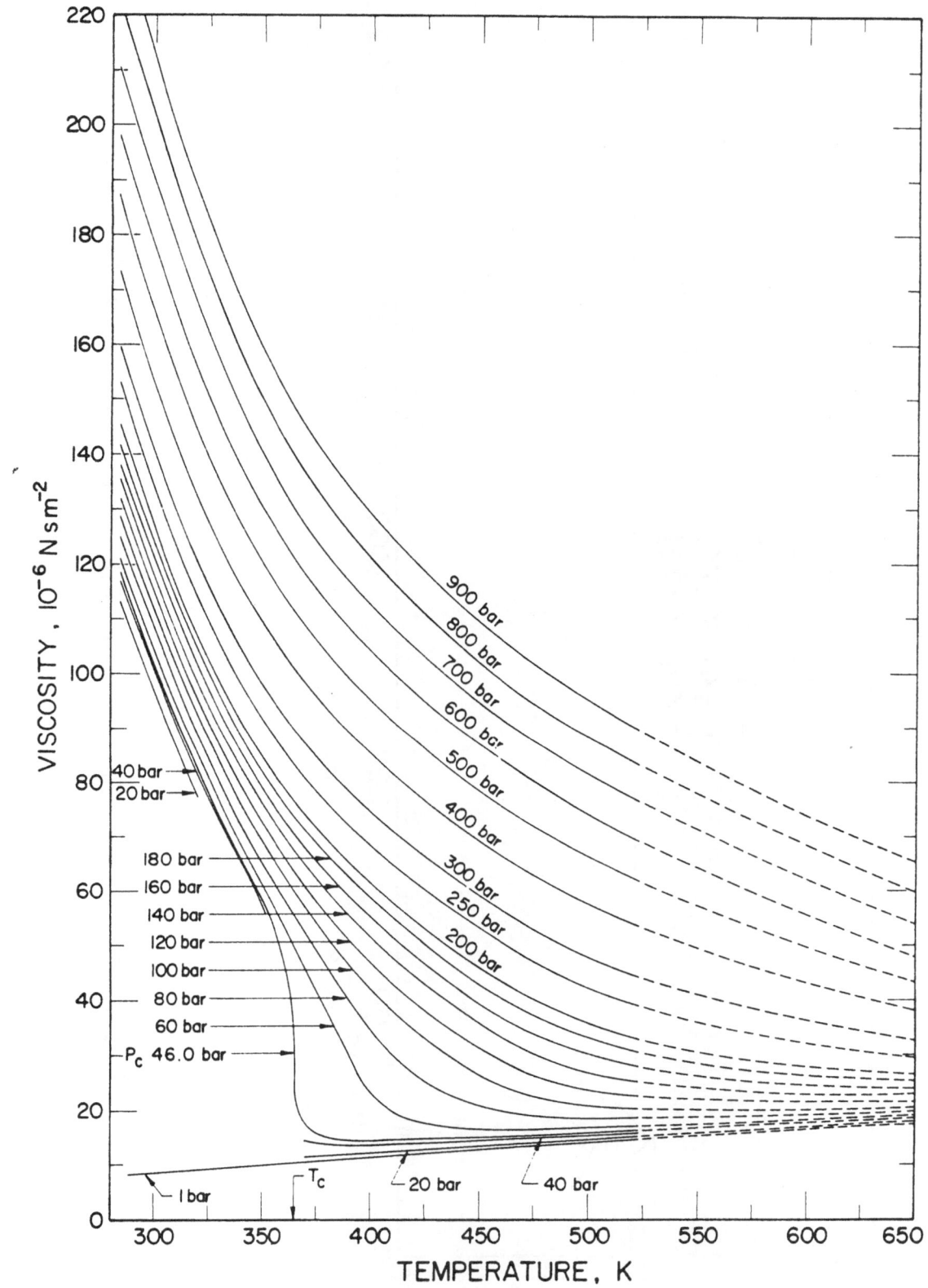

FIGURE 2. VISCOSITY OF PROPYLENE [57].

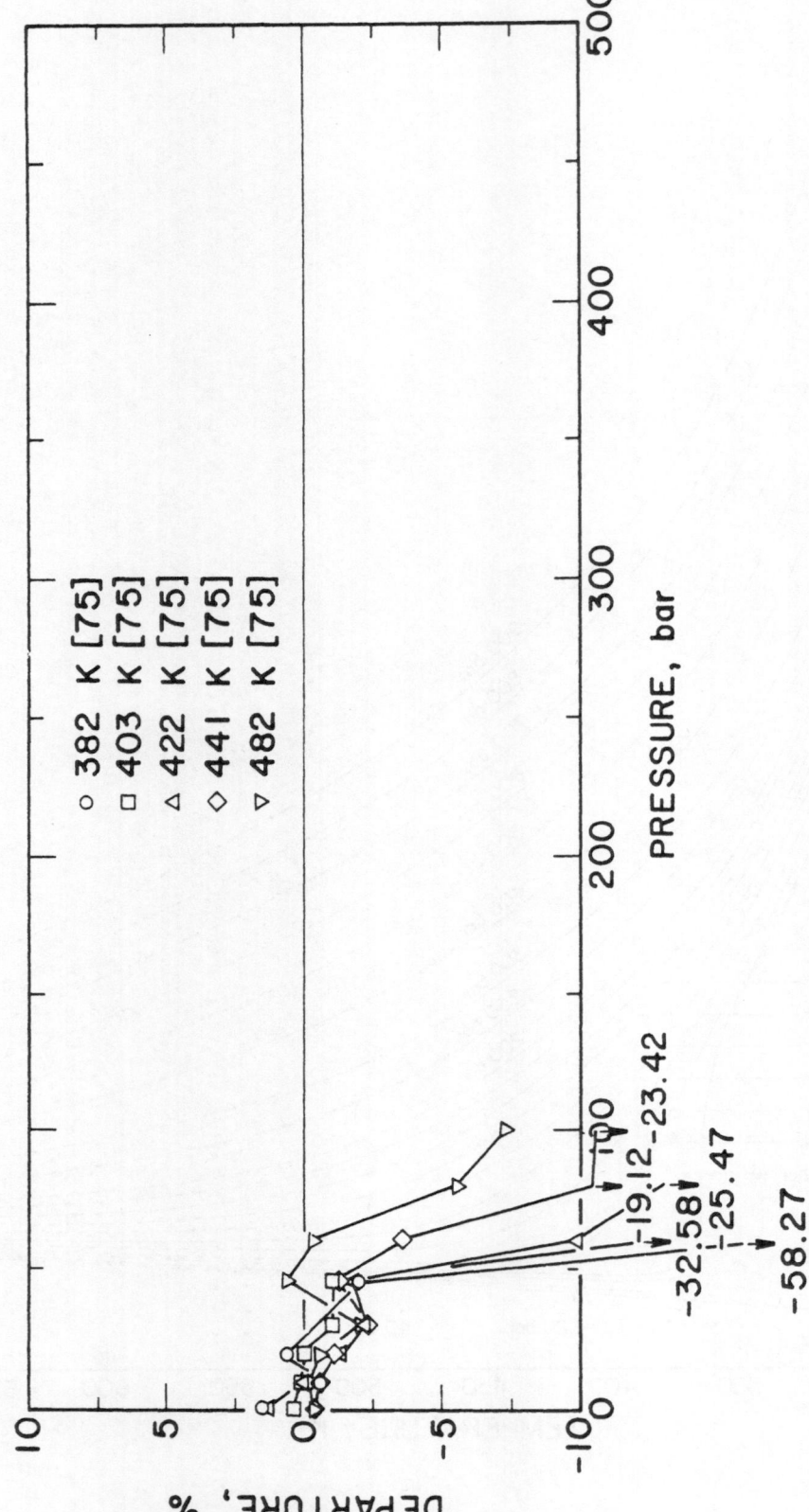

FIGURE 3. DEPARTURE PLOT ON THE VISCOSITY OF PROPYLENE.

TOLUENE

The data of Akhundov et al. [7], obtained by a capillary tube viscosimeter, were chosen to generate the recommended values presented below. The data are plotted as isotherms against pressure in Figures 1A and 1B. In Figures 2A and 2B isobars are plotted against temperature. The recommended values were read from the latter figures. Additional works on the pressure dependence of the viscosity of toluene are listed in the summary table below. However, a comparison with the recommended values, was not possible. Thus, in the absence of further experimental evidence, the authors' statement of uncertainty ± 1.2% is reported here.

ADDITIONAL REFERENCES ON THE VISCOSITY OF TOLUENE

Authors	Year	Ref. No.	Temperature K	Pressure bar	Method
Kor et al.	1972	118	303	1-9807	Ultrasonic
Kozlov et al.	1966	119	293-533	Isochores	Falling cylinder
Bridgman	1926	21	303, 348	1-11768	Falling weight

VISCOSITY OF TOLUENE

$$[\mu, \ 10^{-6} \ N \ s \ m^{-2}]$$

T, K	Pressure, bar													
	1	20	41.1*	60	80	100	120	140	160	180	200	250	300	400
295	588	597	608	617	628	638	648	659	667	678	687	714	737	788
300	554	564	573	582	592	601	611	621	630	639	648	673	696	745
310	491	501	509	517	526	534	543	551	560	567	575	598	618	661
320	437	446	452	459	467	474	483	490	497	504	511	531	549	587
330	393	400	405	412	419	425	433	439	446	452	458	475	491	524
340	357	363	368	374	380	386	393	398	404	410	416	431	446	475
350	327	331	337	342	348	353	360	365	370	376	382	395	409	436
360	301	305	310	315	321	326	331	336	341	347	352	365	378	403
370	278	282	287	292	297	302	307	312	317	323	327	340	352	376
380	258	261	266	271	276	281	286	291	295	301	305	317	329	352
390	-	243	248	252	257	262	267	272	276	281	285	297	309	331
400	-	226	231	235	240	245	249	254	258	263	268	279	290	312
420	-	197	202	206	211	215	219	223	227	231	235	246	256	276
440	-	172	176	180	185	189	193	197	201	205	208	218	228	247
460	-	149	153	158	162	166	170	174	178	182	185	195	204	222
480	-	129	133	138	142	147	151	155	159	163	166	176	184	200
500	-	112	117	122	126	131	135	139	143	147	151	160	168	182
520	-	97.0	103	107	112	116	121	125	129	133	137	146	154	168
540	-	-	89.2	94.3	98.9	103	108	112	116	120	124	133	142	156
550	-	-	82.3	88.0	92.7	97.1	102	106	110	114	118	127	135	151 .

*Critical pressure.

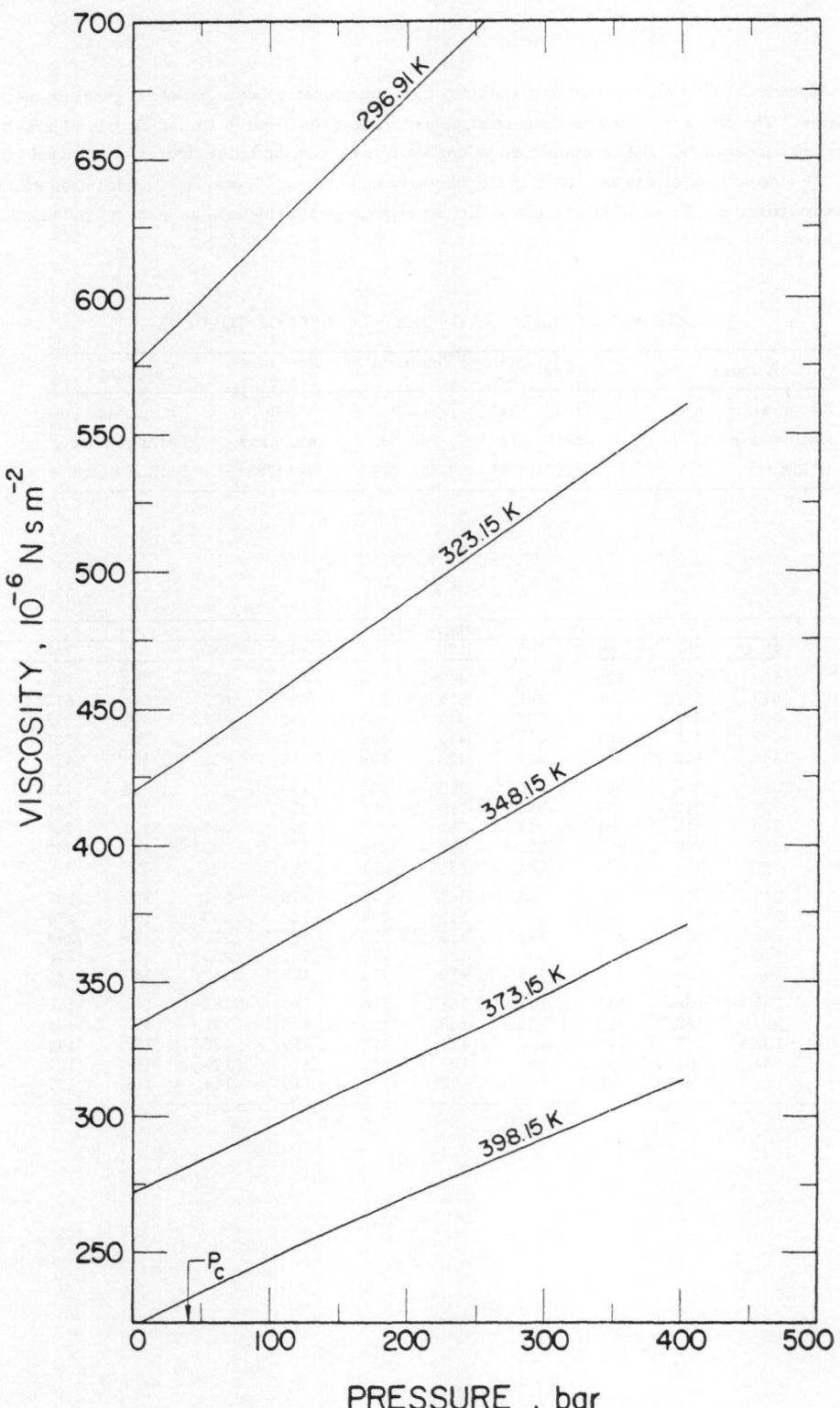

FIGURE I A. VISCOSITY OF TOLUENE [7].

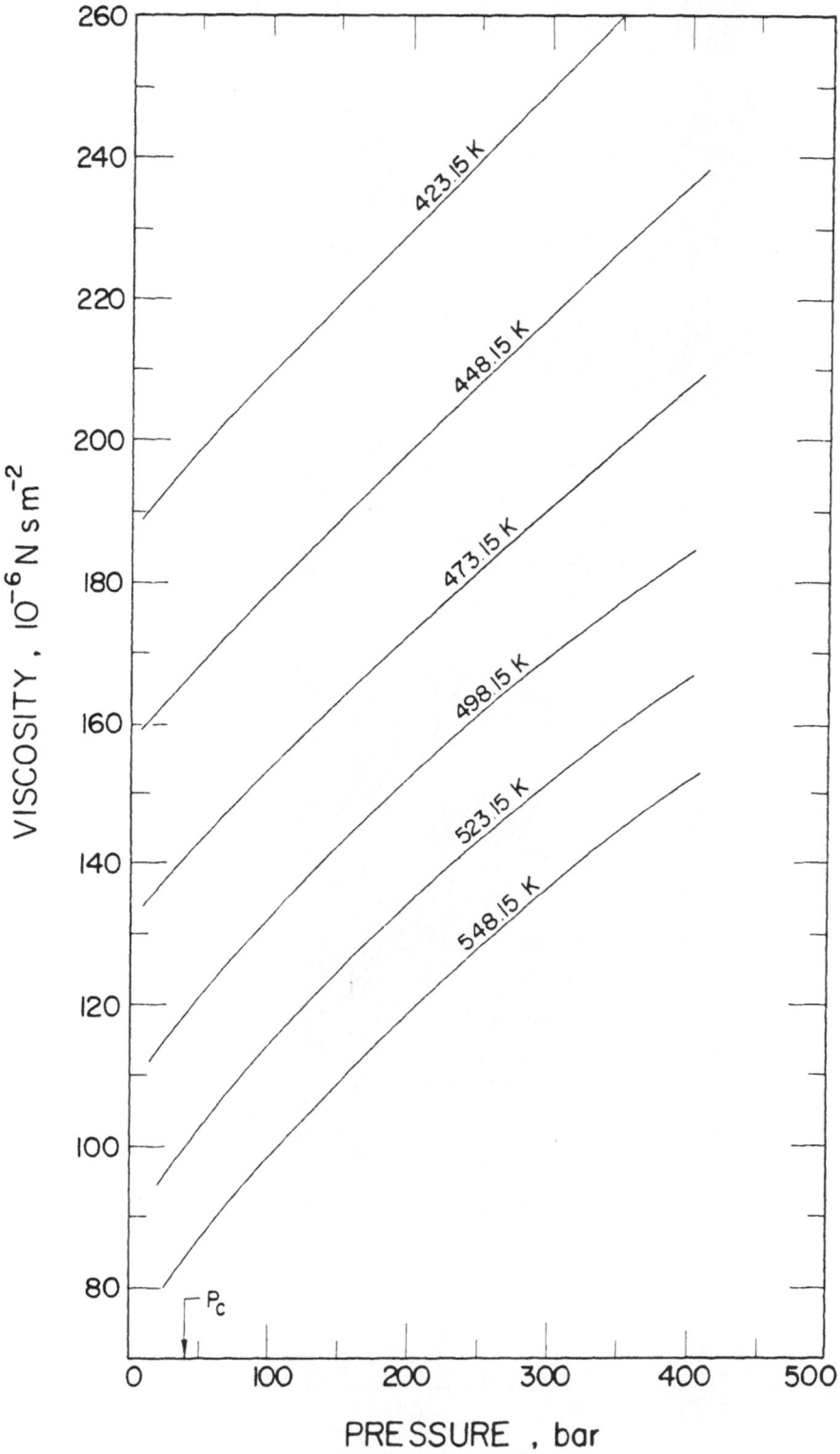

FIGURE IB. VISCOSITY OF TOLUENE [7].

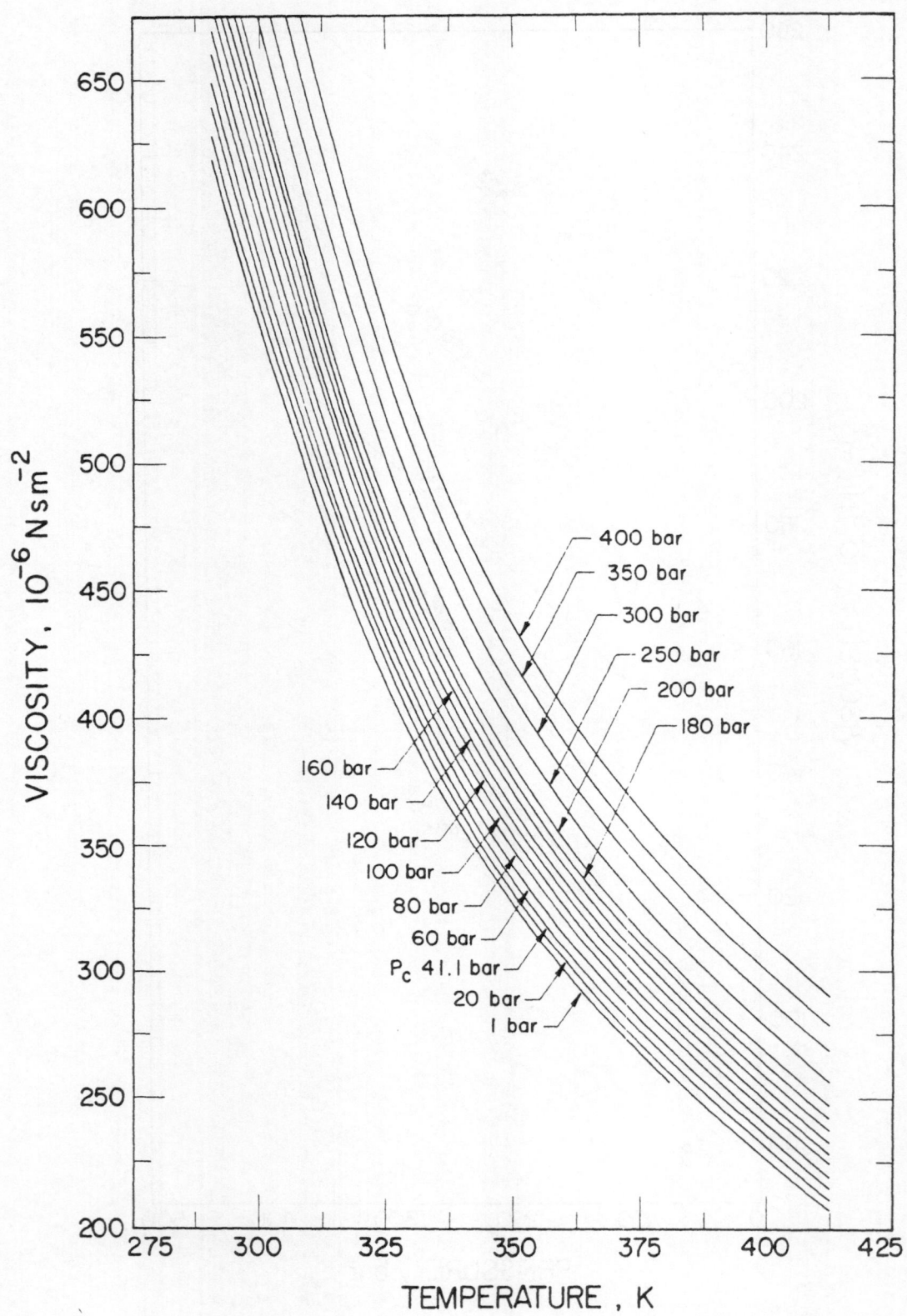

FIGURE 2A. VISCOSITY OF TOLUENE [7].

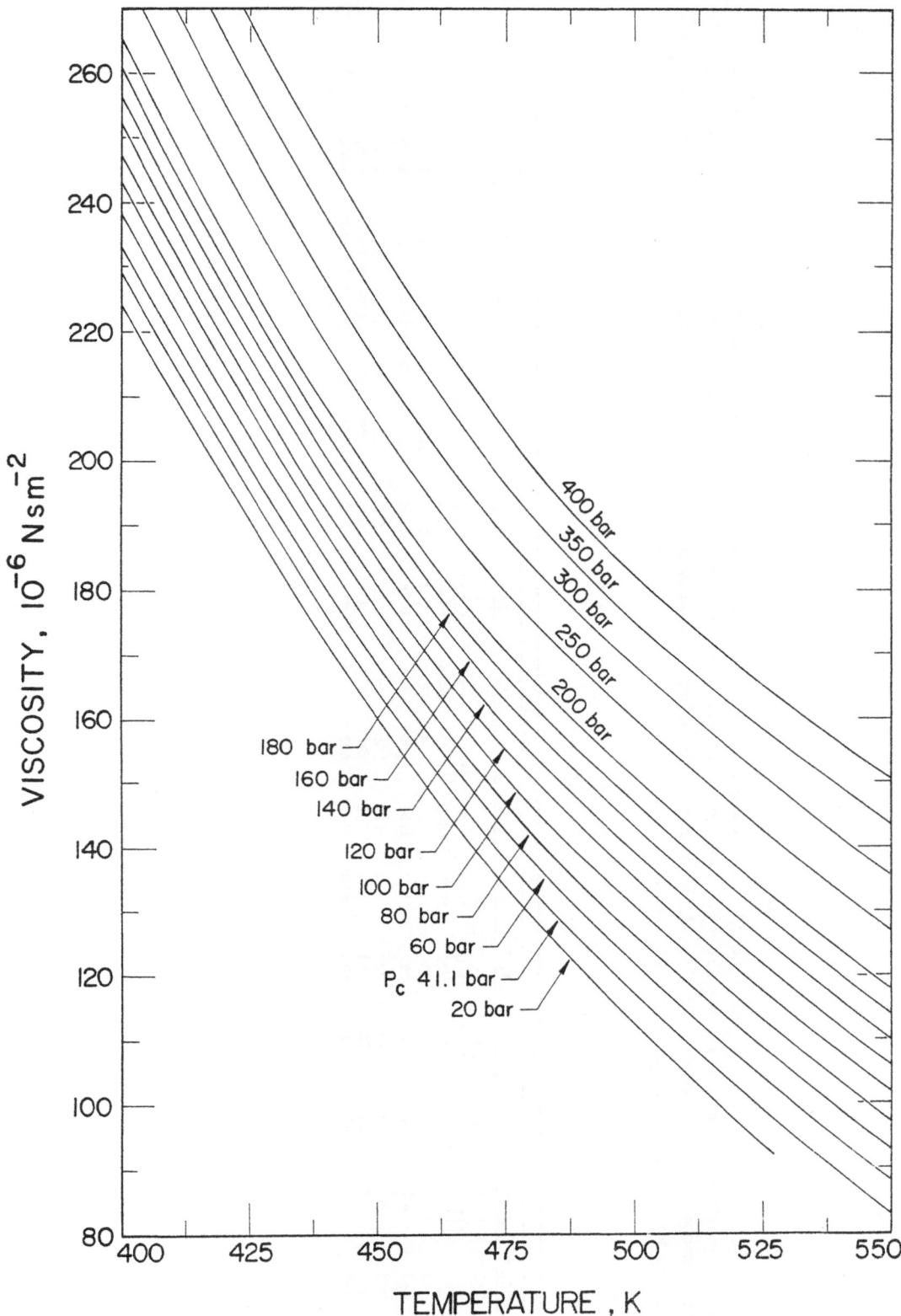

FIGURE 2B. VISCOSITY OF TOLUENE [7].

TRICHLOROTRIFLUOROETHANE (R 113)

Only one set of data by Ivanchenko [89], obtained by a capillary tube viscosimeter, was found in the literature for the pressure dependence of the viscosity of trichlorotri-fluoroethane. The data are plotted as isotherms against pressure in Figure 1, where only minor smoothing adjustments had to be made. In Figure 2, isobars are plotted against temperature. The recommended values, presented below are read from this figure. The authors report an uncertainty of ± 1.5% for their data. In the absence of further experimental evidence, no independent assessment of accuracy may be given at this time.

VISCOSITY OF TRICHLOROTRIFLUOROETHANE (R 113)

$[\mu,\ 10^{-6}\ N\ s\ m^{-2}]$

T, K	Pressure, bar																	
	1	20	34.1*	50	60	80	100	120	140	160	180	200	250	300	350	400	500	600
240	1753	1788	1818	1855	1880	-	-	-	-	-	-	-	-	-	-	-	-	-
250	1440	1480	1508	1540	1558	1588	1626	1668	1714	1758	1792	1830	-	-	-	-	-	-
260	1202	1238	1262	1290	1308	1336	1368	1400	1436	1468	1500	1536	-	-	-	-	-	-
270	1009	1040	1062	1088	1104	1130	1156	1184	1214	1240	1268	1300	1376	1460	1540	1630	1820	2010
280	864	886	903	925	937	960	982	1004	1030	1054	1078	1108	1172	1238	1305	1370	1516	1666
290	744	764	778	796	805	825	844	864	885	907	927	950	1007	1062	1120	1180	1304	1428
300	648	664	677	692	700	718	735	752	772	790	808	826	876	928	978	1028	1134	1244
310	569	582	596	608	616	633	650	664	682	699	714	732	774	820	865	910	1004	1100
320	504	518	530	540	548	564	579	592	608	624	638	654	692	731	772	812	900	987
340	-	416	427	435	443	456	468	480	493	507	520	533	565	599	632	665	737	807
360	-	340	347	356	362	374	386	396	408	420	432	442	472	501	528	558	619	678
380	-	280	287	296	302	312	322	332	344	354	364	375	402	428	452	478	528	580
400	-	232	240	249	255	264	275	286	296	305	315	325	349	373	396	419	464	508
420	-	193	200	210	216	227	237	246	257	266	275	285	307	330	350	371	412	453
440	-	156	165	174	180	192	202	212	222	231	240	250	270	291	310	330	367	406
460	-	-	-	140	147	160	170	180	190	200	208	217	238	256	274	292	328	364
480	-	-	-	109	116	129	140	151	160	170	180	188	208	224	241	257	292	326

*Critical pressure.

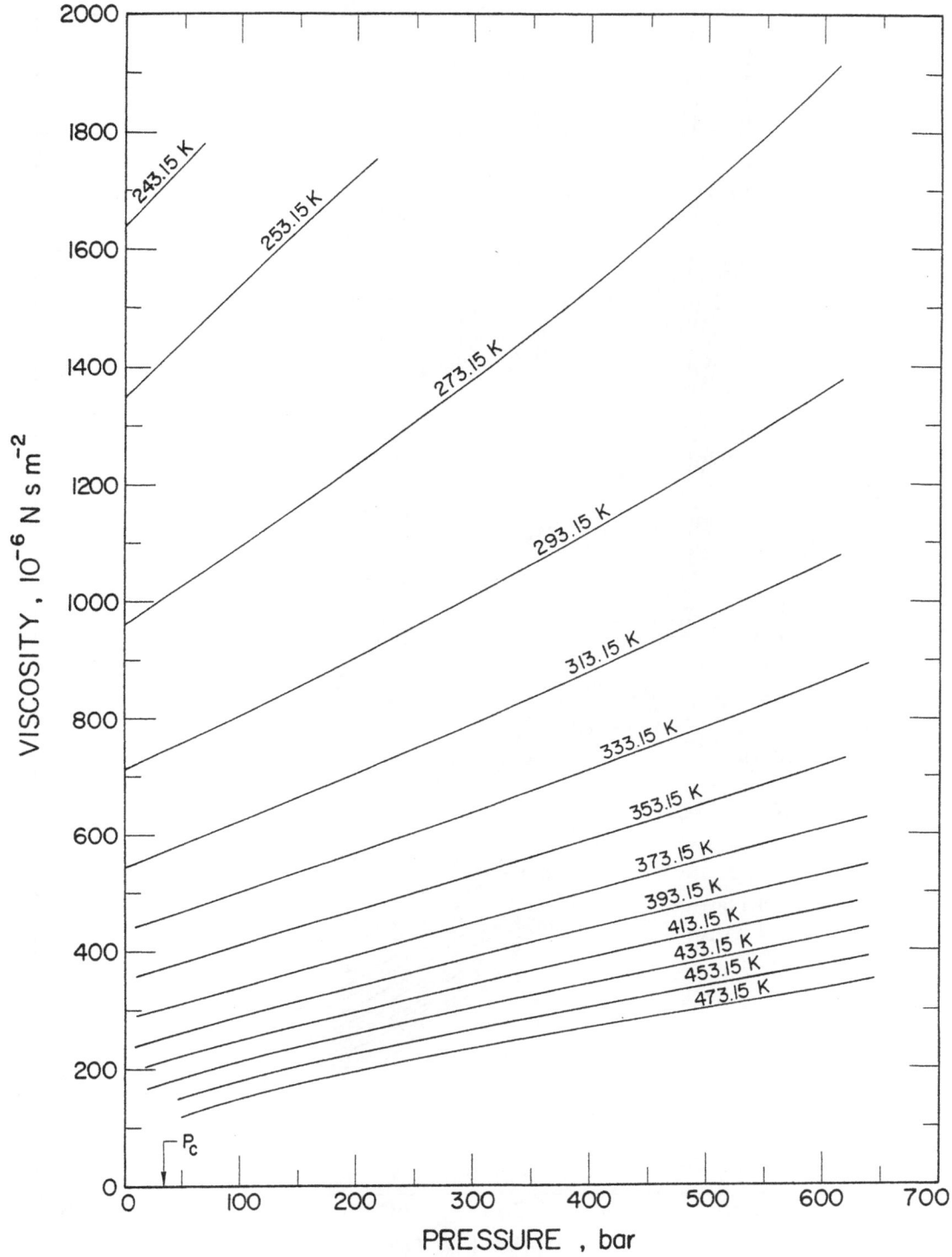

FIGURE 1. VISCOSITY OF R 113 [89].

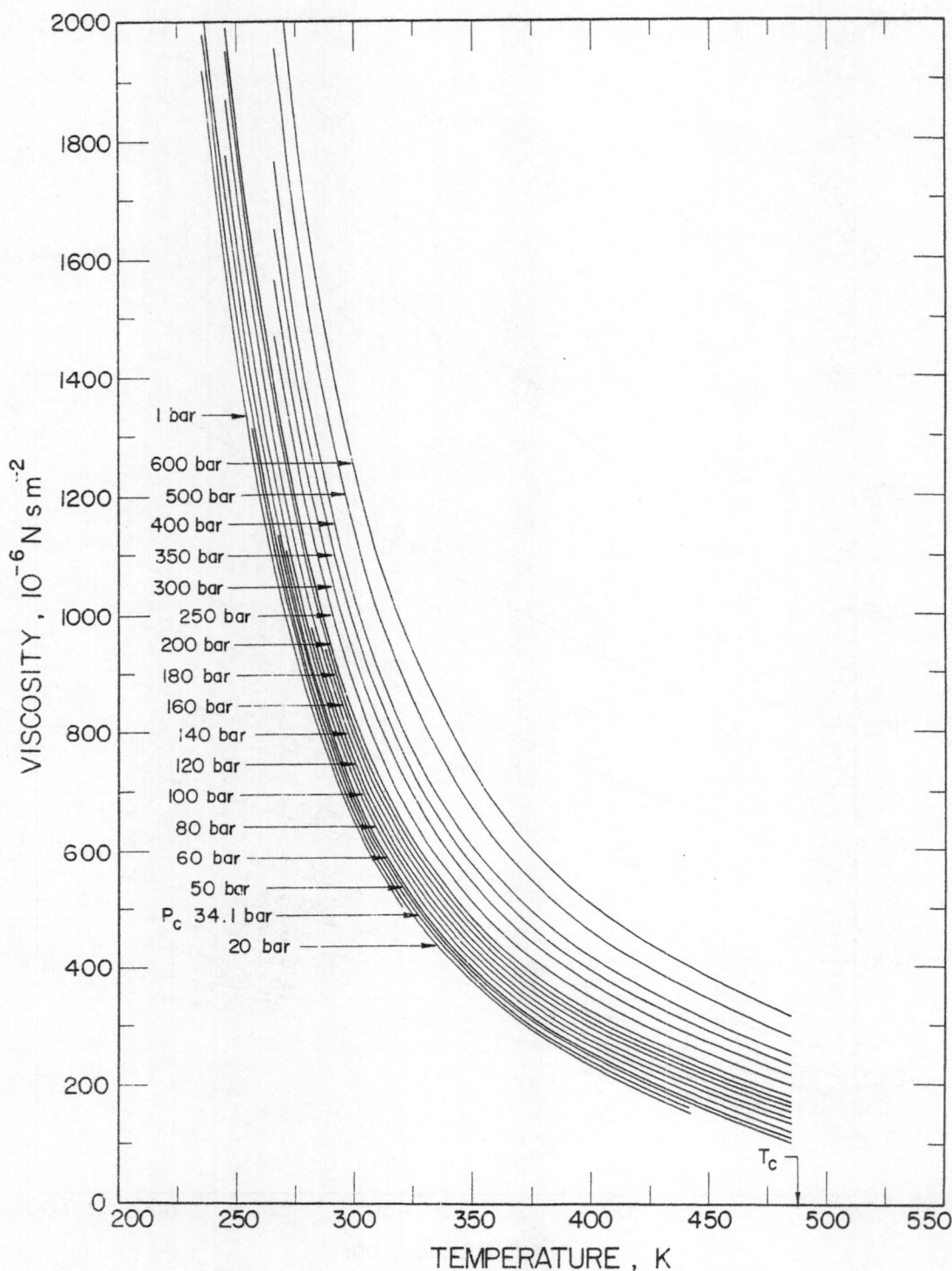

FIGURE 2. VISCOSITY OF R 113 [89].

n-UNDECANE

Two sets of data [72, 163] were found in the literature on the pressure dependence of the viscosity of liquid n-undecane. The data of Rastorguev and Keramidi [163], which were obtained by a capillary viscosimeter were selected to generate the recommended values presented below. The data are shown as isotherms plotted against pressure in Figure 1. Smoothing modifications of maximum 1.1% were applied to the data. A plot of isobars against temperature is shown in Figure 2. The second set of data by Guseinov and Naziev [72] for which no experimental details were reported, covers approximately the same region of fluid states. The agreement between the two sets of data is satisfactory, as seen from Figure 3. The latter deviates from the recommended values by 1.2%.

The uncertainty in the recommended values is estimated to be about ±6%.

VISCOSITY OF n-UNDECANE
$$[\mu,\ 10^{-6}\ N\ s\ m^{-2}]$$

T, K	Pressure, bar												
	1	19.4*	40	60	80	100	150	200	250	300	350	400	500
300	1042	1070	1100	1132	1158	1187	1256	1321	1396	1472	1534	1624	1768
310	908	932	956	982	1006	1032	1092	1154	1218	1285	1340	1403	1529
320	792	816	836	858	880	900	953	1008	1063	1121	1175	1221	1327
330	699	719	736	756	773	792	837	885	933	983	1030	1073	1163
340	620	639	652	669	685	700	741	784	824	867	909	954	1031
350	556	571	584	599	612	627	662	700	735	772	808	853	923
360	499	513	524	538	550	564	596	630	661	693	726	765	830
370	450	463	473	485	497	508	539	570	600	628	658	692	748
380	408	420	429	440	451	462	490	518	544	571	600	628	680
390	372	383	392	401	411	420	448	473	497	522	548	572	621
400	341	351	360	368	377	387	411	434	456	480	502	524	570
420	288	297	304	313	321	329	348	368	389	409	429	449	489
440	243	252	259	266	274	280	299	319	338	356	376	396	432
460	204	212	218	225	232	241	259	279	296	316	334	352	388
480	-	-	-	-	-	210	228	246	262	280	298	315	348
500	-	-	-	-	-	187	203	219	234	250	265	280	311
520	-	-	-	-	-	169	183	197	210	223	236	250	278

*Critical pressure.

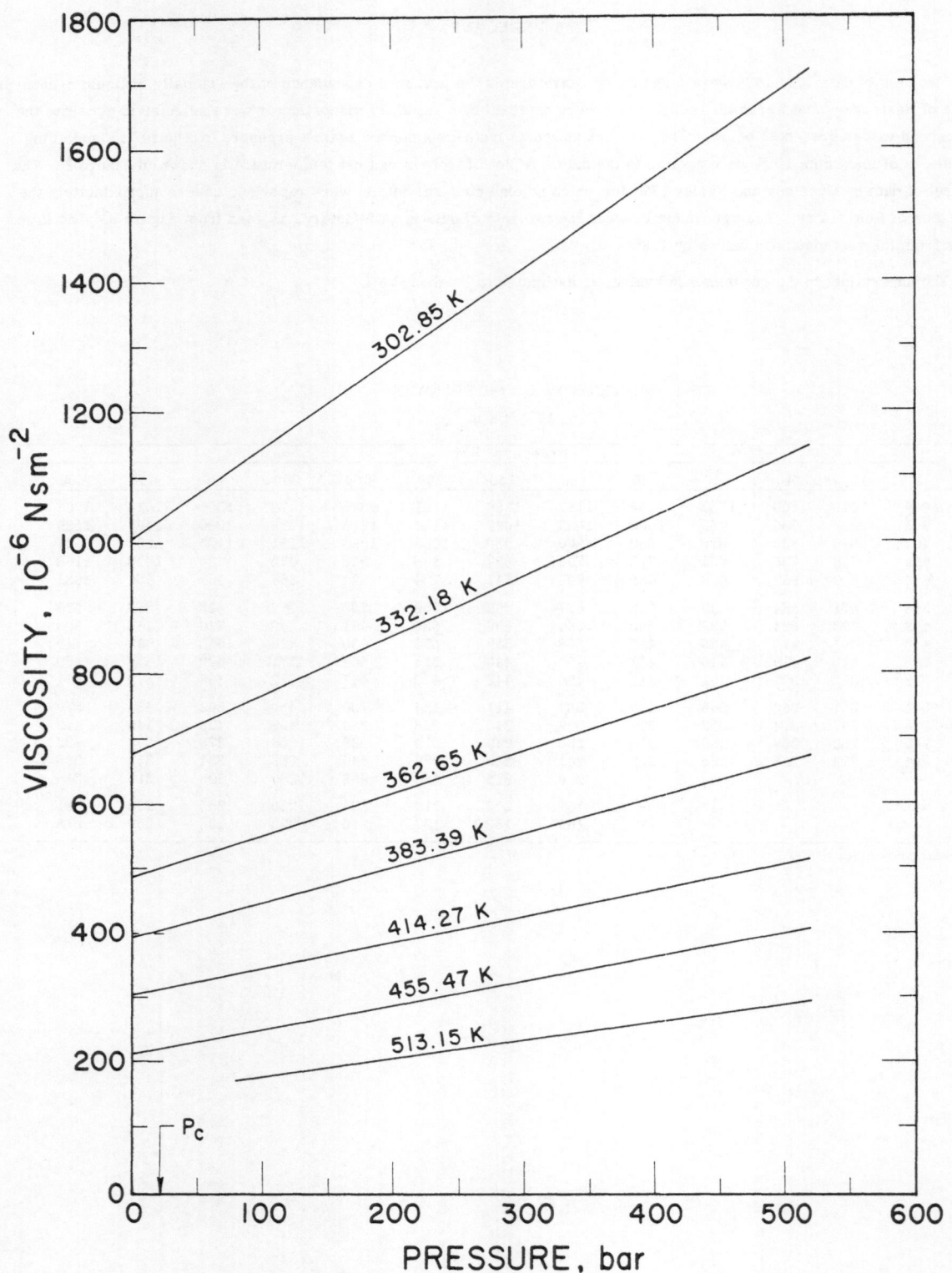

FIGURE I. VISCOSITY OF n-UNDECANE [163].

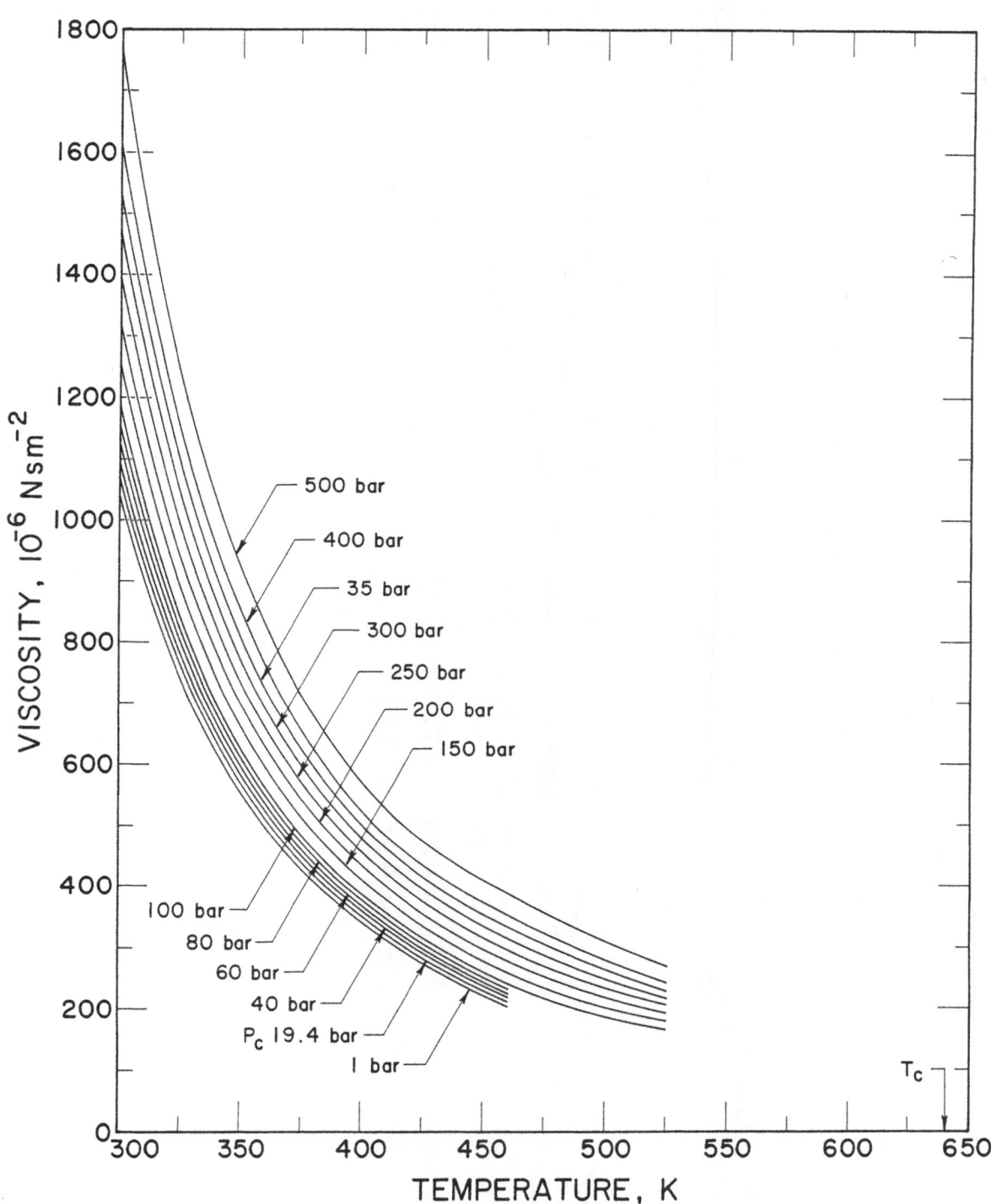

FIGURE 2. VISCOSITY OF n-UNDECANE [163].

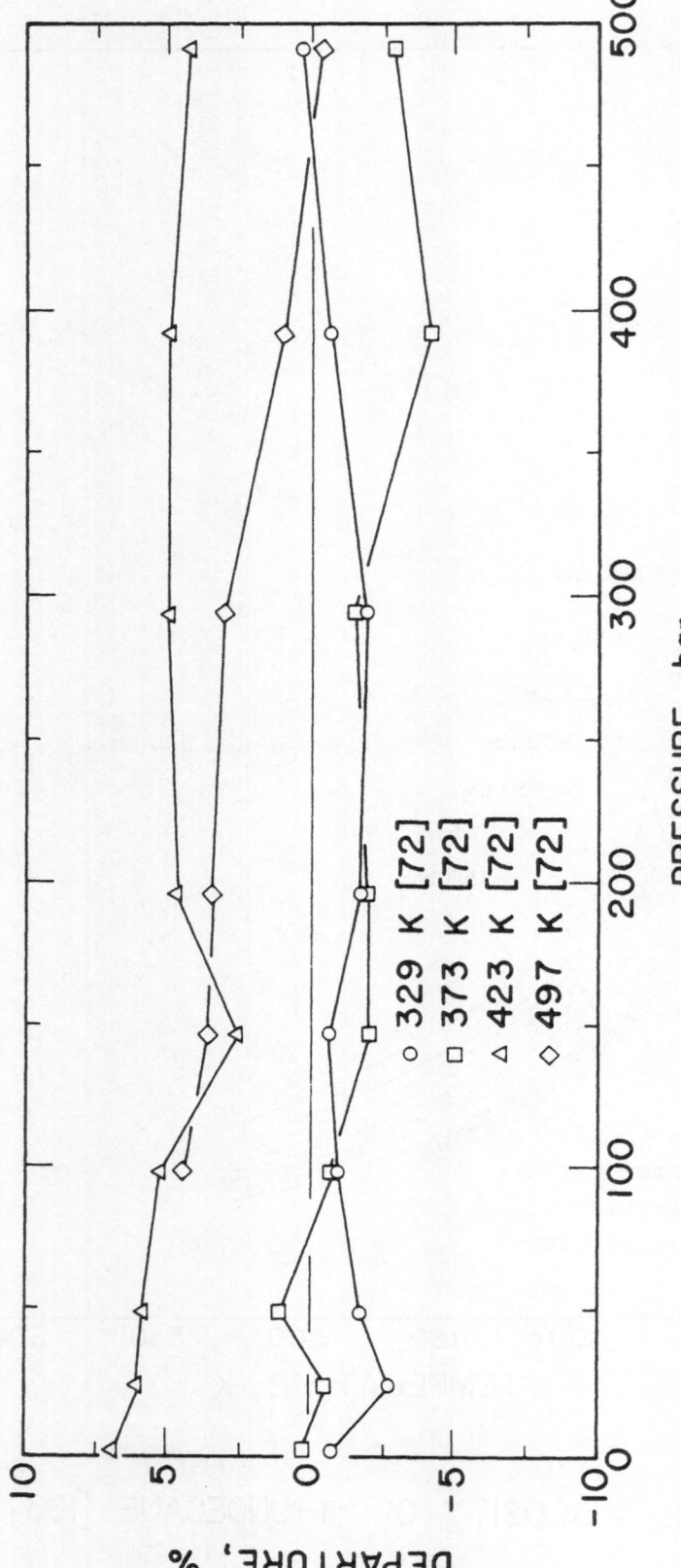

FIGURE 3. DEPARTURE PLOT ON THE VISCOSITY OF *n* – UNDECANE.

XENON

The data of Trappeniers et al. and Van Oosten [196] from the van der Waals laboratory, have been used to generate the recommended values given on the next page. These data, obtained by a capillary tube viscosimeter, are plotted as isotherms against pressure as full lines in Figure 1. Using the P-V-T data of [152], the dilute gas viscosities of [195] and the residual viscosity concept, it was possible to extrapolate from 348 K up to 430 K and 500 bar. The extrapolated values are shown as dashed lines. In Figure 2, isobars are plotted against temperature. Again the extrapolated values are shown as dashed lines.

Additional works on the pressure dependence of the viscosity of xenon are listed in the summary table below. The reliable measurements of Kestin et al. [104] confirm the recommended values for low pressures. However, the data of Reynes et al [169] deviate substantially and systematically as seen from Figure 3. A truly objective conclusion appears to be impossible at this time, but the view is taken here, that the recommended values based on the data of Trappeniers et al. are more reliable, though both sets were obtained by the same experimental method. This judgement is based on the high standards of experimental sophistication at the van der Waals laboratory. The uncertainty in the recommended values is estimated to be $\pm$ 2% in the region of measurements and $\pm$ 4% in the region of extrapolation. This is substantiated by comparison with a set of recommended values by [76], which was derived from the principle of corresponding states using experimental data of argon. These latter values include the liquid region and extend to somewhat higher temperatures. They are recommended for use in the regions not covered here.

ADDITIONAL REFERENCES ON THE VISCOSITY OF XENON

Authors	Year	Ref. No.	Temperature K	Pressure bar	Method	Departure % (no. points)
Strumpf et al.	1974	188	Crit.	Crit.	Torsional Crystal	–
Slyusar et al.	1972	181	161-290	SL	Falling cylinder	–
Boon et al.	1967	17	163-169	SL	Capillary tube	–
Reynes and Thodos	1964	169	323-473	71-830	Capillary tube	-4.82 ± 2.49 (21)
Reynes	1964	167	323-473	71-830	Capillary tube	-4.82 ± 2.49 (21)
Kestin and Leidenfrost	1959	105	298	1-28	Oscillating-disk	1.5 (8)

VISCOSITY OF XENON
[μ, 10^{-6} N s m^{-2}]

T, K									Pressure, bar											
	1	20	30	40	50	55	58.4*	60	65	70	75	80	90	100	150	200	250	300	400	500
300	23.2	24.7	25.9	27.4	30.0	31.8	33.5	34.3	39.4	54.4	67.2	76.2	87.2	94.3	117.0	133.6	147.3	159.2	179.8	198.1
302	23.4	24.9	26.1	27.5	29.9	31.6	33.0	33.8	37.9	46.9	60.7	71.3	83.3	90.7	114.4	131.0	144.4	157.0	177.7	196.0
304	23.5	25.0	26.2	27.5	29.9	31.3	32.8	33.4	36.9	42.0	54.5	66.4	79.4	87.0	111.7	128.9	142.5	154.9	175.6	193.9
306	23.6	25.2	26.3	27.6	29.8	31.1	32.4	33.0	36.3	40.1	48.9	61.6	75.5	83.5	109.1	126.5	140.3	152.6	173.5	191.9
308	23.8	25.3	26.4	27.6	29.8	31.0	32.2	32.8	35.7	39.0	45.5	56.8	71.6	79.9	106.6	124.2	138.0	150.5	171.5	189.7
310	24.0	25.5	26.5	27.7	29.7	30.8	32.0	32.5	35.3	38.1	43.4	51.9	67.8	76.3	104.3	122.0	136.0	148.5	169.4	187.6
312	24.1	25.6	26.6	27.8	29.8	30.8	31.9	32.4	34.8	37.5	41.9	48.0	63.9	72.8	101.9	119.8	133.9	146.5	167.4	185.7
314	24.3	25.8	26.8	27.9	29.8	30.7	31.8	32.3	34.5	37.0	40.6	45.7	60.0	69.5	99.5	117.5	131.8	144.5	165.5	183.8
316	24.4	25.9	26.9	28.0	29.8	30.7	31.7	32.2	34.2	36.5	39.7	44.0	56.0	66.1	97.4	115.5	129.7	142.5	163.5	181.8
318	24.5	26.0	27.0	28.1	29.8	30.7	31.6	32.1	34.0	36.1	39.0	42.6	52.9	63.0	95.0	113.5	127.8	140.5	161.6	179.9
320	24.6	26.2	27.1	28.2	29.9	30.7	31.6	32.1	33.8	35.7	38.4	41.6	50.6	60.0	92.8	111.4	125.9	138.6	159.8	178.0
325	25.0	26.5	27.4	28.5	30.0	30.9	31.6	32.1	33.4	35.0	37.2	39.7	46.3	54.3	87.3	106.5	121.1	133.9	155.0	173.4
330	25.4	26.9	27.7	28.8	30.2	31.0	31.7	32.1	33.2	34.6	36.3	38.4	43.6	50.2	82.1	102.0	116.6	129.4	150.6	168.9
335	25.8	27.2	28.0	29.1	30.4	31.1	31.8	32.2	33.1	34.3	35.7	37.5	42.0	47.4	77.4	97.5	112.3	125.0	146.3	164.6
340	26.1	27.5	28.4	29.4	30.6	31.4	31.9	32.3	33.1	34.2	35.4	37.0	40.9	45.4	73.0	93.5	108.1	121.0	142.2	160.3
345	26.5	27.8	28.7	29.6	30.9	31.5	32.0	32.4	33.1	34.1	35.3	36.6	40.0	44.0	69.0	89.5	104.2	117.0	138.2	156.2
350	26.8	28.1	28.9	29.9	31.1	31.7	32.2	32.4	33.2	34.2	35.2	36.4	39.4	43.0	65.6	85.9	100.6	113.2	134.5	152.3
360	27.5	28.7	29.6	30.4	31.5	32.1	32.5	32.7	33.4	34.2	35.2	36.0	38.5	41.5	60.3	79.0	93.8	106.4	127.4	145.0
370	28.1	29.3	30.1	31.0	32.0	32.5	32.9	33.0	33.6	34.4	35.1	36.0	38.0	40.5	56.5	73.6	88.0	100.2	120.9	138.2
380	28.7	29.9	30.7	31.4	32.5	32.9	33.3	33.4	34.0	34.6	35.3	36.0	37.8	39.9	53.9	69.1	82.9	94.6	115.0	132.0
400	30.0	31.2	31.8	32.5	33.4	33.8	34.1	34.2	34.7	35.2	35.8	36.5	37.8	39.4	50.2	62.9	74.8	85.6	104.9	121.4
420	31.3	32.4	32.9	33.5	34.3	34.7	35.0	35.1	35.6	36.1	36.5	37.0	38.2	39.6	48.2	58.5	69.0	78.8	96.6	112.4
430	31.9	33.0	33.5	34.1	34.9	35.3	35.6	35.7	36.2	36.6	37.0	37.4	38.5	39.8	47.6	56.7	67.0	75.9	92.9	108.3

*Critical pressure.

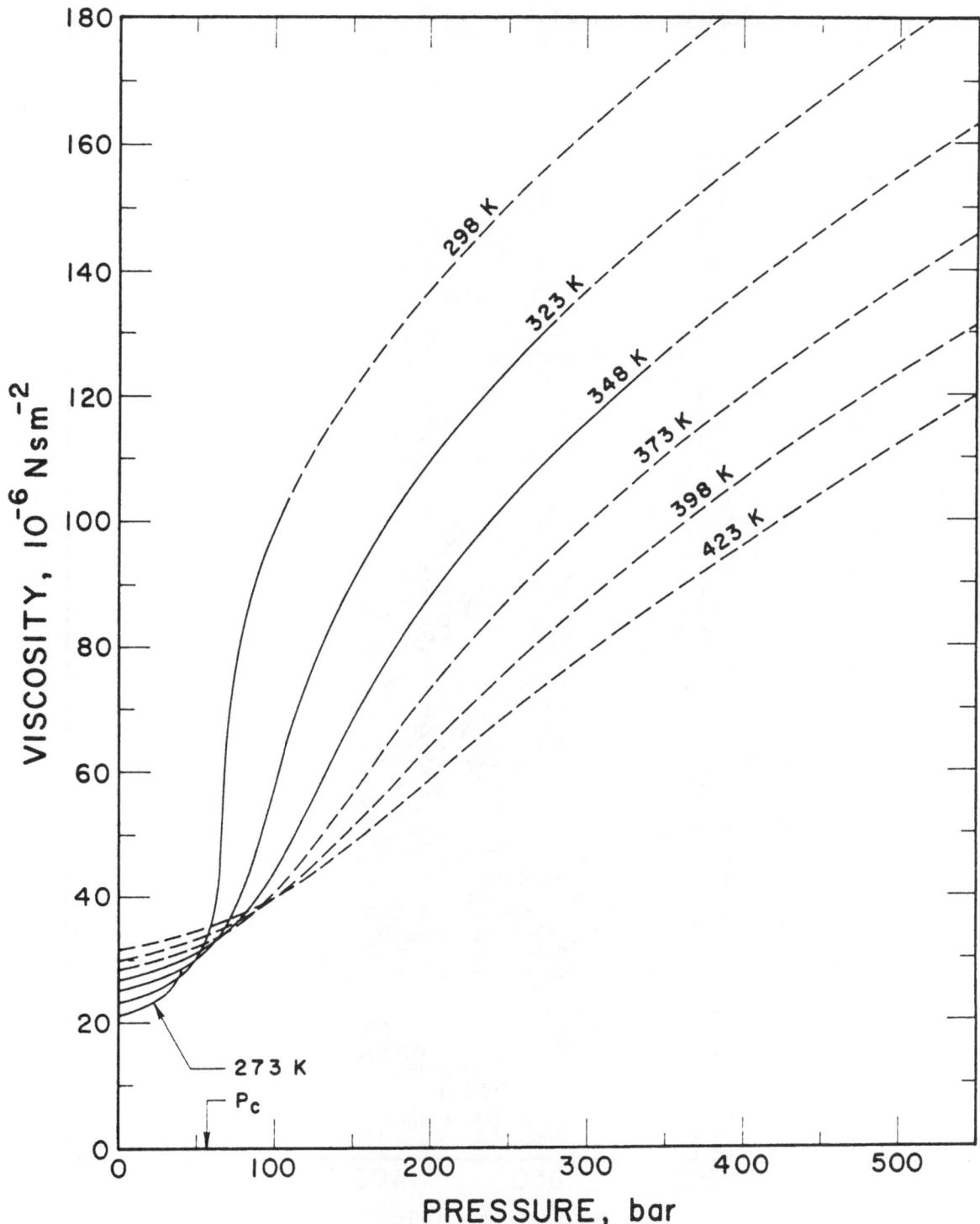

FIGURE 1. VISCOSITY OF XENON [196].

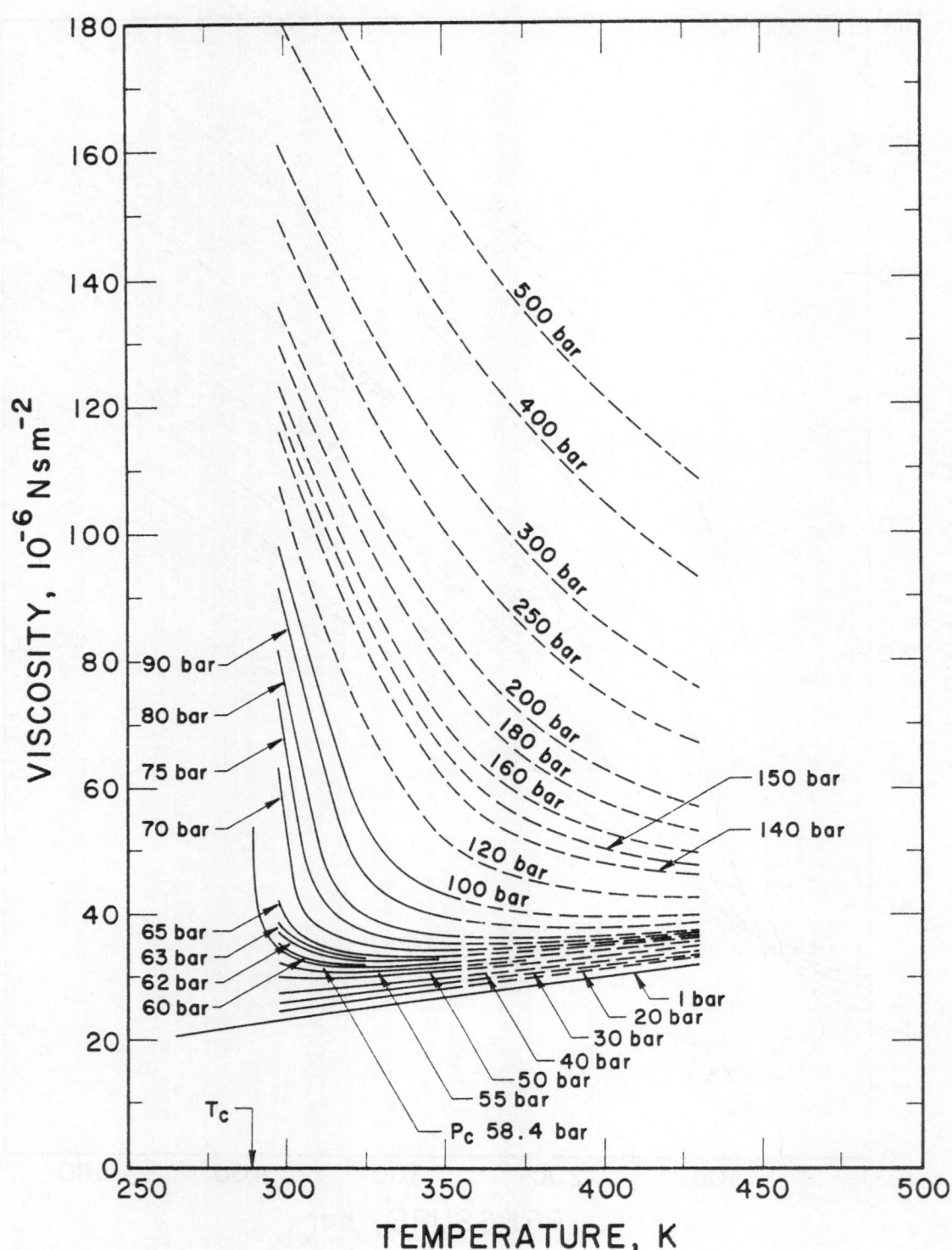

FIGURE 2. VISCOSITY OF XENON [196].

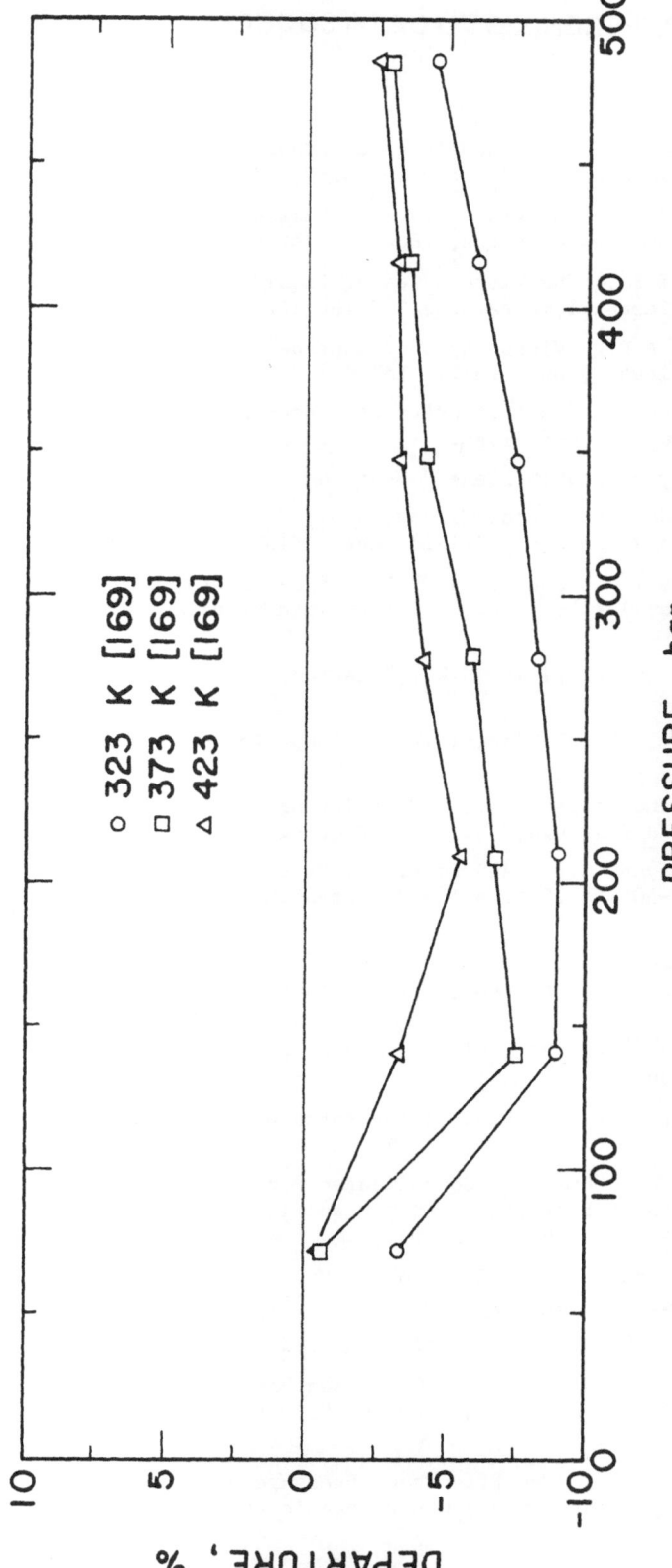

FIGURE 3. DEPARTURE PLOT ON THE VISCOSITY OF XENON.

REFERENCES TO DATA SOURCES

1. Agaev, N.A. and Golubev, I.F., "The Viscosities of Liquid and Gaseous n-Pentane at High Pressures and Different Temperatures," Gazov. Promst., 8(5), 45-50, 1963.

2. Agaev, N.A. and Golubev, I.F., "The Viscosity of n-Hexane in the Liquid and Gaseous State at High Pressures and Different Temperatures," Dokl. Phys. Chem., 151, 635-40, 1963.

3. Agaev, N.A. and Golubev, I.F., "The Viscosities of Liquid and Gaseous n-Heptane and n-Octane at High Pressures and Different Temperatures," Gazov. Promst., 8(7), 50-3, 1963.

4. Agaev, N.A. and Yusibova, A.D., "Viscosity of Isobutane at High Pressures," Foreign Technology Division Rept. FTD-MT-71-1605, 9 pp., 1971. [AD 742 714]

5. Agaev, N.A. and Yusibova, A.D., "The Viscosity of Isooctane at High Pressures and Various Temperatures," Gazov. Promst., 46-8, 1966.

6. Akhundov, T.C., "Viscosity of Ethylbenzene," Neft. Gazov. Promst., 10, 46-74, 1973.

7. Akhundov, T.C., Ismail-Zade, Sh.M., and Tairov, A.D., "Toluene Viscosity at High Pressures and Temperatures," Foreign Technology Division Rept. FTD-HC-23-54-72, 7 pp., 1972. [AD 743 340]

8. Andreev, I.I., Tsederberg, V.N., and Popov, V.N., "Experimental Investigation of the Viscosity of Argon," Teploenergetika, 13(8), 78-81, 1966; English translation: Therm. Eng., 13(8), 111-6, 1966.

9. Arnold, E.W., Liou, D.W., and Eldridge, J.W., "Thermodynamic Properties of Isopentane," J. Chem. Eng. Data, 10(2), 88-92, 1965.

10. Babb, S.E. and Scott, G.J., "Rough Viscosities to 10000 Bars," J. Chem. Phys., 40, 3666-8, 1964.

11. Baron, J.D., Roof, J.G., and Wells, F.W., "Viscosity of Nitrogen, Methane, Ethane, and Propane at Elevated Temperature and Pressure," J. Chem. Eng. Data, 4, 283-8, 1959.

12. Barua, A.K., Afzal, M., Flynn, G.P., and Ross, J., "Viscosity of Hydrogen, Deuterium, Methane, and Carbon Monoxide From -50° to 150°C Below 200 Atmospheres," J. Chem. Phys., 41, 374-8, 1964.

13. Beattie, J.A., Brierley, J.S., and Barriault, R.J., "The Compressibility of Krypton. I. An Equation of State for Krypton and the Weight of a Liter of Krypton," J. Chem. Phys., 20, 1613-5, 1952.

14. Benning, A.F. and Markwood, W.H., "The Viscosities of Freon Refrigerants," Refrigerating Engineering, 4, 243-7, 1939.

15. Bicher, L.B. and Katz, D.L., "Viscosities of the Methane-Propane System," Ind. Eng. Chem., 35, 754-61, 1943.

16. Biles, M.B. and Putnam, J.A., "Use of a Consolidated Porous Medium for Measurement of Flow Rate and Viscosity of Gases at Elevated Pressures and Temperatures," Natl. Advisory Comm. Aeron. Tech. Note, NACA-TN-2783, 1952.

17. Boon, J.P., Legros, J.C., and Thomaes, G., "On the Principle of Corresponding States for the Viscosity of Simple Liquids," Physica, 33, 547-57, 1967.

18. Boon, J.P. and Thomaes, G., "The Viscosity of Liquefied Gases," Physica, 29, 208-14, 1963.

19. Braune, H. and Linke, R., "The Viscosity of Gases and Vapors. III. Influence of the Dipole Moment on the Magnitude of the Sutherland Constant," Z. Phys. Chem. (Leipzig), 148A, 195-215, 1930.

20. Brazier, D.W. and Freeman, G.R., "The Effects of Pressure on the Density, Dielectric Constant, and Viscosity of Several Hydrocarbons and Other Organic Liquids," Can. J. Chem., 47(6), 893-9, 1969.

21. Bridgman, P.W., "The Effect of Pressure on the Viscosity of Forty-Three Pure Liquids," Proc. Am. Acad. Arts Sci., 61, 57-99, 1926.

22. Bridgman, P.W., "Viscosities to 30000 Kg/cm²," Proc. Am. Acad. Arts Sci., 77, 115-28, 1949.

23. Bridgman, P.W., "The Physics of High Pressures," Bell and Sons, London, 1949.

24. Carmichael, L.T., Berry, V.M. and Sage, B.H., "Viscosity of Hydrocarbons. Propane," J. Chem. Eng. Data, 9, 411-5, 1964.

25. Carmichael, L.T., Berry, V., and Sage, B.H., "Viscosity of Hydrocarbons. Methane," J. Chem. Eng. Data, 10, 57-61, 1965.

26. Carmichael, L.T., Berry, V.M., and Sage, B.H., "Viscosity of Hydrocarbons. n-Decane," J. Chem. Eng. Data, 14(1), 27-31, 1969.

27. Carmichael, L.T., Reamer, H.H., and Sage, B.H., "Viscosity of Ammonia at High Pressures," J. Chem. Eng. Data, 8, 400-4, 1963.

28. Carmichael, L.T. and Sage, B.H., "Viscosity of Liquid Ammonia at High Pressures," Ind. Eng. Chem., 44, 2728-32, 1952.

29. Carmichael, L.T. and Sage, B.H., "Viscosity of Ethane at High Pressures," J. Chem. Eng. Data, 8, 94-8, 1963.

30. Carmichael, L.T. and Sage, B.H., "Viscosity of Hydrocarbons. n-Butane," J. Chem. Eng. Data, 8, 612-6, 1963.

31. Carr, N.L., "Viscosity of Gas Mixtures at High Pressures," Illinois Institute of Technology, Chicago, IL, Ph.D. Dissertation, 1952.

32. Chaikovskii, V., Geller, V.Z., and Ivanchenko, S.I., "Viscosity of Dichlorodifluoromethane," Neft. Gazov. Promst., 7, 111-2, 1973.

33. Chierici, G.L. and Paratella, A., "Viscosity Measurements on Carbon Monoxide, Nitrogen up to 900 Atmospheres and Correlation to Mass Diffusion," AIChE J., 15(5), 786-90, 1969.

34. Collings, A.F. and McLaughlin, E., "Torsional Crystal Technique for the Measurement of Viscosities of Liquids at High Pressure," Trans. Faraday Soc., 67, 340-52, 1971.

35. Comings, E.W. and Egly, R.S., "Viscosity of Ethylene and of Carbon Dioxide Under Pressure," Ind. Eng. Chem., 33, 1224-9, 1941.

36. Comings, E.W., Mayland, B.J., and Egly, R.S., "The Viscosity of Gases at High Pressures," Univ. Illinois, Eng. Expt. Sta. Bull., Series No. (354), 68 pp., 1944.

37. D'Ans Lax, Pocketbook for Chemists and Physicists, Volume I, Springer-Verlag, Berlin, 1967.

38. DeBock, A., Grevendonk, W., and Awouters, H., "Pressure Dependence of the Viscosity of Liquid Argon and Liquid Oxygen, Measured by Means of a Torsionally Vibrating Quartz Crystal," Physica, 34, 49-52, 1967.

39. DeBock, A., Grevendonk, W., and Herreman, W., "Shear Viscosity of Liquid Argon," Physica, 37, 227-32, 1967.

40. Diller, D.E., "Measurements of the Viscosity of Parahydrogen," J. Chem. Phys., 42, 2089-100, 1965.

41. DiPippo, R., Kestin, J., and Oguchi, K., "Viscosity of Three Binary Gaseous Mixtures," J. Chem. Phys., 46, 4758-64, 1967.

42. Dolan, J.P., Starling, K.E., Lee, A.L., Eakin, B.E., and Ellington, R.T., "Liquid Gas and Dense Fluid Viscosity of n-Butane," J. Chem. Eng. Data, 8, 396-9, 1963.

43. Eakin, B.E. and Ellington, R.T., "Improved High Pressure Capillary Tube Viscosimeter," Petroleum Trans. (AIME), 216, 85-91, 1959.

44. Eakin, B.E., Starling, K.E., Dolan, J.P., and Ellington, R.T., "Liquid Gas, and Dense Fluid Viscosity of Ethane," J. Chem. Eng. Data, 7, 33-6, 1962.

45. Eisele, E.H., Fontaine, W.E., and Leidenfrost, W., "Measurement of Kinematic Viscosity of the Liquid Phase of a Refrigerant Close to Boiling, at Temperatures and Pressures Other than Ambient, Using a Commercial, Capillary Tube Viscosimeter," 12th Proc. Int. Congr. Refrig., 2, 553-60, 1969.

46. Elverum, G.W. and Doescher, R.N., "Physical Properties of Liquid Fluorine," J. Chem. Phys., 20, 1834-6, 1952.

47. Felsing, W.A. and Watson, G.M., "The Compressibility of Liquid n-Octane," J. Am. Chem. Soc., 64(8), 1822-3, 1942.

48. Filippova, G.P. and Ishkin, I.P., "The Viscosity of Air, Nitrogen, and Argon at Low Temperatures and Pressures to 150 Atmospheres," Inzh.-Fiz. Zh., 4(3), 105-9, 1961.

49. Flynn, G.P., Hanks, R.V., Lemaire, N.A., and Ross, J., "Viscosity of Nitrogen, Helium, Neon, and Argon From -78.5° to 100°C Below 200 Atmospheres," J. Chem. Phys., 38, 154-62, 1963.

50. Förster, S., "Viscosity Measurements in Liquid Neon, Argon, and Nitrogen," Cryogenics, 176-7, 1963.

51. Gallant, R.W., "Physical Properties of Hydrocarbons, Part 20 - Halogenated Methanes," Hydrocarbon Process. Pet. Refiner, 47(1), 135-42, 1968.

52. Geller, V.Z., Ivanchenko, S.I., and Peredrii, V.G., "Coefficients of Dynamic Viscosity and Thermal Conductivity of Difluorochloromethane," Neft. Gazov. Promst., 16(8), 62, 61-5, 1973.

53. Gibbons, R.M., "The Equation of State for Neon Between 27 and 70 K," Cryogenics, 9(8), 251-60, 1969.

54. Giddings, J.G., Kao, J.T.F., and Kobayashi, R., "Development of a High-Pressure Capillary Tube Viscometer and Its Application to Methane, Propane, and Their Mixtures in the Gaseous and Liquid Regions," J. Chem. Phys., 45, 578-86, 1966.

55. Glaser, F. and Gebhardt, F., "Measurements of the Viscosity of Gases and Vapors at High Pressures and High Temperatures," Chem. Ing. Techn., 31, 743-5, 1959.

56. Goldman, K., "Viscosity of N_2 at Low Temperatures and High Pressures," Physica, 29, 510-5, 1963.

57. Golubev, I.F., Viscosity of Gases and Gas Mixtures, a Handbook, Israel Program for Scientific Translations, Jerusalem, 1970.

58. Golubev, I.F. and Gnezdilov, H.E., "Viscosity of Helium and Helium-Hydrogen-Mixtures at Temperatures Between 0 and 250°C and Pressures Up to 500 at," Gazov. Promst., 10(12), 38-42, 1965.

59. Golubev, I. and Kurin, V.I., "Measuring the Viscosity of Gases at Pressures up to 4000 Kgf/cm^2 and Different Temperatures," Therm. Eng., 8, 121-5, 1974.

60. Golubev, I.F. and Shepeleva, R.I., "Viscosity of Hydrogen at Low Temperatures and High Pressures," Gazov. Promst., 11(4), 54-8, 1966.

61. Gonikberg, M.G. and Vereshchagin, L.F., "Measurements of Ethylene Viscosity at Pressures up to 1000 Atm. by the Oscillating Disc Method," Dokl. Akad. Nauk, 55(9), 801-4, 1947.

62. Gonzalez, M.H., Bukacek, R.F., and Lee, A.L., "The Viscosity of Methane," Soc. Petrol. Eng. J., 7, 75-9, 1967.

63. Gonzalez, M.H. and Lee, A.L., "Viscosity of Isobutane," J. Chem. Eng. Data, 11, 357-9, 1966.

64. Goring, G.E. and Eagan, D.P., "An Investigation of the Viscosity of Dry Air at Elevated Pressures and Temperatures Using a Steady-Flow Capillary Viscosimeter," Can. J. Chem. Eng., 49, 25-31, 1971.

65. Gracki, J.A., Flynn, G.P., and Ross, J., "Viscosity of Nitrogen, Helium, Hydrogen, and Argon from -100 to 25°C up to 150-250 Atm," J. Chem. Phys., 51, 3856-63, 1969.

66. Gracki, J.A., Flynn, G.P., and Ross, J., "Viscosity of Nitrogen, Helium, Hydrogen, and Argon from -100 to 25°C up to 150-250 Atm," Purdue Univ. W. Lafayette, Indiana, 28 pp., 1969. [N69-39873] [AD 690 933]

67. Grevendonk, W., Herreman, W., and DeBock, A., "Measurement on the Viscosity of Liquid Nitrogen," Physica, 46, 600-4, 1970.

68. Grevendonk, K.W., Herreman, W., DePesserocy, W., and DeBock, A., "On the Shear Viscosity of Liquid Oxygen," Physica, 40, 207-12, 1968.

69. Groenier, W.S. and Thodos, G., "Pressure-Volume-Temperature Behavior of Ammonia in the Gaseous and Liquid States," J. Chem. Eng. Data, 5, 285-8, 1960.

70. Guseinov, K.D. and Kadscharov, V.T., "The Viscosity of Liquid Propylacetate and Butylacetate at Various Temperatures and Pressures," Nef. Gazov. Promst., 4, 68-78, 1975.

71. Guseinov, S.O., Naziev, Ya.M., and Akhmedov, A.K., "Viscosity of Cyclohexane at High Pressures," Neft. Gazov. Promst., 2, 65-7, 1973.

72. Guseinov, S.O. and Naziev, Ya.M., "The Viscosities of Liquid n-Undecane and n-Heptadecane at High Pressures," Neft. Gazov. Promst., 12, 61-3, 1973.

73. Guseinov, S.O., Naziev, Ya.M., and Ahmedov, A.K., "Viscosity of Liquid Methylcyclohexane and Ethylcyclohexane at High Pressures," Neft. Gazov. Promst., 1, 73-5, 1973.

74. Haepp, H.J., "Measurement of the Viscosity of Carbon Dioxide and Propylene," Ruhr-Univ. Bochum (B.R.D.), Dissertation, 1975.

75. Haepp, H.J., "Measurement of the Viscosity of Carbon Dioxide and Propylene," Waerme Stoffuebertrag., 6, 281-90, 1976.

76. Hanley, H.J.M., McCarty, R.D., and Haynes, W.M., "The Viscosity and Thermal Conductivity Coefficients for Dense Gaseous and Liquid Argon, Krypton, Xenon, Nitrogen, and Oxygen," J. Phys. Chem. Ref. Data, 3(4), 979-1018, 1974.

77. Haynes, W.M., "Viscosity of Gaseous and Liquid Argon," Physica, 67, 440-70, 1973.

78. Haynes, W.M., "Measurements of the Viscosity of Compressed Gaseous and Liquid Fluorine," Physica, 76(1), 1-20, 1974.

79. Heiks, J.R. and Orban, E., "Liquid Viscosities at Elevated Temperatures and Pressures: Viscosity of Benzene from 90° to Its Critical Temperature," J. Phys. Chem., 60, 1025-7, 1956.

80. Hellemans, J., Zink, H., and Van Paemel, O., "The Viscosity of Liquid Nitrogen and Liquid Oxygen Along Isotherms as a Function of Pressure," Physica, 47, 45-57, 1970.

81. Hellemans, J., Zink, H., and Van Paemel, O., "The Viscosity of Liquid Argon and Liquid Methane Along Isotherms as a Function of Pressure," Physica, 46, 395-410, 1970.

82. Herreman, W. and Grevendonk, W., "An Experimental Study on the Shear Viscosity of Liquid Neon," Cryogenics, 14(7), 395-8, 1974.

83. Herreman, W., Grevendonk, W., and DeBock, A., "Shear Viscosity Measurements of Liquid Carbon Dioxide," J. Chem. Phys., 53, 185-9, 1970.

84. Herreman, W., Lattenist, A., Grevendonk, W., and DeBock, A., "Measurements on the Viscosity of Carbon Dioxide," Physica, 52, 489-92, 1971.

85. Huang, E.T.S., Swift, G.W., and Kurata, F., "Viscosities of Methane and Propane at Low Temperatures and High Pressures," AIChE J., 12, 932-6, 1966.

86. Hubbard, R.M. and Brown, G.G., "Viscosity of n-Pentane," Ind. Eng. Chem., 35, 1276-80, 1943.

87. Isakova, N.P. and Oshueva, L.A., "Viscosity of Liquid Methanol," Russ. J. Phys. Chem., 40(5), 607, 1966.

88. Van Itterbeek, A., Hellemans, J., Zink, H., and Van Cauteren, M., "Viscosity of Liquefied Gases at Pressures Between 1 and 100 Atmosphere," Physica, 32, 2171-2, 1966.

89. Ivanchenko, S.I., "Viscosity of Trifluorotrichloroethane," Neft. Gazov. Promst., 15(7), 82 pp., 1972.

90. Iwasaki, H., "Measurement of Viscosities of Gases at High Pressure," Sci. Rep., Tohoku Imp. Univ., 3A, 247-57, 1951.

91. Iwasaki, H., Kestin, J., and Nagashima, A., "Viscosity of Argon-Ammonia Mixtures," J. Chem. Phys., 40, 2988-95, 1964.

92. Iwasaki, H. and Takahashi, M., "Viscosity of Methane," J. Chem. Soc., Japan, Ind. Chem. Sect., 62, 918-21, 1959.

93. Iwasaki, H. and Takahashi, M., "Studies on the Transport Properties of Fluids at High Pressure. I. The Viscosity of Ammonia," Rev. Phys. Chem. Japan, 38, 18-27, 1968.

94. Jobling, A. and Lawrence, A.S.C., "Viscosities of Liquids at Constant Volume," Proc. R. Soc. London, 206A, 257-74, 1951.

95. Kamien, C.Z., "The Viscosity of Several Fluorinated Hydrocarbon Compounds in the Vapor Phase," Purdue Univ., Lafayette, Indiana, M.S. Thesis, 98 pp., 1956.

96. Kamien, C.Z. and Witzell, O.W., "Effect of Pressure and Temperature on the Viscosity of Refrigerants," ASHRAE J., 65, 663-74, 1959.

97. Kao, I.T.F. and Kobayashi, R., "Viscosity of Helium and Nitrogen and Their Mixtures at Low Temperatures and Elevated Pressures," J. Chem. Phys., 47(8), 2836-49, 1967.

98. Karbanov, E.M. and Geller, V.Z., "Viscosity of Trifluorobromomethane," Neft. Gazov. Promst., 11, 22-104, 1974.

99. Kellström, G., "Viscosity of Air at Pressures 1-30 Kg/cm^2," Ark. Mat., Astron. Phys., 27A, 1-15, 1941.

100. Kennedy, G.C., "Pressure-Volume-Temperature Relations in CO_2 at Elevated Temperatures and Pressures," Am. J. Sci., 252, 225-35, 1954.

101. Keramidi, A.S. and Rastorguev, Yu.L., "Viscosity of Hydrocarbons. n-Nonane," Neft. Gazov. Promst., 15(9), 65-8, 1972.

102. Keramidi, A.S. and Rastorguev, Yu.L., "Viscosity of n-Dodecane at High Pressures," Neft. Gazov. Promst., 13(10), 108-14, 1970.

103. Kestin, J. and Leidenfrost, W., "Thermodynamic and Transport Properties of Gases, Liquids and Solids," Symposium, Lafayette, Indiana, 1959 (McGraw-Hill Book Co., Inc., New York, NY, 1959).

104. Kestin, J. and Leidenfrost, W., "The Viscosity of Helium," Physica, 25, 537-55, 1959.

105. Kestin, J. and Leidenfrost, W., "An Absolute Determination of the Viscosity of Eleven Gases Over a Range of Pressures," Physica, 25, 1033-62, 1959.

106. Kestin, J., Paykoc, E., and Sengers, J.V., "On the Density Expansion for Viscosity in Gases," Physica, 54, 1-19, 1971.

107. Kestin, J. and Pilarczyk, K. "Measurement of the Viscosity of Five Gases at Elevated Pressures by the Oscillating-Disk Method," Trans. ASME, 76, 987-99, 1954.

108. Kestin, J. and Wang, H.E., "The Viscosity of Five Gases: A Re-Evaluation," Trans. ASME, 80, 11-7, 1958.

109. Kestin, J. and Whitelaw, J.H., "A Relative Determination of the Viscosity of Several Gases by the Oscillating Disk Method," Physica, 29, 335-56, 1963.

110. Kestin, J. and Whitelaw, J.H., "The Viscosity of Dry and Humid Air," Int. J. Heat Mass Transfer, 7, 1245-55, 1964.

111. Kestin, J., Whitelaw, J.H., and Zien, T.F., "The Viscosity of Carbon Dioxide in the Neighbourhood of the Critical Point," Physica, 30, 161-81, 1964.

112. Kestin, J. and Yata, J., "Viscosity and Diffusion Coefficient of Six Binary Mixtures," J. Chem. Phys., 49, 4780-91, 1968.

113. Khalilov, Kh., "Viscosity of Liquid and Saturated Vapors at High Temperatures and Pressures," Zhetf. T., 9(3), 335-45, 1939.

114. Kiyama, R. and Makita, T., "An Improved Viscosimeter for Compressed Gases and the Viscosity of Oxygen," Rev. Phys. Chem. Japan, 26(2), 70-4, 1956.

115. Kiyama, R. and Makita, T., "A New Simple Viscometer for Compressed Gases and Viscosity of Carbon Dioxide," Rev. Phys. Chem. Japan, 21, 63-8, 1951.

116. Kiyama, R. and Makita, T., "The Viscosity of Carbon Dioxide, Ammonia, Acetylene, Argon and Oxygen Under High Pressures," Rev. Phys. Chem. Japan, 22, 49-58, 1952.

117. Kletskii, A.V., Thermophysical Properties of Freon-22, Israel Program for Scientific Translations, Jerusalem, 1971. [TT-70-50178]

118. Kor, S.K., Singh, B.K., and Rai, G., "Pressure Dependence of Ultrasonic Absorption and Compressional Viscosity Due to Structural Rearrangement in Hexane and Toluene," Nuovo Cimento Soc. Ital. Fis., 12B(2), 205-14, 1972.

119. Kozlov, Yu.V., Yakovlev, V.F., and Malyavin, I.G., "Viscosity of Liquids at Constant Density," Russ. J. Phys. Chem., 40, 1265-6, 1966.

120. Kuloor, N.R., Newitt, D.M., and Bateman, J.S., "Thermodynamic Functions of Gases," Vol. 2, Butterworths, London, 1962.

121. Kurin, V.I. and Golubev, I.F., "The Viscosity of Argon, Air and Carbon Dioxide at Pressures up to 4000 kgf/cm^2 and at Different Temperatures," Teploenergetika, 21(11), 84-5, 1974; English translation: Therm. Eng., 21(11), 125-7, 1974.

122. Kuss, E., "High-Pressure Experiments II: The Viscosity of Compressed Gases," Z. Angew. Phys., 4, 203-7, 1952.

123. Kuss, E., "Viscosity of Compressed Liquids," Z. Angew. Phys., 7, 372-8, 1955.

124. Kuss, E. and Pollmann, P., "Viscosity Pressure Dependence and Degree of Branching of Liquid Alkanes," Z. Phys. Chem. (Frankfurt am Main), 68(3-6), 205-27, 1969.

125. Landolt-Bornstein, Pyrometric Experimental Technique for the Thermodynamic Properties of Homogeneous Materials, Volume IV, Part 4a, Springer-Verlag, Berlin, 1967.

126. Landolt-Bornstein, Numerical Values and Functions, Transportphenomena I, Volume II, Part 5a, Springer-Verlag, Berlin, 1969.

127. Landolt-Bornstein, Numerical Values and Functions, Volume II, Part 2a, Springer-Verlag, 1960.

128. Landolt-Bornstein, Mechanical-Thermal Quantities of State, Volume II, Part 1, Springer-Verlag, Berlin, 1971.

129. Latto, B. and Saunders, M.W., "Viscosity of Nitrogen Gas at Low Temperatures up to High Pressures: A New Appraisal," Can. J. Chem. Eng., 50(3), 765-70, 1972.

130. Latto, B. and Saunders, M.W., "Absolute Viscosity of Air Down to Cryogenic Temperatures and up to High Pressures," J. Mech. Eng. Sci., 15(4), 266-70, 1973.

131. Lee, A.L. and Ellington, R.T., "Viscosity of n-Pentane," J. Chem. Eng. Data, 10, 101-4, 1965.

132. Lee, A.L. and Ellington, R.T., "Viscosity of n-Decane in the Liquid Phase," J. Chem. Eng. Data, 10(4), 346-8, 1965.

133. Lo, H.Y., Carroll, D.L., and Stiel, L.I., "Viscosity of Gaseous Air at Moderate and High Pressures," J. Chem. Eng. Data, 11, 540-4, 1966.

134. Lowry, B.A., Rice, S.A., and Gray, P., "On the Kinetic Theory of Dense Fluids. XVII. The Shear Viscosity," J. Chem. Phys., 40, 3673-83, 1964.

135. Maitland, G.C. and Smith, E.B., "Critical Reassessment of Viscosities of 11 Common Gases," J. Chem. Eng. Data, 17, 150-6, 1972.

136. Makavetskas, R.A., Popov, V.N., and Tsederberg, N.V., "Experimental Study of the Viscosity of Helium and Nitrogen," Teplofiz. Vys. Temp., 1(2), 191-7, 1963.

137. Makita, T., "The Viscosity of Freons Under Pressure," Rev. Phys. Chem., Japan, 24, 74-80, 1954.

138. Makita, T., "The Viscosity of Gases Under High Pressure," Mem. Fac. Ind. Arts., Kyoto Tech. Univ., Sci. and Techn., 4, 19-35, 1955.

139. Makita, T., "The Viscosity of Argon, Nitrogen and Air at Pressures up to 800 kg/cm^2," Rev. Phys. Chem., Japan, 16-21, 1957.

140. Mamedov, A.M., Akhundov, T.S., Ismail-Zade, Sh.M., and Tairov, A.D., "Viscosity of Benzene," Neft. Gazov. Promst., 14(2), 74-6, 1971.

141. Martin, G., Lazarre, F., Salvinien, J., and Viallet, P., "Viscometer Measurements in the Study of Corrosive Gases Under Pressure," J. Chem. Phys., 62(6), 637-42, 1965.

142. Mason, S.G. and Maass, O., "Measurement of Viscosity in the Critical Region. Ethylene," Can. J. Res., 18B, 128-37, 1940.

143. Maxwell, J.B., Data Book on Hydrocarbons, D. van Nostrand Company, Inc., Princeton, New Jersey, 1968.

144. McCarty, R.D., Stewart, R.B., and Timmerhaus, K.D., "p-ρ-T Values for Neon From 27 to 300 K for Pressures to 200 atm Using Corresponding States Theory," NBS Report No. 8726, 1965.

145. Michels, A., Botzen, A., and Schuurman, W., "The Viscosity of Argon at Pressures up to 2000 Atmospheres," Physica, 20, 1141-8, 1954.

146. Michels, A., Botzen, A., and Schuurman, W., "The Viscosity of Carbon Dioxide Between 0°C and 75 °C and at Pressures up to 2000 Atmospheres," Physica, 23, 95-102, 1957.

147. Michels, A. and Gibson, R.O., "The Measurement of the Viscosity of Gases and High Pressures - The Viscosity of Nitrogen to 1000 Atms," Proc. R. Soc. London, 134A, 288-307, 1932.

148. Michels, A., Levelt, J.M., and DeGraaff, W., "Compressibility Isotherms of Argon at Temperatures Between -25°C and -155°C, and at Densities up to 640 Amagat (Pressures up to 1050 Atmospheres)," Physica, 24, 659-71, 1958.

149. Michels, A. and Michels, C., "Isotherms of CO_2 Between 0° and 150° and Pressures From 16 to 250 Atm (Amagat Densities 18-206)," Proc. R. Soc. London, 153A, 201-14, 1935.

150. Michels, A., Michels, C., and Wouters, M., "Isotherms of CO_2 Between 70 and 3000 Atms (Amagat Densities Between 200 and 600)," Proc. R. Soc. London, 153A, 214-24, 1935.

151. Michels, A., Schipper, A.C.J., and Rintoul, W.H., "The Viscosity of Hydrogen and Deuterium at Pressures up to 200 Atm," Physica, 19, 1011-24, 1953.

152. Michels, A., Wassenaar, T., and Louwerse, P., "Isotherms of Xenon at Temperatures Between 0°C and 150°C and at Densities up to 515 Amagats (Pressures up to 2800 Atms)," Physica, 20, 99-106, 1954.

153. Moulton, R.W. and Beuschlein, W.L., "A Study of the Flow of Air in Tubes in the Pressure Range 1 to 300 Atms," Trans. AIChE J., 36, 113-33, 1940.

154. Naldrett, S.N. and Maass, O., "The Viscosity of Carbon Dioxide in the Critical Region," Can. J. Res., 18B, 322-32, 1940.

155. Nasini, A.G. and Pastonesi, G., "Viscosity of Air up to 200 atm," Gazz. Chim. Ital., 63, 821-32, 1933.

156. Naziev, Ya.M., Guseinov, S.O., and Akhmedov, A.K., "Experimental Investigation into the Viscosity of Certain Hydrocarbons at Different Pressures and Temperatures," Neft. Gazov. Promst., 15(6), 65-8, 1972.

157. Naziev, Ya.M., Schuchwerdyew, A.N., and Guseinov, S.O., "Investigation of the Dynamic Viscosity of 1-Heptane and 1-Octene at High Pressures," Chem. Technol. Fuels Oils (Eng. Transl.), 28(12), 736-8, 1976.

158. Neduzhii, I.A. and Khmara, Yu.I., "Experimental Investigation of the Viscosity of Gaseous Ethylene and Propylene," in Thermophysical Properties of Gases and Liquids, 153-7, 1970. [TT-69-55091]

159. Parisot, P.E. and Johnson, E.F., "Liquid Viscosity Above the Normal Boiling Point," J. Chem. Eng. Data, 6, 263-7, 1961.

160. Pavlovich, N.V. and Timrot, D.L., "Experimental Investigation of the Viscosity of Methane," Teploenergetika, 5(8), p. 61, 1958.

161. Phillips, P., "The Viscosity of Carbon Dioxide," Proc. R. Soc. London, 87A, 48-61, 1912.

162. Phillips, T.W. and Murphy, K.P., "Liwuid Viscosity of Halocarbons," J. Chem. Eng. Data, 15(2), 304-7, 1970.

163. Rastorguev, Yu.L. and Keramidi, A.S., "Experimental Study of the Viscosity of n-Undecane Under High Pressures," Neft Gazov. Promst., 14(3), 59-63, 1971.

164. Rastorguev, Yu.L. and Keramidi, A.S., "Viscosity of Hydrocarbons. n-Decane," Neft. Gazov. Promst., 14(5), 59-62, 1971.

165. Reamer, H.H., Cokelet, G., and Sage, B.H., "Viscosity of Fluids at High Pressures," Analyt. Chem., 31, 1422-8, 1959.

166. Reid, R.C., Prausnitz, J.M., and Sherwood, T.K., The Properties of Gases and Liquids, Third Edition, McGraw-Hill, 1977.

167. Reynes, E.G., "The Viscosity of Gases at High Pressures," Northwestern University, Evanston, Illinois, Ph.D. Thesis, 84 pp., 1964. [Univ. Microfilms Publ. No. 64-12331]

168. Reynes, E.G. and Thodos, G., "Viscosity of Helium, Neon and Nitrogen in the Dense Gaseous Region," J. Chem. Eng. Data, 11(2), 137-40, 1966.

169. Reynes, E.G. and Thodos, G., "The Viscosity of Argon, Krypton, and Xenon in the Dense Gaseous Region," Physica, 30, 1529-42, 1964.

170. Robinson, D.W., "The Viscosity of Argon, Helium and Nitrogen at Low Temperatures and High Pressures," Bulletin IIF/IIR Annexe 3, 329-32, 1955.

171. Ross, J.F. and Brown, G.M., "Viscosities of Gases at High Pressures," Ind. Eng. Chem., 49, 2026-33, 1957.

172. Rudenko, N.S. and Shubnikov, L.V., "The Viscosity of Liquid Nitrogen, Carbon Monoxide, Argon and Oxygen as a Function of Temperature," Phys. Z. Sowjetunion, 6, 470-7, 1934; English translation: NASA Technical Translation, NASA-TT-F-11-868, 1-5, 1968. [N68-31-285]

173. Rudenko, N.S. and Slyusar, V.P., "Viscosity of Hydrogen at Constant Density Over the Temperature Range 16.6-300 K," Ukr. Phys. J., 13(6), 656-9, 1968.

174. Sage, B.H. and Lacey, W.N., "Effect of Pressure Upon Viscosity of Methane and Two Natural Gases," Trans. Am. Inst. Min. Metall. Eng., 127, 118-34, 1938.

175. Sage, B.H. and Lacey, W.N., "Viscosity of Hydrocarbon Solutions, Viscosity of Liquid and Gaseous Propane," Ind. Eng. Chem., 30, 829-34, 1938.

176. Sage, B.H., Yale, W.D., and Lacey, W.N., "Effect of Pressure on Viscosity of n-Butane and Isobutane," Ind. Eng. Chem., 31, 223-6, 1939.

177. Sautter, P., "Measurement of the Integral Joule-Thomson Effect of the Refrigerant R12(CF_2Cl_2), and the Calculation of the Thermal and Caloric Quantities of State," University of Stuttgart, Dissertation, 1977.

178. Schröer, E. and Becker, G., "Investigations Above the Critical State. V. Contribution to the Knowledge of Viscosity in the Critical State," Z. Phys. Chem., 173A, 178-97, 1935.

179. Shepeleva, R.I. and Golubev, I.F., "Experimental Measurements of Nitrogen Viscosity at Temperatures of 82 to 276.2 K and Pressures of 1 to 506 x 10^5 N/m^2," Foreign Technology Division, Air Force Systems Command, U.S. Air Force.

180. Shimotake, H. and Thodos, G., "The Viscosity of Ammonia: Experimental Measurements for the Dense Gaseous Phase and a Reduced State Correlation for the Gaseous and Liquid Regions," AIChE J., 9, 68-72, 1963.

181. Slyusar, V.P., Rudenko, N.S., and Tretyakov, V.M., "Experimental Study of Elementary Substance Viscosity Along the Saturation Line and Under Pressure," Ukr. Fiz. Zh., 17(2), 8, 1249-55, 1972.

182. Smith, A.S. and Brown, G.G., "Correlating Fluid Viscosity," Ind. Eng. Chem., 35, 705-11, 1943.

183. Stakelbeck, H., "The Viscosity of Various Freezing Mixtures (Refrigerants)," Z. Ges. Kalteind., 40, 33-40, 1933.

184. Starling, K.E., Eakin, B.E., Dolan, J.P., and Ellington, R.T., "Critical Region Viscosity Behaviour of Ethane, Propane and n-Butane," Progress Int. Res. Thermodyn. Transport Prop., 2nd ASME Symposium on Thermophysical Properties, Princeton, New Jersey, 530-40, 1962.

185. Starling, K.E., Eakin, B.E., and Ellington, R.T., "Liquid, Gas, and Dense-Fluid Viscosity of Propane," AIChE J., 6, 438-42, 1960.

186. Stewart, J.R., "The Viscosity of Natural Gas Components at High Pressures," Institute of Gas Technology, Chicago, Illinois, M.S. Thesis, 1952.

187. Streett, W.B., "Pressure-Volume-Temperature Data for Neon From 80-130 K and Pressures to 2000 atm," J. Chem. Eng. Data, 16(3), 289-92, 1971.

188. Strumpf, H.J., Collings, A.F., and Pings, C.J., "Viscosity of Xenon and Ethane in the Critical Region," J. Chem. Phys., 60, 3109-23, 1974.

189. Swift, G.W., Christy, J.A., and Kurata, F., "Liquid Viscosities of Methane and Propane," AIChE J., 5, 98-102, 1959.

190. Swift, G.W., Lohrenz, J., and Kurata, F., "Liquid Viscosities Above the Normal Boiling Point for Methane, Ethane, Propane, and n-Butane," AIChE J., 6, 415-9, 1960.

191. Timrot, D.L., Serednitskaya, M.A., and Traktueva, S.A., "An Investigation of the Viscosity of Air at Temperatures of 300-570 K and Pressures of 10^5-1.2 x 10^7 Pa by the Oscillating Disc Method," Teploenergetika, 22(3), 84-7, 1975; English translation: Therm. Eng., 22(3), 104-8, 1975.

192. Tjerkstra, H.H., "The Influence of Pressure on the Viscosity of Liquid Helium I," Physica, 18(11), 853-61, 1952.

193. Tkachev, A.G. and Butyrskaya, S.T., "Viscosity of F-22 and Fc-318," Kholod. Tekh., Tr. Nauch. Konf., Leningrad Tekhnol. Inst. Kholod. Prom., SB. Dokl., 227-33, 1970.

194. Tkachev, A.G., Butyrskaya, S.T., and Agaev, N.A., "A Study of the Viscosity of Freons F-22, F-114, F-115, and Fc-318," Heat Transfer-Sov. Res., $\underline{4}$(3), 102-7, 1972.

195. Touloukian, Y.S., Saxena, S.C., and Hestermans, P., "Viscosity," Volume 11 in Thermophysical Properties of Matter - The TPRC Data Series (13 Volumes), Plenum Publishing Corp, New York, New York, 1975.

196. Trappeniers, N.J., Botzen, A., Ten Seldam, C.A., Van den Berg, H.R., and Van Oosten, J., "Corresponding States for the Viscosity of Noble Gases up to High Densities," Physica, $\underline{31}$, 1681-91, 1965.

197. Trappeniers, N.J., Botzen, A., Van den Berg, H.R., and Van Oosten, J., "The Viscosity of Neon Between 25°C and 75°C at Pressures up to 1800 atm. Corresponding States for the Viscosity of the Noble Gases up to High Densities," Physica, $\underline{30}$, 985-96, 1964.

198. Trappeniers, N.J., Botzen, A., Van Oosten, J., and Van den Berg, H.R., "The Viscosity of Krypton Between 25° and 75°C and at Pressures up to 2000 atm," Physica, $\underline{31}$, 945-52, 1965.

199. Trappeniers, N.J., Wassenaar, T., and Wolkers, G.J., "Isotherms and Thermodynamic Properties of Krypton at Temperatures Between 0° and 150°C and at Densities up to 620 Amagat," Physica, $\underline{32}$, 1503-20, 1966.

200. Tsederberg, N.V., Popov, V.N., and Andreev, I.I., "An Experimental Study of the Viscosity of Hydrogen," Therm. Eng., $\underline{12}$(4), 116-8, 1965.

201. Tsederberg, N.V., Popov, V.N., and Panchenko, S.S., "Experimental Investigation of the Viscosity of Helium and Temperatures from 80 to 273 K and Pressures up to 40 MPa," Therm. Eng., $\underline{6}$, 111-4, 1974.

202. Tsui, Chung Yiu, "Viscosity Measurements for Several Fluorinated Hydrocarbon Vapors at Elevated Pressures and Temperatures," Purdue University, Lafayette, Indiana, M.S. Thesis, 95 pp., 1959.

203. Vargaftik, N.B., Tables on the Thermophysical Properties of Liquids and Gases, Second Edition, John Wiley and Sons, 1975.

204. Vassermann, A.A. and Rabinovich, V.A., "Thermophysical Properties of Liquid Air and Its Components," Standards Press, 1968.

205. Vernet, D. and Kniazeff, V., "Determination of Physical Properties of Liquefied Natural Gas," Rev. Inst. Fr. Petrole., $\underline{19}$, 1405-20, 1964.

206. Warburg, E. and v. Babo, L., "Viscosity of Gases and Liquids," Ann. Phys., $\underline{17}$, 390-427, 1882.

207. Wellman, E.J., "Viscosity Determination for Several Fluorinated Hydrocarbon Vapors With a Rolling Ball Viscometer," Purdue University, Lafayette, Indiana, Ph.D. Thesis, 103 pp., 1955. [Univ. Microfilms Publ. UM-13959]

208. Wilbers, O.J., "Viscosity Measurements of Several Hydrocarbon Vapors at Low Temperatures," Purdue University, Lafayette, Indiana, M.S. Thesis, 77 pp., 1961.

209. Yusibova, A.D. and Agaev, N.A., "The Viscosity of Isopentane," Gazov. Promst., $\underline{6}$, 46-7, 1969.

210. Zhdanova, N.F., "Temperature Dependence of Viscosity of Liquid Argon," J. Exp. Theor. Phys. (USSR), $\underline{31}$, 724-5, 1956; English translation: Sov. Phys.-JETP, $\underline{4}$, 749-50, 1957.

MATERIAL INDEX

MATERIAL INDEX

Name	Refrigerant Number	Formula	Physical State				Page
			L	G	SL	SV	
Air	R729	---		G			37
Ammonia	R717	NH_3	L	G	SL	SV	42
Argon	R740	Ar	L	G	SL	SV	46
Benzene		C_5H_5	L				53
Bromotrifluoromethane	R13B1	$CBrF_3$	L	G		SV	59
i-Butane	R600a	$i-C_4H_{10}$	L	G	SL	SV	62
n-Butane	R600	$n-C_4H_{10}$	L	G	SL	SV	67
Butylacetate		$C_6H_{12}O_2$	L				72
Carbon Dioxide	R744	CO_2	L	G	SL	SV	75
Carbon Monoxide		CO		G			80
Chlorodifluoromethane	R22	$CHClF_2$	L		SL		83
Cyclohexane		C_5H_{12}	L				87
n-Decane		$n-C_{10}H_{22}$	L				92
Dichlorodifluoromethane	R12	CCl_2F_2	L	G	SL		96
n-Dodecane		$n-C_{12}H_{26}$	L				99
Ethane	R170	C_2H_5	L	G	SL		102
Ethanol		C_2H_5O	L	G	SL		107
Ethylbenzene		C_8H_{10}	L				110
Ethylcyclohexane		C_8H_{15}	L				113
Ethylene	R1150	C_2H_4		G			116
Fluorine	R738	F_2	L	G	SL		122
Helium	R704	He		G			126
n-Heptane		$n-C_7H_{15}$	L	G	SL	SV	130
n-Heptene		$n-C_7H_{14}$	L		SL		135
n-Hexane		$n-C_5H_{14}$	L	G	SL	SV	138
n-Hexene		$n-C_5H_{12}$	L				142
Hydrogen	R702	H_2	L	G	SL	SV	145
p-Hydrogen	R702	$p-H_2$	L	G	SL		149
Krypton	R784	Kr		G			155
Methane	R50	CH_4	L	G	SL	SV	160
Methanol		CH_4O	L	G	SL	SV	167
Methylcyclohexane		C_7H_{14}	L		SL		170
Neon	R720	Ne	L	G	SL	SV	173
Nitrogen	R728	N_2	L	G	SL	SV	179

Name	Refrigerant Number	Formula	Physical State				Page
			L	G	SL	SV	
n-Nonane		$n\text{-}C_9H_{20}$	L				188
i-Octane		$i\text{-}C_8H_{18}$	L		SL		191
n-Octane		$n\text{-}C_8H_{18}$	L	G	SL	SV	194
n-Octene		$n\text{-}C_8H_{15}$	L		SL		198
Oxygen	R732	O_2	L	G	SL	SV	201
i-Pentane		$i\text{-}C_5H_{12}$	L	G	SL		208
n-Pentane		$n\text{-}C_5H_{12}$	L	G	SL	SV	214
Propane	R290	C_3H_8	L	G	SL	SV	219
i-Propanol		$i\text{-}C_3H_8O$	L	G	SL	SV	225
n-Propanol		$n\text{-}C_3H_8O$	L	G	SL		228
Propylacetate		$C_5H_{10}O_2$	L				231
Propylene	R1270	C_3H_6	L	G			234
Toluene		C_7H_8	L				239
Trichlorotrifluorethane	R113	$C_2Cl_3F_3$	L				244
n-Undecane		$n\text{-}C_{11}H_{24}$	L				247
Xenon		Xe		G			251